System Level ESD Protection

Vladislav Vashchenko · Mirko Scholz

System Level ESD Protection

 Springer

Vladislav Vashchenko
Maxim Integrated Corp.
Palo Alto, CA
USA

Mirko Scholz
IMEC
Heverlee
Belgium

ISBN 978-3-319-34892-6   ISBN 978-3-319-03221-4   (eBook)
DOI 10.1007/978-3-319-03221-4
Springer Cham Heidelberg New York Dordrecht London

To our families

Preface

Subject and Purpose of This Book

Over the last decade a significant paradigm shift in systems and analog IC design was initiated with new market demands and the availability of emerging technologies. The rapid evolution of the handheld and mobile device market segment, dramatic increase of electronic content in automotive products, and the substantial progress in industrial and medical applications created a new need for on-chip protection against system level ESD stresses. The expansion of this trend is not limited to only the ESD specification. Compliance with other, system level electromagnetic compatibility (EMC) standards might also be required. It can include surge protection, EMI compatibility, overvoltage protection (OVP), and even added ability to withstand longer electrical pulses specific to defibrillator equipment or lightning strikes.

This new trend in combination with high level system on chip (SoC) and system in package (SiP) integration, a substantial increase in data rates, electronic system portability, lower power consumption, and lower operation voltages has initiated intensive research and design activities. As a result, an understanding of the need for development of the new test methodologies, adaptation of the system level test standards and procedures toward the IC components, and ESD IP design on test chips has significantly changed the ESD development landscape and the overall research and development investment today. As a result, the engineering of new high current capable ESD on-chip devices and more advanced transient voltage suppressors (TVS) with precise electrical characteristics has emerged.

Essentially, a new on-chip system level protection ESD design culture has been formed. It includes physical design of the high current capable devices taking into account a variety of latchup and transient latchup scenarios, the print circuit board (PCB) and future systems design, and the understanding of the correlation of the characteristics of different devices, stress types, and setups. As an ultimate goal, this activity is now targeting a new system-IC co-design approach. The creation of this *new system level ESD design culture* is getting more and more acknowledged not only by the authors of this book, but by many leading experts in the ESD, IC, and system design community. So far, however, the overall understanding of the on-chip system level design is greatly dispersed across multiple papers, tutorials, white papers, and IC product application notes.

With its main purpose, this book represents a first attempt to organize, to structure, to simplify, and to bring to the reader the understanding of the major aspects of this system level on-chip ESD design culture. The authors pursue this attempt at the level of the most possible logical and simple way of understanding that requires no specific previous knowledge which makes it compatible for a broad audience.

During the work on this book the authors have made a joint effort to combine and summarize their research and industrial design experience, accumulated in the field, to bring the understanding to the next level. The material of this book is composed to present the on-chip system level ESD design within the scope of five chapters. They are logically focused on the introduction of the system level on chip design principles, presentation of the major test methods, on-chip ESD design solutions taking into account latchup phenomena, and finally on bringing the outline for an IC-system co-design approach.

Although many original research papers are referenced across the book chapters, the overall purpose of this book is not to deliver a review of the most up-to-date publications or standards in the field. Instead, the challenging task, as targeted by the authors, was to find and bring under each aspect or design step a logical emphasis of the basic physical principles behind the solution, design, or methodology. This is done to enable readers to apply the in-depth physical understanding generated by reading of the book material toward the solution of their own specific system and chip ESD design problems. The authors expect that creativity and innovation required for solving future system level ESD design problems will be substantially supported by this book. In particular, the readers will see the advantage of physical design approach supported by a mixed-mode device-circuit simulation methodology that uses parameterized devices, circuits, and processes.

The authors carry on a genuine hope that this book will be found useful not only by dedicated ESD design practitioners, but also by a broad audience of IC and system designers, application and product engineers.

The Book Structure

The book has a five-chapter structure. The first introductory Chap. 1 defines major principles and methodologies for the on-chip ESD design. Chapter 2 focuses on ESD test standards and methods. Chapter 3 describes the device and clamp level solutions for the on-chip system level ESD protection, which is extended in Chap. 4 toward the remaining aspects of the chip design, latchup, and transient-induced latchup. Finally, Chap. 5 leverages the previous chapter's knowledge outlining the new chip-system co-design approach.

The introductory Chap. 1 is an important step to understand the material of the following chapters. The chapter sets the terminology that is used throughout the book. It starts from the basic understanding of an ESD event as a transfer of energy between two connected objects and continues with the description of the on-chip ESD protection strategies as well as the differentiation of on-chip and off-chip ESD protection approaches. A more detailed specification of the understanding for the system level ESD pulses, standards, and test methods is further summarized in Chap. 2. The material in Chap. 1 outlines an understanding that the demand of new age electronics creates a significant design paradigm shift both in on-chip system level design with integrated system level ESD protection devices and off-chip PCB design. New precision Si TVS solutions provide now electrical characteristics for significantly more accurate voltage clamping in comparison to the commonly used polymer or zinc-oxide multilayer varistors under very low capacitive load. A more detailed device level description for Si TVS devices is given in Chap. 3, while the understanding of the on-chip and off-chip co-design aspects with Silicon TVS is discussed in Chap. 5. An important aspect to pass certain system level tests is related not only to power-off, but also to power-on stress conditions. Thus, those pins need to be protected for an order of magnitude higher current level than the standard component specifications Charge Device Model (CDM), Machine Model (MM), and Human Body Model (HBM). Chapter 1 emphasizes that the ESD "solutions" for the system are no longer a simple choice of a suppressor component to be connected at the system port. An effective solution requires the application of a design methodology that takes into account the layout of the circuit board, the electrical characteristics of the suppressor, and the ESD characteristics of the IC itself. The chapter also brings up a number of design aspects that should be taken into account during the challenging design of a functional and reliable product. An elevating complexity of the required design solutions requires innovation and novel approaches.

The last two sections of Chap. 1 introduce two key simulation methodologies to support physical ESD design relying on the ESD compact modeling and the new Technology CAD approach based on the parameterized mixed-mode analysis with the DECIMM tool. The approach enables an automated mixed-mode device-circuit analysis with both parameterized process profiles and device templates.

After the main terminology and the direction of the book is set, Chap. 2 brings the background for the physical aspects of key test methodologies and their applications for the on-chip ESD system level design at each development stage. Focus is made on the understanding of the ESD gun testing on board level, followed by the package and wafer level test methods toward an approach for a more effective on-chip design. This chapter focuses first of all on system level tests like the commonly used IEC 61000-4-2 and ISO 10605 standards. This material is followed by the explanation of the key methodological approach of Human Metal Model (HMM) testing, which is the first component-level emulation of system level ESD stress. For completeness, the chapter covers the understanding of other practically used test methods for the on-chip design as the transmission line pulse methodology (TLP), ESD waveform capturing and analysis, and the physical bases behind the correlation factors for different pulses, device types, and test conditions. The testing methodologies presented in this chapter are widely applied in the remaining chapters of the book.

On-chip ESD solutions and the process technology aspects are discussed in Chap. 3. The chapter targets a structured and logical understanding of rather cross-disciplinary subjects required for the successful design of pins protected with system level requirements. These topics include principles of ESD device operation in the breakdown state, under injection and conductivity modulation conditions, clamp layout design, process technology options, the safe-operation area (SOA), the self-protection capability of standard devices, and the co-design of the on-chip ESD network and analog internal circuit blocks. Thus, the chapter challenges the structural presentation of the high-level introductory material for device design, the ESD devices, and clamp design principles specific for rather advanced system level on-chip protection. The major challenge, covered here is the high voltage (HV) system level ESD cell design focused on achieving the proper pulsed-type independent width scaling for high current performance, taking advantage from available process technology features. The nonlinear effects impacting the performance of the structure due to current crowding, improperly balanced layout, the multifinger turn-on effect, or an undedicated "sneak" current path formed in a particular cell and product layout are discussed. The chapter is based on a substantial physical experimental verification and validation of the final solutions, TCAD mixed-mode analysis of the ESD devices, clamp circuits, and analog peripheral circuit blocks based on the new approach enabled by the DECIMM^{TM} tool [19, 20]. The chapter concludes by a discussion of the critical topic related to process capability for ESD solution design and confidence of ESD protection window targeting.

In wafer-level packaging (or micro surface mounting devices) design, the flip chip bonding bumps is evenly distributed on the top of the entire active layout area. At a high current from a system level ESD event, the injection from an analog circuit clamp area can disturb the operation of many more connected active devices. A "sneak" latchup current path can form deep inside the layout of the internal circuit components. Therefore, clamp latchup isolation is an important part of the design. These latchup phenomena are summarized in Chap. 4 as a logical continuation of the previous chapter's material. The key point of Chap. 4 is that the chip-level integration of a validated standalone ESD clamp for system level requirements is not a simple problem. Application specifications and chip functionality need to be thoroughly taken into account to avoid clamp interaction with internal circuit blocks during both system level ESD stress and normal operation. In high injection conditions induced by the system level ESD current, parasitic devices capable of supporting the conductivity modulation regime may also turn on. From this perspective, a system level ESD event in power-on conditions can conceptually be treated as similar to latchup phenomena. Here, differentiation is done between three major latchup scenarios that represent physically different phenomena. The *conventional CMOS Latch-up* is represented as a high-current turn-on of a parasitic SCR formed by a PMOS-NMOS inverter pair as a result of current injection in the form of two subcases: an I/O buffer with internal injection and a core circuit with injection from a remote injector. The *HV Nepi-to-Nepi Latch-up*, the turn-on of the parasitic n-p-n structure in a high current state, is a result of the current injection in one of the pockets. The understanding of the electron injection from a low-side pocket and the hole injection from a high-side pocket are explained. Finally, the *Transient induced Latch-up* is introduced as a combination of the physical phenomena where the ESD clamp turns on as a result of a short-term voltage overstress. Thus the chapter covers many important industrial design aspects that make a bridge for on-chip integration of validated standalone ESD clamps, taking into account the chip functionality, internal circuit blocks interaction, during both system level ESD stress and normal operation. The latter is demonstrated on the example of a highly integrated smart-power IC with a CAN transceiver.

The final Chap. 5 combines the tools and methodologies of the previous chapters to develop an effective and robust ESD co-design for system level IC pins and systems. The chosen approach combines transient device characterization with test boards and on-wafer setups, and device and circuit simulations. The combination of simulations and on-wafer characterization enables the design and verification of system level ESD protection solutions during an early stage of the IC design and even long before a final system is designed or built. The first section introduces the available off-chip ESD protection devices. This is followed by a study of the available simulation tools for the ESD protection design. Methodologies and examples for the device and circuit modeling are provided. The simulation models are used for two system level ESD design methodologies: datasheet-based design and co-design. The required input for each methodology is discussed together with the advantages and disadvantages of each methodology. Through several case studies, recommendations for the design of ESD protection structures for system level IC pins are provided. The chapter concludes with a comparison, benchmarking, and discussion of the introduced design methodologies. Simulations and transient device characterization are essential for the analysis and development of ESD protection solutions for system level IC pins. ESD device models can be simplified to reduce the modeling effort and the simulation time of large protection networks. Mixed-mode ESD simulations enable higher accuracy or the simulation of devices with a complex behavior during ESD stress. Highly accurate FEM models allow the simulation and extraction of the device behavior in the transient domain. SPICE and compact models can be added to the mixed-mode simulation setup to enable circuit-like simulations. The presented simulation approaches are applied to different case studies to analyze and benchmark system level ESD design concepts. The chapter introduces the *datasheet-based system design* with TVS on-PCB components as a methodology which is based on designs experience or the request of higher ESD protection level by the IC supplier. The application of the datasheet-based approach leads in many cases either to an overdesign on the system side and/or costly overdesigned ESD protection solution on IC-level. In some applications this approach can create additional challenges to design simultaneously an ESD robust and functional system. As a result many more off-chip components than necessary might be added.

The alternative methodology is an *IC-system co-design* approach which involves obtaining the pulsed device characteristics. TLP testing and TLP I-V curves are used to provide information about the on-chip and off-chip ESD protection to the system designer. The material in this chapter is greatly based on case studies which combine TLP testing results with transient simulations required to identify the transient behavior of the on-chip and off-chip ESD protection devices during system level ESD stress. An important extension of the *IC-system co-design* methodology is achieved by adding extensive HMM characterization data, obtained from the external IC pins to the design flow. The information is used to verify if the residual currents into the on-chip ESD protection do not exceed the standalone HMM failure level. It is concluded that the co-design approach will stay as a challenge for future SiP and SoC due to the continued scaling of CMOS technologies, novel device concepts like 3D transistors (multi-gate-FET, FINFET), as well as new integration principles like 2.5 and 3D integration which are defined with new backend of line structures like through-silicon-vias.

Acknowledgments

The authors are grateful for numerous discussions and support of their many colleagues for the last two decades from EOS/ESD Association, Industry and University Research Groups. They would like to recognize an exceptional contribution of Dr. Andrei Shibkov from Angstrom Design Automation for DECIMM^{TM} simulation support and implementation of the new features that, in particular, enabled the desired composition of this book, especially the automated simulation of the latchup (Chap. 4) and the revolutionary process capability index simulation study for ESD devices (Sect. 3.5).

Another special appreciation from both authors is for Dr. Augusto Tazzoli for his dedicated effort to complete a detailed high quality review of the entire manuscript with multiple valuable comments made and technical discussions that have significantly improved the material of this book.

In addition, Dr. Vladislav Vashchenko appreciates Yana Vashchenko for her work on copyediting the material of Chaps. 3 and 4. He appreciates his many colleagues from Maxim Integrated Corp. in ESD filed Joseph Sheu, Todd Mitchell, Dr. Slavica Malobabic, Blerina Aliaj, Dimitrios Kontos, Dr. Ali Rezvani, and Ph. D.

student Yunfeng Xi who impacted the composition of this book by many stimulating discussions during the work on projects related to the system level ESD subject. He is deeply thankful for the multiple discussions of ESD subjects during the past decade with his colleagues in the field: Dr. Dimitri Linten and Dr. Geert Hellings from imec, Dr. Misha Khazhinsky, Jeremy Smith and Anirudh Oberoi from Silicon Labs, Prof. Elyse Rosenbaum, from the University of Illinois UC, Dr. Markus Mergens from QPX GmbH, Dr. Vess Vassilev from Novorel, Dr. David Tremouilles from LAAS, Dr. Harald Gossner from Intel, and Prof. Juin Liou from UCF.

Dr. Mirko Scholz would like to gratefully recognize his current and former colleagues in the imec ESD team for numerous discussions in the daily ESD work. This includes especially Dr. Shih-Hung Chen, Dr. Dimitri Linten, Dr. Steven Thijs, Dr. Geert Hellings, Roman Boschke, and Dr. Alessio Griffoni. He also wants to recognize the direct and indirect support of many co-workers in the different imec departments and groups. He wants to acknowledge the former National Semiconductor ESD team (now Texas Instruments), especially Dr. Ann Concannon, Dr. Antonio Gallerano, and Dr. David Lafonteese for the collaboration on various component and system level ESD topics in the previous years. He thanks HANWA/Japan for the joint development collaboration on ESD testers which enabled many of the new and advanced ESD testing methodologies which are partly also discussed in this book. He further wants to acknowledge past and current members of the ESDA working group 5.6 for the work on the HMM test method. And, last but not least, he acknowledges his former colleagues and co-researchers from the ELEC department at the Faculty of Engineering at the Vrije Universitaet in Brussels.

Contents

Chapter 1
System Level ESD Design

1.1 Understanding of ESD Events

1.1.1 IC and System-Level ESD Stress

An ESD event represents the transfer of energy between two connected objects with different electrostatic potentials until the potentials become equal or the connection is removed. The "connection" assumes the current path provided by any media including air. An ESD event results in a decaying current pulse proportional to the level of the electrostatic potential difference and the rise time and current level determined by the impedance of the connection. The transfer occurs either through contact or via an ionized ambient discharge (a spark). This transfer is modeled in various standard circuit models for testing the compliance of device to corresponding passing level. Typically, these models use a capacitor charged to a given ESD pulse voltage and a network which acts as a current-limiting resistor (or ambient air condition) accompanied by inductive and capacitive loads to control the pulse rise time and waveform parameters. The specifications of the ESD pulses, standards and test methods are summarized in Chap. 2.

ESD transient current pulses have $\sim$1–200 ns duration with the rise times from a few hundreds picoseconds to a few tenths of nanoseconds and current amplitudes from $\sim$1 to over 50 A. Unless special protection measures are taken, at certain critical amplitude the ESD transient current pulses may directly impact the reliability of both the systems and components. This can result in a reduction of the manufacturing yield or loss of consumer products.

In real life, ESD discharges can strike the system or devices during their product lifetime by the end user, during manufacturing and assembly or maintenance, as a result of a discharge after triboelectric charging, interfacing with other charged systems or devices in form of mechanical connection of the conducting surfaces or by air discharge. For example connecting charged cables or connectors to a system input and output port can create a discharge current that must be dissipated inside a system through a dedicated ESD current path (Fig. 1.1).

V. Vashchenko and M. Scholz, *System Level ESD Protection*,
DOI: 10.1007/978-3-319-03221-4_1, © Springer International Publishing Switzerland 2014

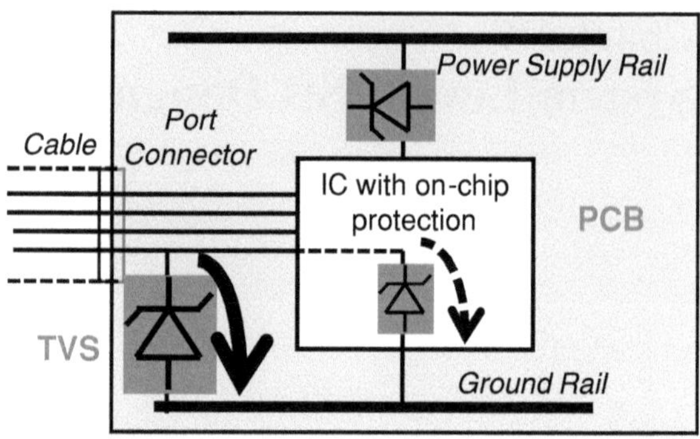

Fig. 1.1 ESD current path in a system with internal IC

The clamping voltage and residual current realized in an IC are functions of the breakdown voltage and the dynamic resistance of the protection device. The clamping voltage can be extrapolated by multiplying the current on dynamic on-resistance of the circuit involved in the ESD current path.

To guarantee the reliability and certify of the systems both the component and system-level ESD qualification tests are carried out based on a set of corresponding standardization documents. This qualification tests conduction is an undividable part of both the IC and system design and production. Passing these tests is a requirement to obtain different product certifications, for example the CE mark in Europe, necessary for the introduction of products in the consumer market.

Meantime the ESD protection strategy itself is based on a rather simple approach. It consists of an implementation of a dedicated current path for the discharge by means of either on-chip protection structures in case of IC components or/and embedded into a system ESD protection network. The network protects the system itself in addition to other isolation measures. This protection network is usually composed using active and passive on-chip and off-chip on board components. Altogether the components and interconnects essentially represent a pulsed power network circuit that remains passive during normal operation, but becomes active during ESD pulse providing the discharge current path. The ESD pulse protection network activation is typically realized by a combination of the rise time and overvoltage detection. In this case on-chip and off-chip ESD protection network components turn on if a critical voltage at the IC pins or system ports exceeds a certain threshold value.

1.1.2 Trends in the IC Component and System ESD Design

The level of ESD that an end-user can generate and introduce into an electronic system product during its lifetime is much more severe than the typical level generated in a controlled manufacturing environment. Moreover, since the introduction of the S20.20 procedure [1] in year 1993, the ESD pulse levels detected in

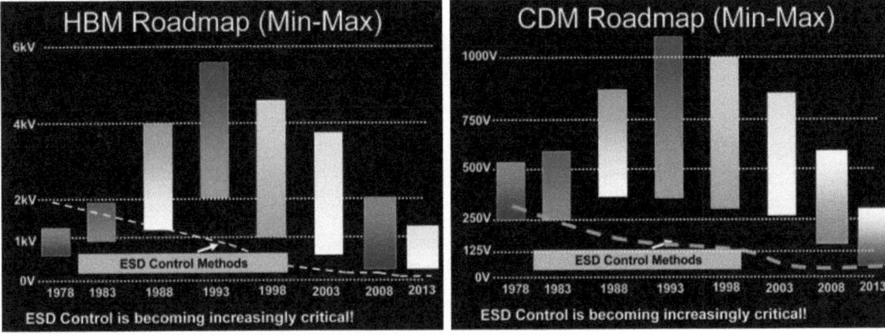

Fig. 1.2 ESD pulse roadmap suggested by the EOS/ESD Association for HBM and CDM stress

Fig. 1.3 Impact of different
ESD control methods on the
voltage of a person in a
manufacturing environment:
a no ESD control, **b** ESD
flooring installed and **c** ESD
flooring installed and ESD
footwear used data from [2]

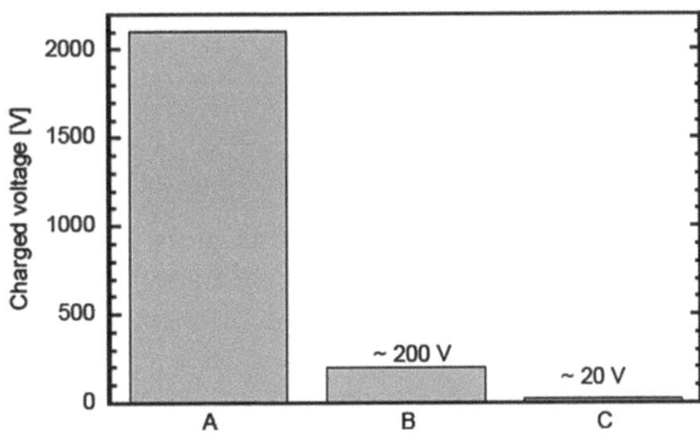

real ESD controlled environments have a continuously reduced trend. As a result, the EOS/ESD Association (ESDA) has proposed a corresponding reduction of the qualification requirements (Fig. 1.2). This initiative has not been fully accepted by the industry yet and the standard corporate requirements from IC manufacturers and particular product specifications are often dictated by customers. They typically include 2 kV HBM and 100–200 V MM ESD pulses for the component-level. However, the fact of the significant reduction of the ESD stress event amplitudes in the ESD protected areas hardly causes any doubts and it is widely used at least for the passing level waiver justification.

ESD control programs significantly reduce the accumulation of charges in a manufacturing place. For example in case "A" (Fig. 1.3) the voltage on the person is measured when no specific ESD control method is in place. It can reach over 2 kV. If the manufacturing place is equipped with an ESD floor the electrostatic induced voltage is reduced down to ~200 V level ("B", Fig. 1.3). If the operators wear ESD footwear the walking on the ESD floor generates only ~20 V ("C", Fig. 1.3). Thus the voltages generated by the person can be reduced by two orders of magnitude by implementation of the ESD control programs.

An opposite trend over recent two decades was observed for the system-level protection where a number of new standards have been introduced to support the

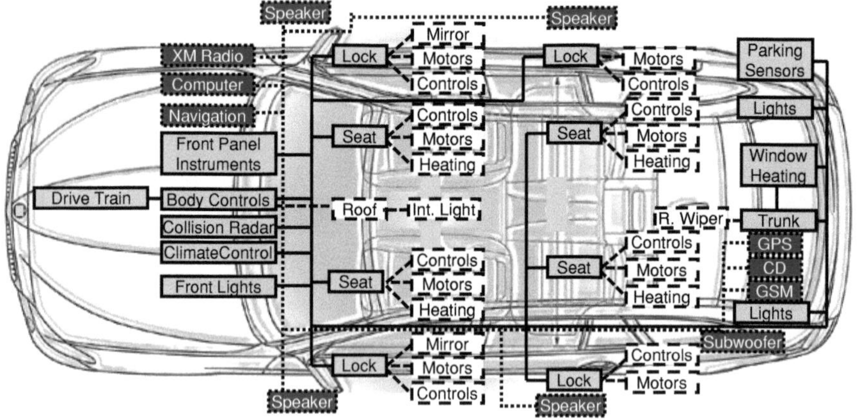

Fig. 1.4 Illustration of the robust system-level design approach for automotive products

elevated ESD protection requirements of systems. One of the typical examples is to address requirement for zero defect part per million failure rate in automotive applications (Fig. 1.4). In addition to the system level ESD pulses a number of new for IC manufacturing standards have been propagated down to the component level. This includes power surge specifications (Chap. 2). Thus a major paradigm shift had occurred when these, originally system-level standards, have been propagated down to the requirements of the passing level of selected pins of IC components.

To some extent the emphasis on ESD protection has been shifted from the chip-level ESD robustness to the system robustness. In the past the original attempt to address this was made by intuitive requirements of just a higher passing level of the standard component-level ESD tests. The IC manufacturers were requested to supply IC with some pins protected up to for example 8 kV or even 15 kV HBM. Even today such an unreasonably high (comparing to Fig. 1.2) components HBM stress level can be found in datasheets of some new product listed together with the system level passing data. However, the passing level for the specific system-level standard pulses became a dominant approach.

In general the component passing level for CDM, MM and HBM standard pulses at the chip level do not guarantee a system-level added robustness of the IC. The discharge through the IC during chip-level ESD qualification is applied in power-off conditions. In general the resultant ESD current path is not the same during a system-level ESD test in a powered-up system. Therefore the capability of IC to withstand at least a fraction of the system-level ESD current needs to be verified. This should be done for the IC pins directly interfacing with the system ports unless a board-level ESD protection network guarantees the appropriate limitation of the current into the IC pin.

The original purpose of the chip level verification is to guarantee the reliability of the chip itself during the manufacturing processes and assembly of the system itself, rather than during the life cycle of the system in the field. A system might experience rather high failure rates in the field and unlikely will pass certification unless the system is specifically designed to dissipate any ESD discharge by an intentional implementation of a system-level ESD protection network.

The most optimal design approach today to implement ESD robust system-level design involves both the IC component and the system board protection levels. On-chip ESD protection network is typically composed of ESD clamps, diodes, self-protected power devices and robust metal busses and interconnects to conduct the component-level ESD currents. The system-level ESD protection networks rely on passive components in addition to optional transient voltage suppressors (TVS). The major strategies to combine both networks will be reviewed in the following two sections.

There are two important trends leading to fundamental changes in ESD protection strategies at the system design level—increased susceptibility of system-level ICs to ESD with migration toward more advanced process technology nodes and increasingly stringent signal integrity requirements as data rates continuously increase [3].

The traditional approach to deal with higher data rates is to reduce the capacitive load in the transmission line. This includes the capacitance of the ESD protection device. However the capacitance of the ESD device is proportional to the device active regions width. A reduction of the ESD protection capability of the device is an expected undesired side effect. As a result, system designers become under pressure to make tradeoffs between their system reliability and signal integrity or involve matching networks and ESD protection co-design strategies.

For certain signal integrity requirements the traditional system-level ESD approach cannot deliver suitable protection solutions due to the limitation in the semiconductor process technology used. Therefore an off-chip ESD protection design becomes unavoidable. In general the clamping voltages and residual current levels that were acceptable in previous generation of application specific ICs are no longer suitable for the new generation circuits implemented in the scaled semiconductor processes. Adoption of high speed data interfaces USB, HDMI and Display Ports adds to the complexity of maintaining the signal integrity under robust ESD protection. Similar trends can be observed in RF antennas, automotive networks, medical, industrial and even emerging servers for cloud computing applications.

The changes of the application environments directly contribute to the ESD vulnerability. For example various handheld and mobile devices are now being used in everyday harsh and uncontrolled environments with real life ESD events generating system-level ESD pulses up to 30 kV, power surge and EMI. The devices must regularly withstand multiple cable connections and disconnections to the system ports, accumulation by the isolated portable device triboelectric charges in different application conditions like for example in the pocket of a running or biking person.

With traditional ESD architectures, an inverse relationship between robust ESD protection and low parasitic capacitance negatively impacts the signal integrity. This results in an increasing difficulty of maintaining capacitance and impedance matching. On the other hand, the combination of multiple specifications, standards and testing criteria are often quite cumbersome. Identifying which ESD protection approach and device provides the best results is rather difficult due to unclear voltage clamping waveform realized by the off-chip devices under system-level ESD discharge conditions and by the on-chip devices at the residual current propagated to the IC pin.

1.2 On-Chip ESD Protection Strategies

The original role of an on-chip ESD protection is to guarantee that an IC is capable of withstanding the ESD stress through its entire manufacturing process in an ESD Protected Area (EPA). This includes both chip foundry and system assembly. Respectively the qualification of each chip includes requirements for certain passing level for the pulses defined by the various standards. The component level standards have been established to verify the immunity of the IC's in the controlled ESD environment. The major standards include: The Charged Device Model (CDM), Machine Model (MM) and Human Body Model (HBM) (Fig. 1.5).

CDM testing verifies the susceptibility of the IC to a discharge event from the charged package to a grounded metal object. This pulse has a fast rise time $\sim$ 100 ps, very short 1–5 ns, and its amplitude is proportional to the IC component size. The IC package can accumulate charges at a given electrical potential. The typical test levels for CDM are between 250 and 750 V.

The MM ESD event physically represents the discharge of pre-charged metal tools that are suddenly connected to an IC pin. It is assumed that the IC has connection to the discharge ground path. Due to inductance of the metal tool the pulse has an oscillatory waveform. It is expected that in an EPA the metal tools unlikely can be charged up to a too large electrical potential due to their low resistance and if proper grounding procedures are in place. Therefore typical test level for MM is either 100 or 200 V. The 200 V pulse produces $\sim$ 3 A peak current.

The HBM standard physically represents the discharge of a charged person to an IC pin with another IC pin grounded. It is agreed that the electrostatic potential generated by person within EPA is between 0.5 and 2 kV. The generated current pulse is limited by a 1.5 kΩ discharge resistance which results in a peak current of $\sim$ 1.33 A for a stress level of 2 kV.

These IC component test voltage levels (Fig. 1.2) are only meaningful in a controlled manufacturing environment within an EPA, where precautions are taken to ensure that static electricity levels on personnel and equipment are minimized. System-level stress pulses at the same pre-charge voltage result in significantly

Fig. 1.5 Comparison of the idealized waveforms for major IC component and IEC system-level pulses

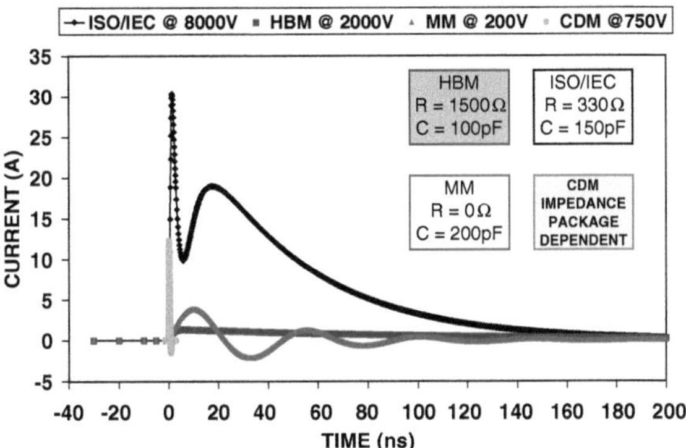

Fig. 1.6 Comparison of the total charge for major IC component and IEC system-level pulses

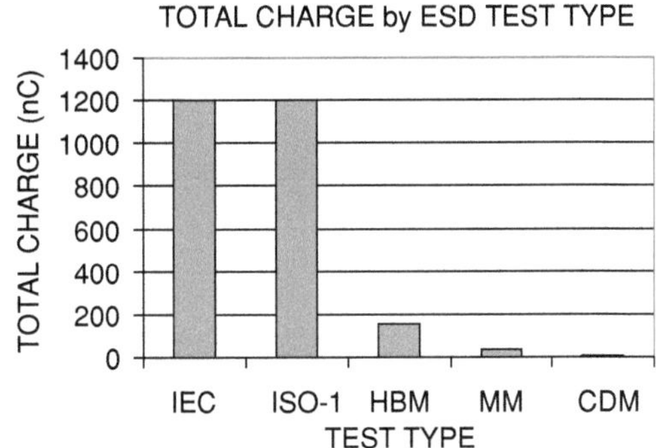

larger pulsed current amplitudes (Fig. 1.5) and total charges (Fig. 1.6). The system-level standards are discussed in details in Chap. 2.

On-chip ESD protection network is represented by an embedded power circuits that provide different reversible current paths for each IC pin-to-pin combination. At the discharge through these current paths an appropriate voltage limitation is realized in order to prevent any irreversible changes of the protected internal circuit. In spite of the high variety of analog circuits, the network can be constructed based on rather common simple principles. Typically networks are differentiated as the local clamp based or as the rail-based network. The first may also include self-protection capability of the on-chip active circuit components.

The local protection network relies on a dedicated ESD protection clamp attached either between the protected IC pin and IC ground or between two IC pins. In this case the current path is expected to be realized for each pin to pin combination through the local clamp. The alternative rail-based network delivers the protection of the same power domain IC pins through the diodes connected to the ESDPLUS and ESDMINUS rails with core clamp connected. The current path between two pins is realized through the diode and the core clamp network. Both approaches have advantages and drawbacks (Table 1.1).

Table 1.1 Comparison of two major ESD protection strategies

	Local clamp network	Rail-based network
Advantages:	Independent of bus resistance	Area efficient
	Local to every pin	Highly process portable
	Suitable for multiple pin count	SPICE-based design
	Does not require pad ring	Optimal for small pin count IC
	SoC and system-level design	
Drawbacks	More area intensive	Performance based on
	Highly process sensitive	bus resistance
	Difficult to simulate and account for "sneak" current path	Distributed ESD network requires complex chip level implementation and verification

Design of the ESD protection network includes an appropriate selection of ESD clamps developed for a given process as an intellectual property (IP) and compatible with the process design kit (PDK), design metallization routing adequate to conduct the large ESD currents, and accounting for the alternative current path(s) throughout the internal blocks of the analog circuitry. The choice of the local or rail-based clamps for different power domains is an application-specific task. Due to interaction of analog I/O blocks with the active internal circuitry both mixed-mode simulations and circuit ESD simulations with compact modeling tools are used to provide an important design verification and optimization.

Overall on-chip ESD protection network strategies are described in details in [5]. The major function of the network is to react on high transient voltages by enabling a discharge current path. Depending on the internal circuit specifications, both transient-triggered and voltage-reference ESD clamps can be used to implement the on-chip protection network. Design of the protection clamps for the component-level specifications is broadly discussed in [5]. The specific of ESD clamp design for system-level pins is discussed in Chap. 3.

The term *ESD pad ring* is often used in digital circuit design or in small pin count analog circuits where the periphery of the chip is used for I/O and power supply pads, ESD clamps and I/O circuit placement. Any ESD pad ring includes pads and a distributed ESD protection circuit created near the pads from a combination of cells. ESD protection network components can be embedded in the I/O cells too. In most cases, the ESD pad ring can be decoupled from the internal circuitry and then practically reused to support different internal circuit blocks.

For analog circuits, understanding of the ESD pad ring design is not always as straightforward as for digital circuits. The complexity is the result of a greater level of interaction of the current path with the internal power circuit. Often, an analog circuit pin can interface with many connected active devices. Thus inside of the internal circuit some, especially power, components may see a significant residual ESD stress.

In the case of micro SMD or wafer level flip chip packaging design the bump pads are arranged in form of ball grid array on top of the whole silicon die area. In this flip-chip package the implementation of the ESD pad ring becomes impractical.

Although ESD network generally may not be decoupled from the internal analog integrated circuit the ESD current path and voltage limitation can be still analyzed to a certain level using hierarchical and mixed-mode simulation methods.

Depending on the internal circuit design, the absolute maximum voltage of the internal components may significantly vary due to different coupling of the control electrodes of the output devices. From this prospective, power analog circuits and ESD protection networks often may require custom co-design, especially when system-level requirements are specified for certain IC pins.

1.2.1 Rail-Based ESD Protection Network

To pass the ESD qualification each IC pin must rely on either self-protected by the internal power devices or must be connected to the ESD protection network. Local ESD protection is independent of bus resistance, as well as of other network components. It provides a relatively easy accountable local voltage waveform at every pin. But it is not an optimal space-saving solution and more process-sensitive especially in case of high voltage tolerant analog IC pins.

In general a rail-based ESD protection network can be composed of diodes and so-called core clamps connected between ESD rails. In principle ESD rails can share the power supply busses or be independent from them. The core clamps can have either snapback or non-snapback characteristics. However when snapback clamps are used the total accumulated voltage drop might become excessive to limit the voltage at the pin at the appropriate level. Therefore the approach has a limited use, for example for amplifiers with large ESD protection windows.

Rail-based protection can be implemented with the core clamp acting either as a power and core clamp or just as a core clamp. In the first case the ESDPLUS and ESDMINUS rails are the same as ground and power supply, for example VDD and VSS (Fig. 1.7a). For the core clamp functionality the power supply can be connected similarly to the I/O through the rail diode pair (Fig. 1.7b). In this case, if an active clamp is used, a charge "bleeding" resistor must be used from the ESDPLUS rail to any other pad in order to avoid the internal charge accumulation and corresponding disabling of the active clamp as a result of multiple ESD pulses.

A rail-based ESD network can be identified by the presence of ESD busses (rails) with the shared core clamp and the ESD diodes connecting the other pins to the rails unless the diode function is accomplished by the body diodes of the appropriately large internal devices connected to the pin. ESD diodes can connect input/output (I/O) pins or other control pins, including power supply pins. The power supply pins connected to the rails typically share the metallization routing. However, in principle, the network can be organized with ESDPLUS and ESDMINUS busses that separate the power supply and ground for the domain.

ESD protection can be embedded in the input/output (I/O) cells. Under this approach, a set of ESD and latch-up rules must be taken into account to provide a

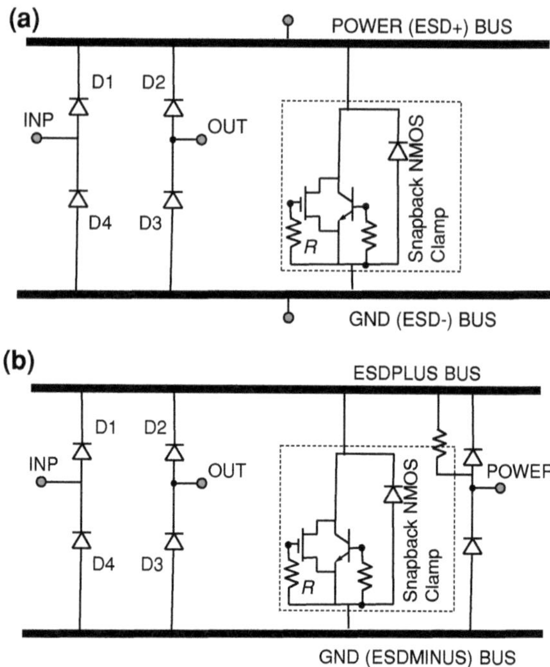

Fig. 1.7 Rail-bases ESD protection network with snapback clamp as a power clamp (**a**) and as a core clamp (**b**)

non-conflicting operation of ESD elements and I/O cell components. The latch-up physics and related design aspects are discussed in Chap. 4.

One of the major advantages of the active clamp network design is a direct applicability of circuit analysis for ESD pulse operation with regular active device compact models. In the case of the rail-based network, there are several steps in the complex chip verification. They usually include: verification that every IC pin has ESD protection; analysis of the dedicated ESD discharge paths and pin-to-pin voltage drop; extraction and verification of interconnect robustness for the ESD current density level (in contacts, metal and vias); verification of ESD cells types, their dc voltage and pin signal tolerance as well as compliance with ESD design guidelines.

In rail-based ESD protection, the ESD diodes are expected to conduct pulsed ESD current only in the forward bias mode and have high reverse breakdown voltage protected by the ESD network. To avoid interference with the major ESD current path, the voltage tolerance of the ESD diodes is an important parameter. An appropriate breakdown voltage of the rail-based protected domain—above the power supply level—should be provided to avoid ESD diode burnout due to avalanche breakdown mode and to minimize direct impact on circuit functionality during normal operation.

An example of the simplified 4-pin rail-based circuit (Fig. 1.7a) combines the protection of the input (INP), output (OUT), POWER, and ground (GND) pins. Additional pins can be added to the network by reusing similar diode connections to the I/O pins. For each pin-to-pin combination the network delivers an ESD current path through the forward-biased ESD diodes, corresponding parts of the rails and the core power clamp. Both the high-voltage current path and the reverse path through the clamp diode are realized in the power clamp. For example, at positive ESD zap of INP versus OUT, the current path is formed through the upper diode D1, the upper part of ESDPLUS bus, then through the snapback clamp, the corresponding part of the ESDMINUS bus, and the forward-bias diode D4.

Rail resistance should also be taken into account at ESD current level, the total voltage drop on the circuit and the voltage drop at each pin relative to the ground can be estimated for the clamp voltage drop V_C, forward-biased diodes voltage drop V_D and the contribution of bus resistances R_{ESD+} and R_{ESD-} depending on the current path (Table 1.2). Certainly more complex multiple pin count cases require CAD automation tools for the automatic extraction of multiple current path resistances [6].

In the distributed network the clamp components can be shared. For example the distributed core active clamp includes multiple instances of NMOS arrays with shared RC-timer (Fig. 1.8). The advantageous low on-state resistance of the distributed network of RC-controlled power NMOS devices delivers rather low clamping voltages across the pad ring area that is proportional to the number of distributed cells used.

If the system-level requirements are important the chip-level protection with active clamps becomes rather challenging since active clamps are disabled when the power supply voltage is present. To limit the voltage at the input pins, anti-parallel diode clamps with a number of diodes that corresponds to the desired differential signal level are used to protect the differential BJT input pins (Fig. 1.9).

1.2.2 Local Clamp Network and Two Stage Protection

The rail-based network for IC inputs and outputs is usually inefficient in case of system-level ESD stress due to the high voltage drop accumulated by the network. Under power-on conditions the active clamp is disabled and thus may not sufficiently react on the ESD pulse. Other cases preventing the active clamp approach involve the protection of the pads with a voltage tolerance above the rating of the available active devices, for example in the case of amplifiers realized in a 5 V semiconductor process with high common mode 65 V tolerant differential input pins connected to thin film input resistors [7]. Additional examples include circuit pins with dual-direction high voltage tolerance. Protection solutions for such pins may not tolerate the lower diode-to-ground rail and thus require also an exclusion of the upper diode-to-power-supply rail. Therefore the local clamp approach represents the only practical solution (Fig. 1.10). In the local ESD protection network approach every pin is protected by a dedicated local ESD clamp that delivers a current path either from pin to the ground or directly another pin.

Table 1.2 Rail-based peak voltage at peak ESD current

ESD zap combination	Total voltage drop on zap pins	INP voltage versus ESD–	OUT voltage versus ESD+	POWER voltage versus ESD–
INP to ESD+	V_D	V_D	0	0
PWR to ESD–	V_D	$\sim V_D/2$	$\sim V_D/2$	V_D
ESD– to INP	V_D	V_D	0	0
PWR to ESD–	V_C	$V_C/2$	$V_C/2$	V_C
INP to OUT	$2V_D + I_H(R_{E-} + R_{E+}) + V_C$	$V_D + I_H(R_{E-} + R_{E+}) + V_C$	V_D	$\sim V_D$
INP to ESD–	$V_D + I_H(R_{E-} + R_{E+}) + V_C$	$V_D + I_H(R_{E-} + R_{E+}) + V_C$	$(I_H(R_{E-} + R_{E+}) + V_C)/2$	$I_H(R_{E-} + R_{E+}) + V_C$

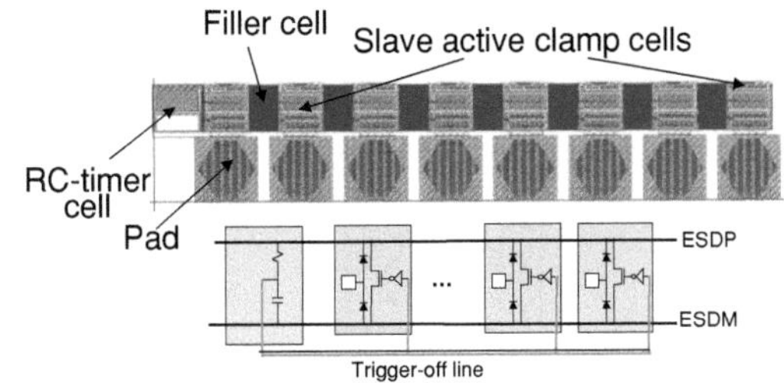

Fig. 1.8 Example of layout and schematic views for 3 V ESD domain active clamp network composed from ESD library cells

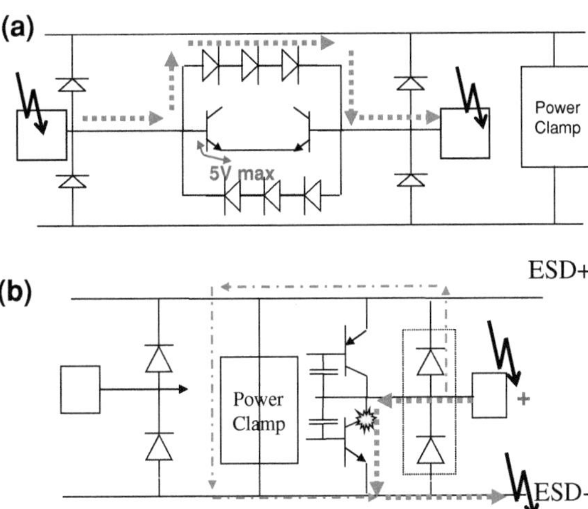

Fig. 1.9 Example of rail-based bipolar differential input (**a**) and output protection (**b**) with indicated ESD current path

A two-stage on-chip ESD protection principle is applicable for local clamp protection. In general the two-stage network consists of a combination of two ESD clamps separated by a resistive current path or a delay component. The resistor-based two-stage network (Fig. 1.11a) combines the primary clamp that limits the operational voltage to V_P, while the smaller secondary clamp limits the voltage at the internal node to V_I. The second stage resistor R_S is selected depending on the second stage clamp current I_{C2}: $R_S = (V_P - V_I)/I_{C2}$. The two stage clamp is very efficient in filtering the short pulse overstress which is typical for CDM events due to the $R_S C_I$ time constant. C_I is the equivalent capacitance of the internal node. The use of secondary clamp becomes optional if the internal device can support an adequate pulsed current level.

The two-stage protection against system-level ESD pulses is similarly applied both on-chip and on board. On PCB the advantage lies in the high capacitance and inductances of discrete components, while transient voltage suppressor may function as a primary protection stage.

An example of the two stage network is a first stage snapback clamp separated by a resistor from a second stage and a voltage limiting avalanche diode (Fig. 1.11b) or two stage diodes with a matching resistor (Fig. 1.11c). The primary

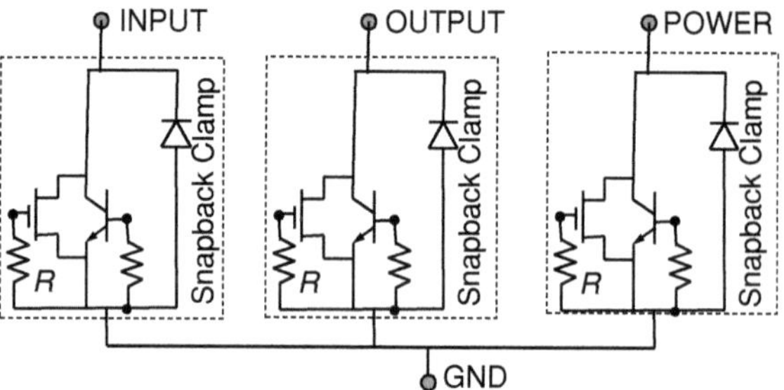

Fig. 1.10 Example of a 4-pin local clamp based protection network

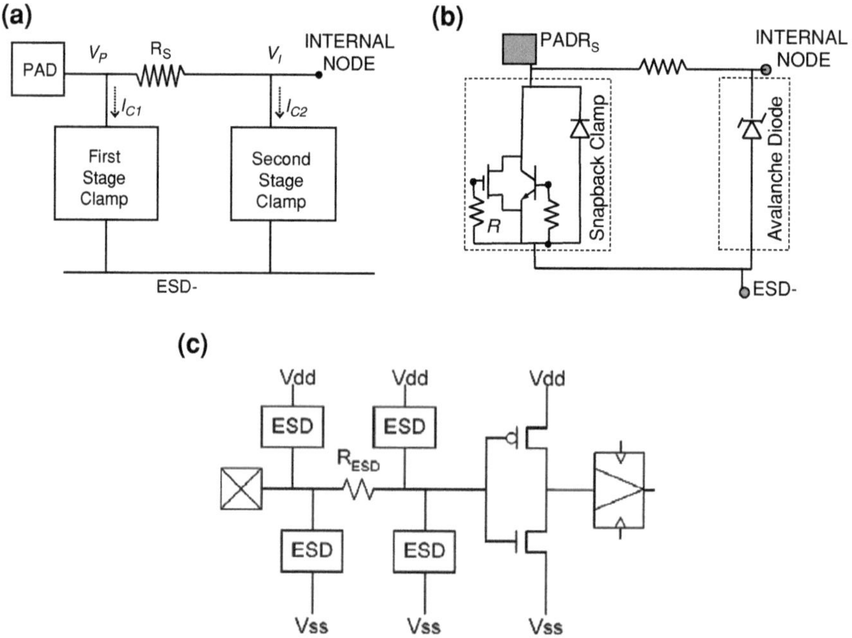

Fig. 1.11 Two-stage ESD protection for CMOS input (**a**), the local based network with the second stage avalanche diode (**b**) and ESD protection for a high-speed receiver with a two-stage matching resistor (**c**)

clamp directly interfaces with the input PAD. It limits the voltage only to certain extent and conducts the most of the ESD current. However the primary camp can be designed to target the clamping voltage waveform only approximately. This separated function is addressed by a more precise, but less robust secondary clamp. The result of the separation of the current and voltage limiting function between the local protection stages is an overall decrease of parasitic capacitance, leakage and noise.

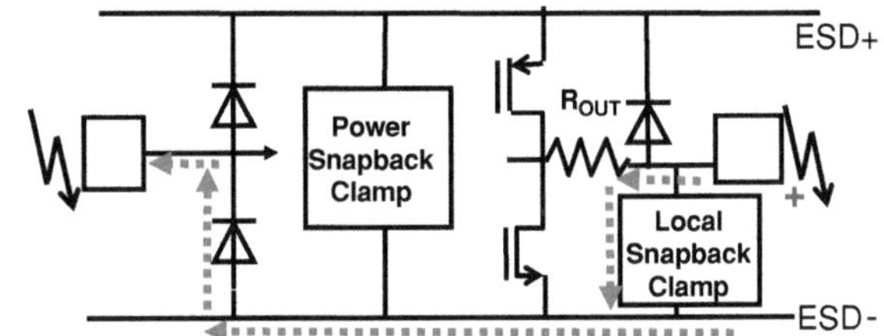

Fig. 1.12 Simplified circuit diagram of an NMOS output pad with a current-limiting transistor and an additional local clamp for ESD protection

An additional advantage can be taken from implementation of non-linear resistors used in the stage. For example it can be implemented as a saturation resistor (R_S Fig. 1.11a) that provides an additional voltage drop starting from some saturation current level.

The principle of the two-stage network is widely used across both the component and the system levels protection for inputs and outputs. In case of outputs the internal circuit power arrays may provide an appropriate matching current for the second stage protection even in cases where the stage separation is only accomplished with metallization resistors (Fig. 1.12). The complexity of local ESD protection is related to generally unknown conditions of the control electrodes of the devices interfacing with the pad during ESD pulse. As a result the triggering voltage of the internal devices can become comparable to the triggering voltage characteristics of the ESD clamp.

Under power-on system-level test conditions the local snapback clamp needs to be chosen by taking into account the potential transient latch-up issue. For example various silicon controlled rectifier (SCR) devices (Chap. 3) generally produce a relatively low holding voltage below the power supply level and thus provide a latch-up risk.

The two-stage principle can be implemented also at the device level. An example of a compact solution is represented by a three-terminal diode structure (Fig. 1.13a, b). It incorporates two-sided diodes that form a three-terminal structure. One side of the diode is used for power ESD operation (a power diode connected to the I/O pad) and the other side just holds the lower potential in the internal circuitry (a potential diode connected to the internal circuitry). Both internal diodes share the same well, thus creating a two-stage ESD protection circuit with a corresponding "built-in" internal well-resistor. An important feature of the diode structure is the internal potential sharing during ESD operation.

The ESD protection of the I/O pin can be realized using a corresponding circuit presented in Fig. 1.13c. Operation of the two-stage components can be analyzed using numerical simulations of the device cross-sections (Fig. 1.13d) demonstrating over two time's lower voltage at the internal node.

The two stage approach can be effectively used to resolve the issues in the entire ESD pulse time domain. For example it can be used to limit the residual voltage at the internal after the snapback device triggering off. This can be illustrated by an "Erase" pin protection case in an EEPROM memory module with external programming.

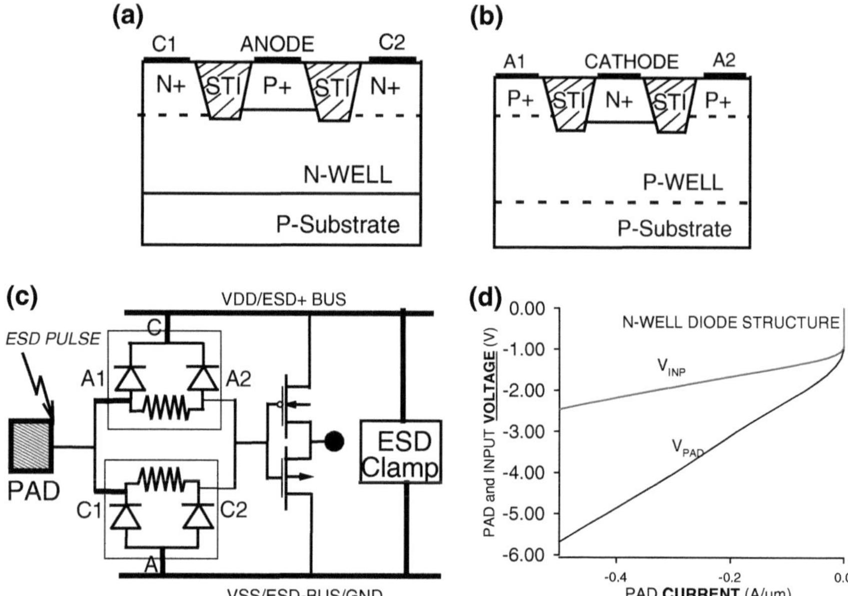

Fig. 1.13 Three-terminal diode structures for nWell (**a**) and pWell (**b**) diodes with device level 2-stage protection principle implementation, corresponding rail-based network (**c**) and calculated voltage drop the pad at the internal node for the nWell diode as a function of the pad current (**d**)

The failure was observed in form of a reprogramming of the memory cells as a result of ESD stress due to combination of the high residual voltages at the erase pin with other nodes potential at the remaining ESD stress steps after ~ 800 ns from the beginning of the pulse. The measured waveform at the original SCR-based erase pin protection delivered different load dependent residual voltages (Fig. 1.14). With the small load with only one EEPROM cell the reprograming as a result of ESD stress was observed in comparison to the case of higher load with 48 EEPROM cells.

To reduce the residual voltage the original HV SCR based circuit (Fig. 1.15a) has been modified with a second stage component and additional feedback loop (Fig. 1.15b). An enable module has been added to disable the second stage clamping during normal operation.

The local on-chip protection is not only limited by the local clamping of the peripheral pins. In addition to the goal of achieving a good ESD performance the local clamps can be included in the internal circuit. One of the most typical examples of the impact of ESD stress on the internal pins in high voltage circuits is the internal voltage regulator of low side control circuits, logic and drivers for switching devices. Because of the relatively low impedance of internal regulators, a second stage resistor often cannot be used in order to realize the two-stage protection approach. As a result, the regulators can exhibit increased sensitivity to

Fig. 1.14 Voltage and current waveform for the EEPROM with single cell and 48 cell arrays measured at 2 kV HBM

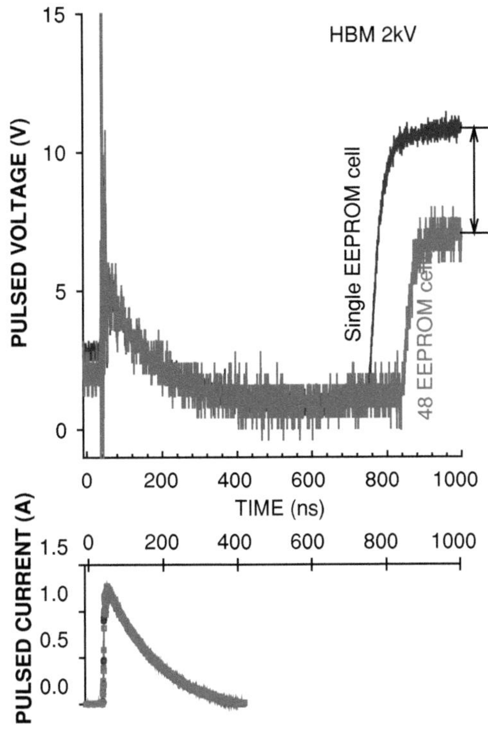

Fig. 1.15 Original (**a**) and modified (**b**) EEPROM Erase pin protection clamp circuit to reduce the residual voltage after SCR turn-off

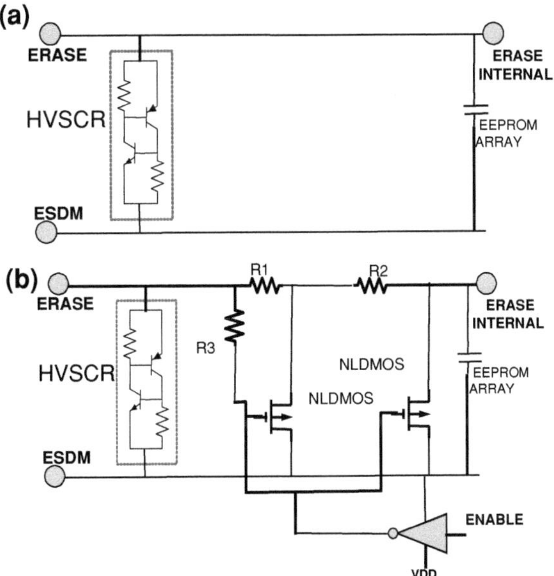

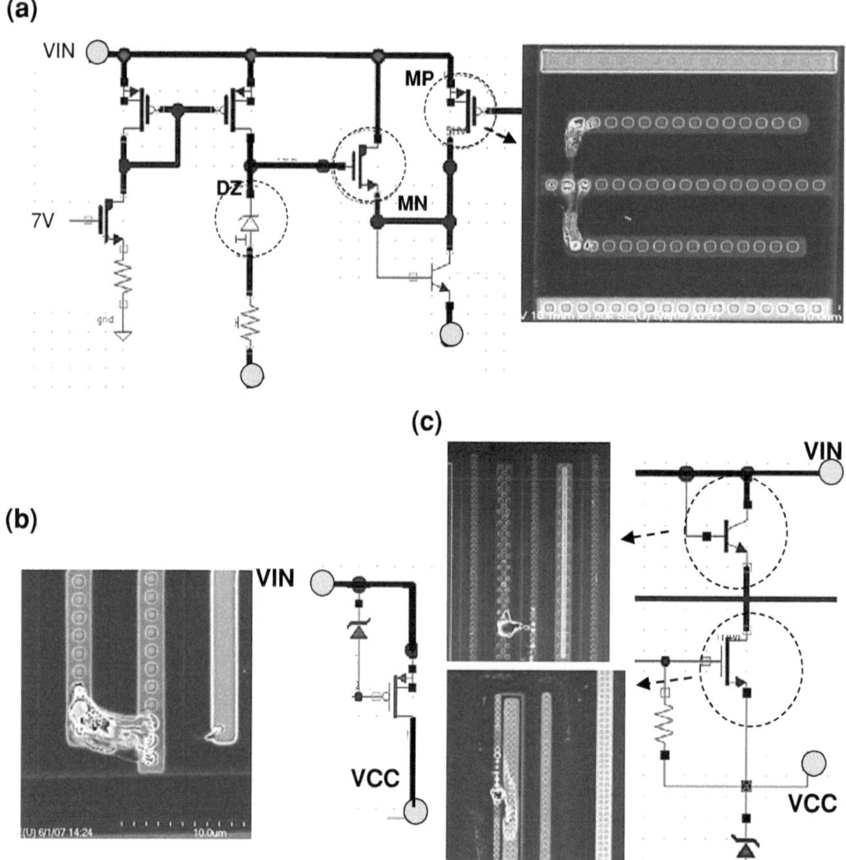

Fig. 1.16 Examples of the FA results for PMOS (**a**, **b**) and NPN BJT based VCC regulators (**c**) [5]

ESD damage and therefore must be designed to withstand certain ESD current levels. Usually, the internally generated VDD voltage node requires an internal voltage clamp capable of handling high ESD current. Such a clamp is applied even if the VDD node is not connected to the external pin at all.

The most typical locations of failure in internal voltage regulator designs are the high-side high-power components. In the high-side PMOS-based regulator the high-voltage PMOS fails (Fig. 1.16a). In this case, the gate clamping diode may not provide a substantial improvement (Fig. 1.16b). A similar ESD current failure path was observed in bipolar and stacked regulators. In this case, due to the domino effect, damage in both stacked components might be observed (Fig. 1.16c). ESD protection of the high-voltage PMOS used in VCC regulators (Fig. 1.16) is based upon the expected ESD path formation between the VIN and CBOOT pins.

Fig. 1.17 Simplified circuit of the DC–DC voltage regulators with three possible ESD current path scenarios for VIN to CBOOT ESD zap

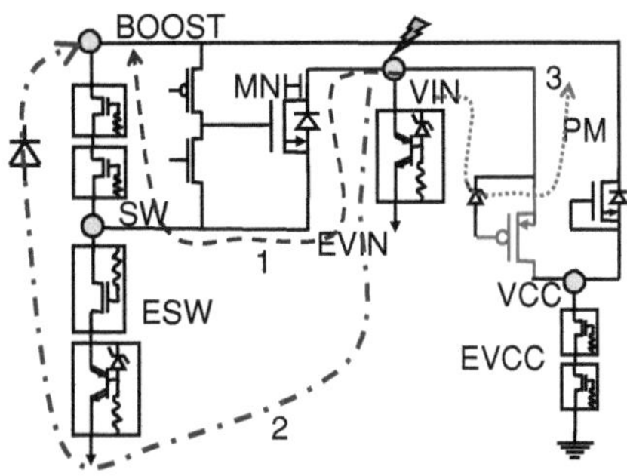

In general, there are three current paths that could be realized for ESD pulse current (Fig. 1.17). The first current path is provided by the large switching NLDMOS controlled by the driver. The NLDMOS turns on for a short period of time due to the drain-gate coupling that may be sufficient enough to discharge the ESD current. This current is directed to the switch pin SW, followed by the reverse current path through the BOOST-SW clamps. The clamp's reverse path ESD diode provides this remaining current path. This current path is reversible in properly designed ESD protection clamps. The second alternative current path is realized through the high-voltage ESD clamp to the power ground, followed by the ESD diode between the BOOST and ground. This current path is also reversible.

Finally, the third, and generally irreversible, current path is provided by the high-voltage PMOS and other stacked components of the regulator (for example, the NPN diode shown in Fig. 1.17) directly to BOOST pin. If this current path is dominant, the PMOS is exposed to failure. This current path explains the scenario demonstrated by Fig. 1.16a, b.

At low level of array coupling and a small ESD protection window, ESD failures are related to the internal voltage regulators and they require product-specific measures to overcome the triggering voltage variation in the HV ESD protection VIN clamp, dependence of the NLDMOS array self-protection capabilities upon layout and internal circuit driver design, variation of the pulsed SOA of high-voltage PMOS and NLDMOS devices.

The specific way to improve ESD protection depends on the dominant ESD current path realized in the particular product circuit. Perhaps, the first, most logical solution is to improve ESD protection by reducing the VIN clamping voltage to a lower level, thus limiting the voltage drop on the PMOS device. However, this adjustment of the parameters of the high-voltage ESD protection clamp might not be an option in case of relatively low safe-operating area (SOA) margins of the power-optimized components. In this case, the most robust way is to implement ESD oriented co-design of the VCC regulator, which provides a much higher pulsed absolute maximum voltage for the circuit pins. Design measures may include stacked or oversized components or application of less power-efficient components with higher voltage tolerance, if allowed by the process. To protect the low side driver an additional internal power clamp may be required.

1.2.3 Multiple Voltage Domains

When the analog circuit has different voltage domains, the ESD protection network can be constructed using several approaches that depend on circuit specifications. The most typical example of a multiple voltage domain circuit is in the combination of the digital, analog, and power domains. Since the ESD protection of a chip is required to withstand each pin-to-pin combination, all domains should be connected into a network.

Consider an example of analog and power domains. The inter-domain connection in the network is usually realized between the power and analog grounds. The simple reason for this arrangement is that the power supply voltages cannot be shared. To avoid cross-talk due to the potential drop on the ground busses, the grounds can be connected using a back-to-back diode clamp. This clamp decouples the ground bus interference on the voltage drop on the forward-bias diodes and, at the same time, will provide a low voltage drop at ESD current level (Fig. 1.18). In this case, the stress on the pins from the different domains will include the additional voltage drop on the diode. A similar approach is used in the case of a two-voltage domain with a distributed active clamp network.

The case of back-drive compatible pins is one of the examples of when a rail-based ESD protection network with diode clamps is not possible. The pin must be independent from power and thus the diode-to-ESD+ rail is not suitable.

An example of multiple voltage domains with level shifts used in LCD display applications is presented in Fig. 1.19. This application is designed to operate at a substrate voltage of -8 V with power supply pins of -8, -6, 0, 6, and 12 V.

The analog circuit is completed using 3.3 and 6 V isolated CMOS devices. The ESD protection network combines five blocks with separate ESD clamps: distributed active clamp (AC) blocks, with a clamp width of $W \sim 500$ μm, $Lg = 0.8$ μm, and an RC timer with $RC = 4$ μs; 6 V NMOS snapback clamp; dual-direction ESD devices for 3.6 V inputs and dual-direction ESD devices for 20 V inputs (level shifter).

The ± 6 V-tolerant input pin V_O is protected by the dual-direction DIAC ESD clamp (Fig. 1.19) or by the substrate-isolated stacked back-to-back snapback NMOS clamp. The parasitic resistance and parasitic diodes in the internal circuit (Fig. 1.19, dotted lines) can provide the additional voltage drop and the alternative current path, respectively. These parasitic components of the network should be accounted for.

1.3 Off-Chip ESD Protection Strategies

Passing the component-level low energy ESD (CDM, MM, HBM) standard ESD tests do not guarantee withstanding of high energy ESD transients introduced by the system-level ESD pulses up to 30 kV in the field application that could impact real products reliability. The most common test methodology to validate the ESD

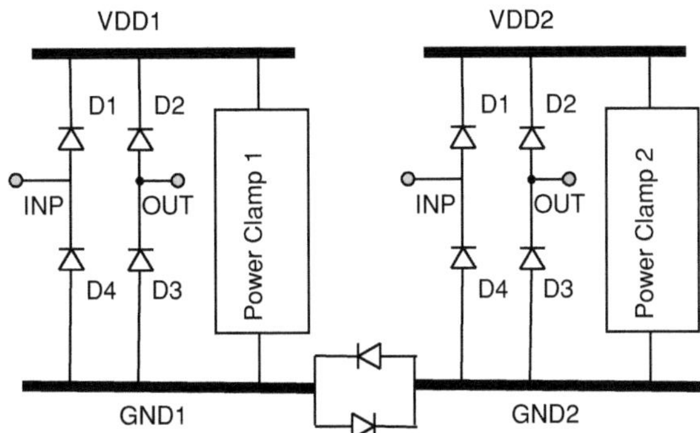

Fig. 1.18 ESD protection network for multiple voltage domains with the antiparallel diode clamp

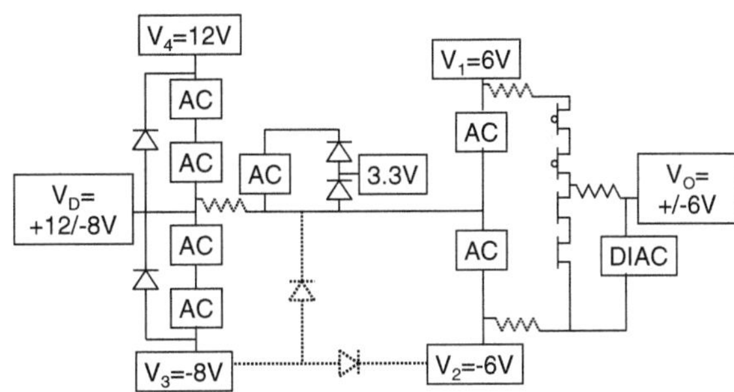

Fig. 1.19 ESD protection network for LCD driver using DIAC protection for output

immunity (susceptibility) of the products is IEC 61000-4-2 [8]. It defines the ESD pulse waveform, voltage/current values, test conditions and procedures. The standard is discussed in detail in the following Chap. 2.

The trend towards smaller form factors of the systems is propagated to the IC product specifications and further supported by scaling of the semiconductor processes down to the advanced technology nodes both for active devices and interconnects. The implementation of the power optimized high-speed circuits that brings overall more susceptibility and damages at already low ESD stress levels. Under those conditions the on-chip protection design for system-level ESD stress becomes increasingly challenging. For many cases the off-chip protection becomes a more cost effective approach. A detailed off-chip protection overview is described in Chap. 5. In this introductory section some material towards an overall off-chip ESD protection strategy is provided to help further understanding of the on-chip and off-chip ESD protection aspects discussed in the following Chaps. 2 3 and 4.

1.3.1 Trend Toward High Level Integration: SoC and SiP

The trend towards high level of the system functional blocks integration on-chip resulted in removing the barrier between the components and systems. System-

Fig. 1.20 Illustration of building blocks in a system on a chip

on-chip (SoC) and system-in-package (SiP) designs now can combine a variety of analog and digital circuit blocks that directly interface with system ports. The trend is supported by emerging manufacturing options like the 2.5D and 3D packaging techniques. This creates a significant paradigm shift and in particular results in the definition of system-levels ESD passing specifications for selected IC pins.

This trend requires the implementation and integration of different ESD design strategies for RF, high-voltage analog, and low voltage digital circuits on the same die or in the same IC package. While the majority of the SoC or SiP IC pins are still required to pass only component-level ESD specifications, the presence of the pins with system-level requirements significantly impacts the overall ESD design strategy. This is mainly because of the possible interaction of the IC blocks.

Many of the SoCs are used in handheld and mobile devices like smart phones and tablet computer. Next to the central processing unit (CPU) and memory controller interfaces like USB and flash memory, the power management, display and sensor ICs and the radio for wireless communication are integrated in one IC (Fig. 1.20). The challenge of predicting the system-level ESD robustness of an IC before it is mounted in a system is combined with the trend to minimize the number of (off-chip) components in an application board. As a result IC suppliers are expected to deliver system-level ESD robustness without a detailed knowledge of the final application board, block or system design. At the same time the system designers expect a more robust component (IC)-level ESD protection to ensure higher system-level ESD robustness.

1.3.2 ESD Voltage Suppression

One of the major approaches for the off-chip protection is based on discrete ESD pulse clamping components that act similarly to the on-chip protection clamps. The off-chip protection strategy is significantly enhanced by the opportunity to use different PCB components in the ESD protection network. Thus, in general the off-chip ESD protection strategy is based on a similar to on-chip two- or multi-stage distributed networks. The primary port protection usually relies on active discrete

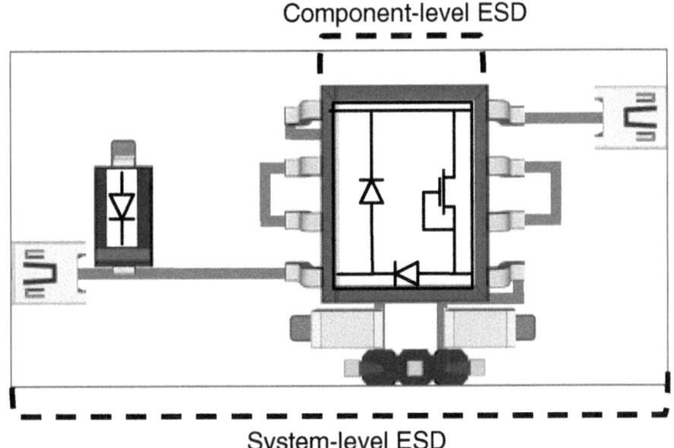

Fig. 1.21 Concept of system-level and component-level ESD protection design: example of an application board with IC, decoupling network, off-chip ESD protection and connectors

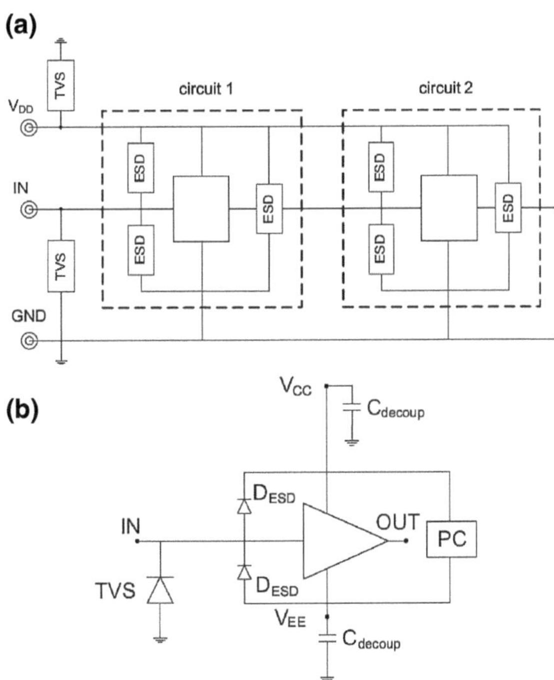

Fig. 1.22 Illustration of the system-level ESD protection design concept for TVS system-level off-chip ESD protection of two circuits (**a**) and a buffer amplifier (**b**)

ESD protection components, while the secondary network components are represented by either on PCB passive or on-chip active components (Fig. 1.21).

Off-chip protection can involve numerous options including isolation and filtering circuits and transient voltage suppressor (TVS) components. The variety of the suppression components includes multilayer varistors, silicon diodes or polymer-based suppressors.

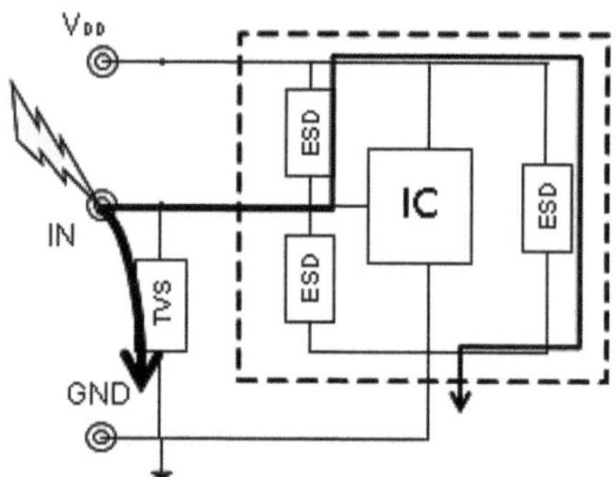

Fig. 1.23 Illustration for the current path in case of ESD stress at the system-level pin

System-level ESD protection is only implemented on the external IC pins that typically include input, output and supply pins of ICs which are connected to a system port. The internal pins usually do not require any additional off-chip ESD protection. The ESD protection concept for the external IC pins integrates at the system-level discrete off-chip ESD protection devices placed from the input and supply connector to ground for the *circuit 1*, while *circuit 2* remains not connected to any system port (Fig 1.22a).

An example of the system-level ESD protection of a buffer amplifier for RF applications is shown in Fig. 1.22b. It uses on-chip ESD protection diodes "DESD" a power clamp "PC" and an off-chip TVS. The RF input of the amplifier is protected on system-level with a TVS diode. The supply pins *VCC* positive and *VEE* negative are protected using the board-level supply network with decoupling capacitors. The amplifiers output pin is considered as an internal pin that requires no system-level ESD protection device.

Suppression components protect the circuit by clamping the ESD voltage to a safe level with respect to the protected circuit. Connected in parallel with the signal lines (Fig. 1.23), the suppressors clamp the ESD voltage and shunt the majority of the ESD current away from the protected chip to the appropriate reference bus.

The intrinsic electrical characteristics of a TVS that must be considered are not only the size, pin out, the leakage current, but must include the clamping voltage waveform. Depending on the TVS the voltage waveform at the internal IC pin can be substantially different and directly impact the passing level (Fig. 1.24).

Intentionally or unintentionally the end-result of the off-chip protection is related to operation of both the off-chip and on-chip ESD protection components in the same network. In the simplest case an additional isolation resistor between off-chip and on-chip protection part of network can be added to limit the current into the on-chip ESD protection (Fig 1.25a). A more efficient design example for multi-stage ESD protection combines TVS diode, decoupling capacitor (*Cx*) and ferrite bead (FB) (Fig. 1.25b). In this network the TVS component diverts to the ground only part of the ESD current while the passive components in the decoupling network act as a filter for the remaining residual current.

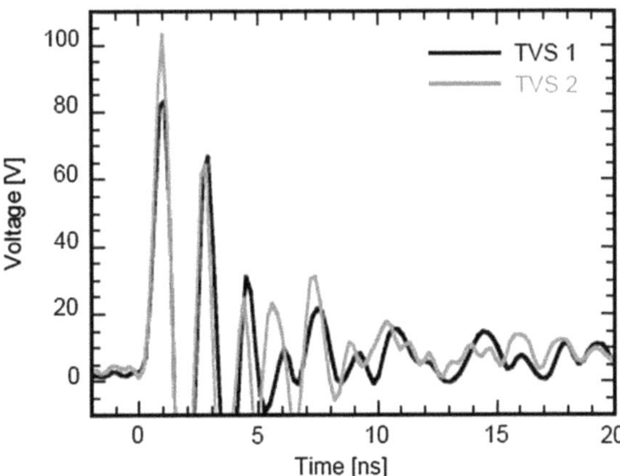

Fig. 1.24 Voltage waveform during IEC 61000-4-2 stress of two different TVS diodes

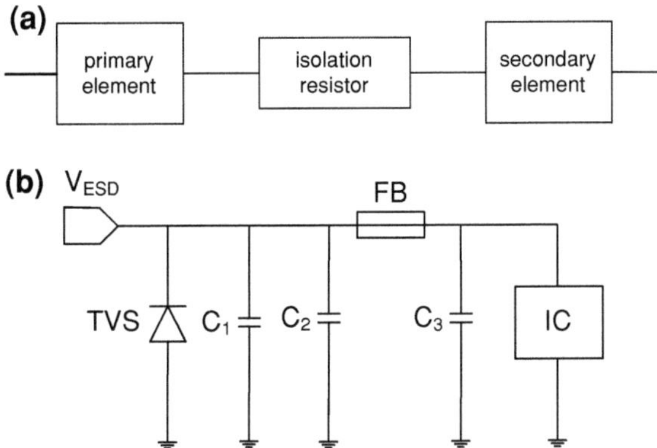

Fig. 1.25 Illustration of the multi-stage ESD protection concept for the off-chip with the second stage resistor (**a**) and ferrite bead (**b**)

From cost and performance perspective the number of ESD protection components on board of system-level must be reduced. Thus, a logical approach is the on-chip implementation of system-level ESD protection structures (Fig. 1.26). The trade-off between the off- and on- chip ESD protection strategies is related to the required on-chip area versus the cost of the PCB components assembly.

The latter is the major cost component since typically the cost of the discrete components itself is rather low. An ESD clamp with an intrinsic current capability of 5 mA/μm and passing 2 kV HBM in assumption of linear width scaling must be designed at least 13 times wider to be able to pass 8 kV IEC 61000-4-2 stress. In addition to the larger ESD clamp size also the metal routing needs to be improved to carry safely the high system-level ESD stress current to the protection device. The larger device and metal area goes along with a significant increase of the overall parasitic capacitance. This aspect is discussed more in the following section.

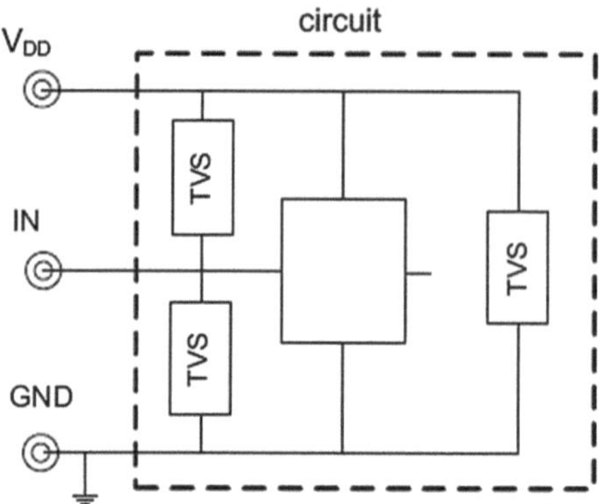

Fig. 1.26 Illustration for the implementation of system-level ESD protection on-chip

In more details the on-chip design for the system-level pins from the protection devices perspective is described in Chap. 3. The system-level ESD protection for external IC pins with on-chip and off-chip protection devices is addressed in the Chap. 5.

1.3.3 Capacitance and Signal Integrity

In the past ESD energy has been traditionally shunted away from the IC pins using devices with equivalent capacitances of ~ 10–100 pF. Due to the trend towards higher signal frequencies this parasitic capacitance level unfortunately introduces a significant signal distortion. At the same time modern high speed data ports can only tolerate a very limited increase of the line capacitance allocated for ESD protection devices.

Implementation of the integrated on-chip ESD protection devices with a pulsed current carrying capability of over 30 Amps and less than 1 pF equivalent capacitance often is simply incompatible with the integrated process technology. Indeed, to achieve a low equivalent capacitance the blocking junction requires the incorporation of a rather large depletion region. For example to achieve a system-level ESD robustness for an asymmetrical p-n junction with the 0.1 pF capacitance and 5 V breakdown, the diode requires an isolated n-region of ~ 8 μm with the doping level of ~ 1–15 cm^{-3} (Fig. 1.27).

This combination of parameters is hard to find in integrated processes, especially, if substrate isolation is required. Therefore for high data rate system-level protection the main off-chip protection approach includes a discrete off-chip suppressor with optimized characteristics based on its vertical device architecture.

For low signal frequencies the overall implementation of the protection with the suppressor can utilize a separation between the signal frequency domain and any undesired frequencies domains like EMI and ESD transient pulses. Thus the TVS capacitance still enables filtering by forming a low band pass filter with the PCB

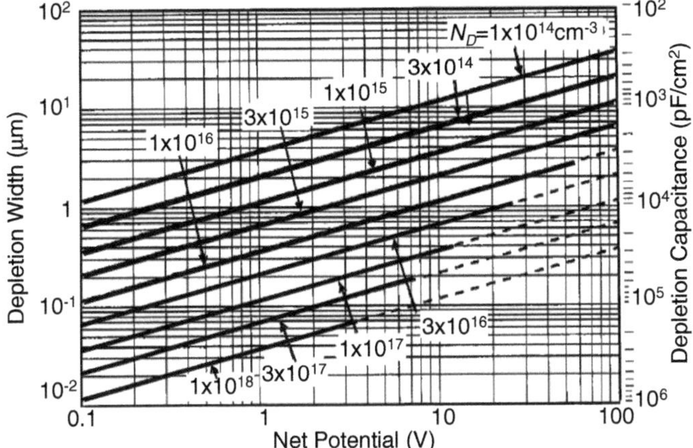

Fig. 1.27 Depletion-layer width and depletion-layer capacitance per unit area as a function of net potential for one-sided abrupt junctions in Si; doping N is from the lightly doped side; *dashed lines* represent breakdown conditions [9]

trace inductance. In this case the suppressor provides both the clamping functions for transient suppression and simultaneously EMI noise filtering against unwanted, high-frequency signals that couple into the protected data line. For example the headset terminals on a cell phone operate at relatively low audio frequency, while the ESD and the cell phone operating frequencies (0.8–1.9 GHz) are much higher. For these applications a high capacitance TVS can be a more appropriate choice to provide system-level ESD protection by the filtering out the radiated cell phone signals from the headset lines.

However, the advantages of low pass filtering are vanishing when the signal data rate is high. The high-speed data lines include the USB2.0, IEEE 1394, Gigabit Ethernet, and Infiniband protocols, etc. The data rates of all these protocols exceed 100 Mbps to guarantee a high throughput for video, audio, and data signals. The corresponding wide bandwidth of the transmission line cannot tolerate high TVS capacitances used for filtering, since the side-effect is a filtering of the data signals themselves or distortion of data waveforms causing system inoperability.

The distortion is realized in form of rounding leading and trailing edges of high/low state transitions due to slower rise and fall times. The slower rise/fall times introduce timing issues in the circuit functionality mixing the expected "high" and "low" states at the specific times. With the signal components degradation, the ability of the circuit to recognize the intended information is decreased making it much harder to maintain the data signal integrity. Therefore over recent years a number of low parasitic capacitance TVS components with the equivalent

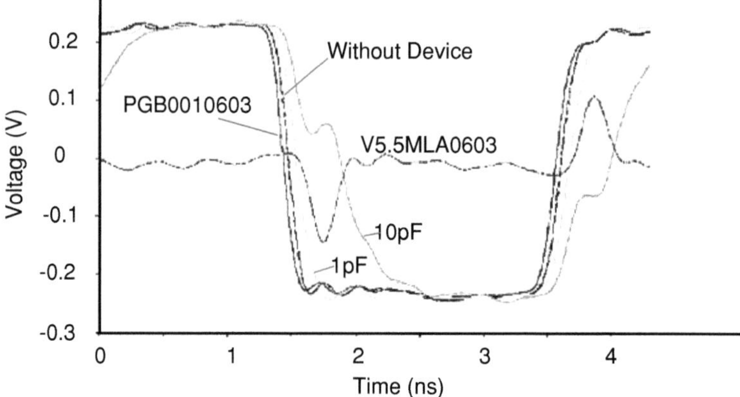

Fig. 1.28 Signal response waveforms of different devices and capacitors at USB 2.0 High Speed at peak to peak voltage amplitude 480 mV and signal frequency at 480 Mbit/s from data [10]

capacitance $\sim$0.1 pF have been developed and released by many industry leading TVS production companies.

An example of the signal distortion by the discrete active and passive components in case TVS devices and ceramic capacitors is presented in Fig. 1.28 for the suppressor with 0.05 pF, ML ceramic capacitor of 1 pF, ML ceramic capacitor of 10 pF and multilayer varistor of 660 pF [10].

For the 480 Mb/s data rate the 660 pF varistor capacitance causes a distortion that completely prevents the reaching of the signal operating voltage. Even the capacitance value of 10 pF is high enough to cause substantial distortion to the waveform, the decrease in the amount of level time and significant changes in the leading and trailing edge shapes. The capacitance value of 1 pF shows a small amount of edge distortion, while the 0.05 pF capacitance value allows the data waveform to pass without distortion. The rise time (10/90 %) data for each capacitance value for the bit rate of 480 Mbps demonstrate the corresponding effect [10]. For the measurement setup with either no devices or 0.05 pF Suppressor 10/90 % rise time was at 225 ps, while ceramic SMD capacitors of 1 and 10 pF impacted the rise time up to 275 and 526 ps, respectively. Thus the capacitance characteristic of an ESD suppressor is extremely important when protecting high speed data lines to maintain the integrity of high-speed data signals.

1.3.4 ESD Suppressor Considerations for Off-Chip Network

In addition to the two major factors, the clamping voltage waveforms and the capacitance realized in the off-chip ESD protection network with transient voltage suppressors (TVS), a number of other important considerations must be taken into account.

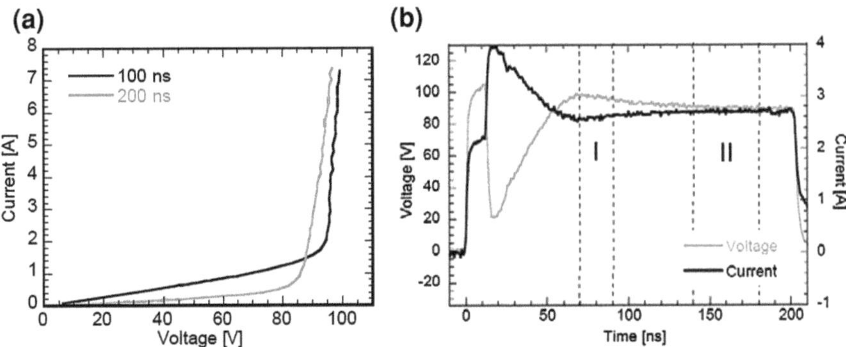

Fig. 1.29 TLP I-V characteristics (**a**) and 200 ns TLP voltage and current waveforms of varistor S10k40 for the stress level 120 V with marked measurements windows I and II (**b**)

Some aspects of TVS device design are discussed in Chap. 3 and their application for co-design in Chap. 5. The majority of TVS components can be subdivided into three categories: polymer, varistor, and semiconductor diode suppressors. The polymer suppressor diodes are based upon tunneling and breakdown conductivity modulation effects between conductive nanoparticles introduced into the polymer. They provide an advantage in extremely low parasitic capacitance, but cannot limit voltage at low level. The varistors (or voltage variable resistors) have different designs. Most of them are zinc oxide metal-oxide varistors. The varistor operation principle is based upon the combination of thermo-ionic emission and electron tunneling. The varistors are usually effective to protect against very strong ESD events—for example street lights against lightning strike. For the semiconductor systems the disadvantage of varistors is their high capacitance and low amount of zaps they can withstand. Similar to polymer suppressors the varistors clamping voltage is usually rather high.

For example varistors are made of metal electrodes and Zinc-oxide (ZnO) ceramic layers. ZnO grains on the ceramic layer form diode-like junctions which allow the flow of current only in one direction. During normal circuit operation the varistor is off. The capacitance of a varistor is typically rather high due to the large amount of junctions. The varistor starts conducting if during ESD stress the breakdown voltage of the varistor is exceeded. The current conduction spreads uniformly throughout the varistor which allows a very high robustness also to surge currents. According to TLP characteristics the positive feedback in the device does not provide S-shape I–V characteristics (Fig. 1.29a) [5].

The clamping voltage in the I–V characteristics of the varistor varies with the TLP stress duration is due relatively slow turn-on (Fig. 1.29b). After a TLP stress of 100 ns the varistor is not reaching the quasi-static state. The large capacitance varistors have a rather poor long-term reliability. After a few ESD zaps varistors degradation can be noticed in form of an increased off-state leakage current.

Another alternative solution—spark gap is usually composed by two electrodes with a certain distance to each other. This "gap" is filled with air or another gas that allows sparking between the two electrodes. When the (usually very high) breakdown voltage is exceeded, the gas in between the electrodes becomes ionized creating an arc discharge with low resistance. In addition to the high breakdown voltage the spark gap needs a certain time for the ionization of the air or gas used. This delays the breakdown or triggering of the spark gap. Ones triggered, spark gaps show a "snapback-like" behavior with a very low clamping voltage. The use of spark gaps and varistors can be justified mainly for high voltage application with several hundreds of volts tolerance.

Similar characteristics are delivered by polymer voltage suppressor (PVS) devices. The principle design is somewhat similar to a spark gap with the discharge realization in a polymeric material instead of air. The polymer material has a very low dielectric constant which results in a capacitance below 200 fF for a device in SMD0603 form factor. However the main drawback is related to high triggering and clamping voltages above 100 V.

Passive board-level components were historically used as primary current path components in system-level ESD protection. Two commonly used devices are capacitors and ferrite beads. Both devices are able to either conduct or block the ESD current by filtering.

A capacitor in parallel to an IC pin conducts large part of the ESD current under very low clamping voltage. However, all discrete capacitors have metal electrodes and pins with a parasitic inductance. An inductance of a few nH may be present depending on the length of the connector pins and the design of the capacitor case. The capacitor parasitic can cause an unwanted voltage drop across the ESD device during the fast rising edge of ESD currents. These voltage peaks must be taken into account when using a capacitor as an ESD protection device in parallel to an IC.

Ferrite beads are passive devices that represent a functionality similar to a band pass filter. Depending on the ferrite bead model, either the lower frequency or the higher frequency components of the ESD current are filtered. Ferrite beads are used in series to the protected pin. Depending on the model, a ferrite bead can suffer from saturation at large stress currents thus limiting its suitability for ESD protection.

Table 1.3 summarizes the electrical parameters of the devices that can be used for off-chip system-level ESD protection with the parameters: R_{ON}: on-resistance, V_{BD}: DC breakdown voltage (at current $I = 1$ mA), V_{clamp}: clamping voltage after 30 ns @ 8 kV IEC 61000-4-2, t_{ON}: turn-on time, C: at 1 MHz. TVS diodes trigger at low breakdown voltages and are available with comparable low parasitic capacitance. This makes them the most suitable candidates for the system-level ESD protection for low-voltage and high-speed applications. For analog and high-voltage applications high-breakdown TVS diodes, varistors and capacitors represent the most suitable off-chip protection devices.

When a TVS is chosen for appropriate suppression level and other electrical characteristics including a good match with the system circuit parameters, the remaining design steps include the choice of the on-board TVS location for most

Table 1.3 Overview on electrical parameter of common system-level ESD protection devices

	R_{ON} (Ω)	V_{BD} (V)	V_{clamp} (V)	C (pF)	t_{ON} (ns)
TVS diodes	<1	>3	>10	>2	<1
	1 … 1.5	>3	>10	<2	<1
Varistors	>20	>30	>100	>2	<40
	>20	>50	>200	<2	<40
Polymers	<1	>250	>200	<0.1	>5
Capacitors	≪1			Full range	<1

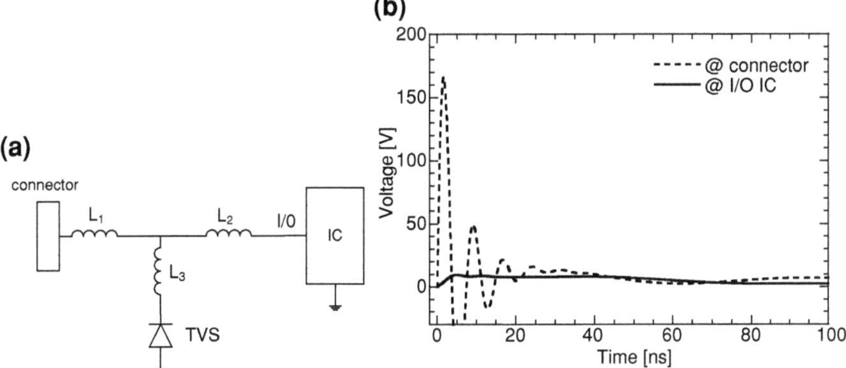

Fig. 1.30 Simplified block diagram for protection with TVS on PCB (**a**) and corresponding voltage waveforms measured at TVS and I/O nodes (**b**)

optimal performance to take advantage of the parasitic inductance of the PCB traces.

Similar to capacitance, low frequency signals will be unaffected by the inductance that is presented by the board traces. However, at high frequencies, the inductance will be present as an impedance component that can affect signal integrity: $R_L = \omega L$. Even a relatively short trace inductance can provide rather high impedance if the frequency is high enough. Therefore the more distance is between the ESD suppressor and the protected chip pin on PCB the lower is the voltage at the pin. In the diagram (Fig. 1.30a) the inductance *L1* physically represents a connection between the port and the ESD suppressor; *L2*—between the ESD suppressor and the I/O pin of the chip, *L3*—between the I/O line and the ESD suppressor. This strategy provides a significant suppression of the voltage waveform at the IC pin (Fig. 1.30b).

The inductance *L2* provides attenuation of the ESD pulse voltage and current as the energy is stored and dissipated in the electromagnetic field around the board trace. The attenuation of the ESD pulse propagates down the board trace. Thus the TVS placement location is favorable to be installed at a connector which is usually one entry point for ESD transients.

In the case (Fig. 1.30) a TVS is chosen with a rather high voltage overshoot and clamping voltage. The waveform at the suppressor for the 1 kV TLP pulse at the connector has the measured peak voltage ~ 350 V and the holding voltage of 75 V. It is clear that a voltage of 650 V at a PCB trace may create a substantial coupling at high frequency. If a 3-inch long trace ($L2$) is connected to the ESD suppressor site and the input pad for the IC the attenuation of the overshoot peak voltage is ~ 6 times and "clamping" voltage is reduced from ~ 60 V to approximately 25 V. The reduction corresponds to the frequency dependent impedance of the PCB line $L2$.

By increasing the trace length between the ESD suppressor and the chip the stress at the IC pin can be dramatically reduced. The ESD suppressor should be located directly behind the connector. It should be the first board-level component that the ESD transient encounters.

Extrapolating this empirical analysis the optimal location can be derived as follows:

(i) Inside connectors to the system shielding (chassis),
(ii) At the point where circuit board traces interact with the connector pins,
(iii) On the circuit board immediately behind the connector,
(iv) On robust, unprotected lines that may efficiently couple to I/O lines,
(v) Before a series resistive element on a data line,
(vi) Before a fan-out point on a data line,
(vii) Near the IC pin.

Another placement consideration is the distance from the board data signal trace to the TVS itself represented by the inductance $L3$ (Fig. 1.29). This trace inductance should be minimized to reduce the voltage drop in the signal line cross-point of the circuit where all inductors are connected. If the $L3$ is high the TVS becomes isolated from the signal line by the inductance impedance. Thus the ideal placement for the TVS solder pad is on top of the data line trace.

Respectively the chassis or frame ground is expected to work as ESDMINUS bus. To avoid the coupling the ESD stress signal with the data signal two different current paths implemented by design is the most optimal approach. By referencing the TVS to the chassis ground, unintentional noise effects and ground bounce can be reduced to maintain signal integrity.

The most useful suppressors that started to dominate over recent year for portable semiconductor systems are based on Si material. The design is somewhat similar to other on-chip ESD devices combining the precise breakdown characteristics and high performance with multiport integration. However most of the design advantages can be taken from the discrete approach using vertical device architecture supported by the dedicated process technology.

The comparison of the ESD protection devices using only data sheet parameters with ESD level ratings and cost is not always optimal. For example, device "A" may have in its datasheet a level of IEC 61000-4-2 robustness ~ 5 kV, while device "B" can guarantee ~ 20 kV. However this range represents only the level of survivability of the TVS devices rather than a diode clamping waveform characteristic. If device "A" provides a more appropriate ESD operation waveform the system design can be substantially better rather than when using device "B". At the same PCB trace connection the waveforms realized at the TVS connection will determine the magnitude of the secondary current through the IC pin.

Unless provided in the datasheet, the clamping voltage and residual current estimation in the ESD time domain may not be simple. The clamping voltage parameter in the suppressor datasheets, if it is present at all, might be misleading, since it can only indicate the clamping voltage after the voltage overshoot at the beginning of the ESD pulse. At the same time the residual current can be only calculated based on the PCB trace design.

Polymer based TVS seem to be attractive for high-frequency applications due to their low capacitance of 0.05–1.0 pF. However, they usually have triggering voltages that are significantly higher than the clamping voltage. A typical polymeric suppressor breakdown is around 300–500 V with a clamping voltage after the snap back of up to 150 V. In addition polymer suppressor may have a very long recovery time up to a few hours of even a day to transit to the high-impedance state after the ESD pulse.

Varistor type suppressors are low cost, but have too high triggering and clamping voltages, as well as a high resistance. Typical low capacitance varistor TVS have the clamping voltages range 150–500 V and a dynamic resistance of above 20 Ω. Another side effect is the possible degradation with multiple stresses. Most varistors can reversibly operate for 10–20 zaps only.

The most optimal method used for systems ESD protection is based upon the semiconductor TVS diodes. ESD protection diodes are characterized by low clamping voltages, low resistance, fast turn-on times, and good ESD reliability. In general, semiconductor diodes offer the best ESD protection, and now available with capacitances below 0.1–1 pF to guarantee a good signal integrity.

Current industry practice is to publish clamping voltages based on a pulse with an 8 μs rise time and duration of 20 μs. This is the specification for the power surge rather than ESD event. Most datasheets document clamping voltage using a 1 A pulse and sometimes a higher current pulse as well. This pulse is not equivalent to a fast transient ESD system level pulse with below 1 ns rise time and duration of 100–150 ns. The TVS clamping voltage during a level 4 IEC 61000-4-2 pulse with a peak current of $\sim$30 Amps is in general different from the microsecond time domain at 1–3 A surge current provided in the TVS datasheet. Thus the IC-system co-design methodologies involve additional pulsed characterization (Chap. 5)

In general, semiconductor diodes deliver the lowest peak clamping voltages, while suppressors and polymers have significantly higher clamping voltage characteristics. Typically the low voltage semiconductor TVS diodes are rated to clamp the voltage at the level of 8–15 V range for 8 kV IEC 61000-4-2 stress. This is achieved in significant contrast with varistor and polymer suppressors that can have clamping voltages in the range of 150–500 V. Consequently the secondary current through the IC pin depends on the dynamic resistance of the entire off-chip network with the voltage limited by a TVS placed at the system port.

There are various options available for ESD protection with Si suppressor devices: SCR diodes, TVS Avalanche diodes and punch-through BJT devices. The board-level ESD protection design varies from system to system. Factors that influence the design include board layout, the ESD capabilities of the IC, and the physical ability of ESD transients to get on the data lines. Empirical testing can also be done to determine the system's susceptibility. If it is determined that a supplemental ESD protection is desired, the next step is to identify the appropriate suppressor. There are many characteristics that should be considered including: capacitance, peak and clamping level, leakage current and the system operating voltage.

Another consideration is the number of lines to be protected. This is determined by the system's data protocol. For example, USB buses have two data lines, RS 485 uses two lines per differential pair, 10/100BaseT Ethernet uses four lines, etc. In cases where multiple data lines are protected, it may be desirable to use a multi-port suppressor to save board space and assembling costs.

Recognizing that a radically new approach is required to meet today's ESD requirements a number of companies has introduced the new Si TVS architecture getting significantly better ESD performance than traditional off-chip ESD protection devices.

1.4 Network Simulation with ESD Compact Models

A significant progress in the predictability of system-level ESD design can be achieved by circuit simulation when ESD compact models are available for the on-chip and off-chip ESD devices. Typically ESD compact models are not provided in a standard process design kit. However they can be developed using standard device models and sub-circuit components to model snapback regimes with avalanche breakdown and injection.

The sub-circuit components for ESD compact models include parasitic BJT structures, an avalanche current source, and saturation resistors. Development of the accurate snapback model is generally not an easy task. Due to complicated extraction methods for ESD current conditions, the construction of the scalable model requires advanced expertise in the field. One of the most helpful ways of developing a snapback model is to use the physical process and device simulation to generate the reference data for model extraction. Such an approach provides a significant acceleration of the extraction procedures under reasonable accuracy.

The generated compact models can be used for both ESD and functional simulations of large circuits, using standard circuit simulation tools such as HSPICE and Spectre, as well as for mixed-mode simulators, for example using DECIMM^{TM} (see next section).

A number of original studies have been done to support the snapback model development [11–15] including the snapback ESD compact model for a low-voltage NMOS [15]. This section presents an understanding of a general approach for custom ESD compact development principles to support the on-chip and off-chip ESD protection co-design. The following section is addressing the same goal based on the alternative approach of mixed-mode device-circuit simulation.

1.4.1 ESD Compact Model for LV Devices

The simplest ESD compact model example is represented by the snapback NMOS device. The cross-section and the equivalent circuit representing the low-voltage NMOS device combines the standard NMOS represented by a non-snapback N-MOSFET circuit model, calibrated for a particular technology node, the parasitic n-p-n BJT represented by the transistor compact model and the avalanche current source (Fig. 1.31a). The bipolar device requires extraction of the current gain, forward ideality factor, reverse ideality factor, equivalent high-current collector resistor, and equivalent high-current emitter resistor [15].

The current-controlled voltage source *VRSUB* (Fig. 1.31b) represents the substrate resistance modulation analytically implemented as

$$V_{R_{SUB}} = \frac{I_{SUB}(L + Ls + SWS)^{alfa\_Rsub}}{a\_Rsub * W - b\_Rsub * I_{SUB}}, \tag{1.1}$$

where *L*, *Ls*, *SWS* and *W* are geometrical structure parameters (Fig. 1.31), and *a_Rsub*, *b_Rsub*, *alfa_Rsub* are extracted from the experimental data model parameters [15]. The avalanche breakdown current source component *Iavc* is described by:

$$I_{avc} = (M_b - 1)I_c + (M_{ch} - 1)I_{ch}, \tag{1.2}$$

where the current components I_c and I_{ch}, and M_b and M_{ch} (Fig. 1.31) are the avalanche multiplication factors due to the bulk and channel regions with a high electric field in the vicinity of the drain-substrate junction depletion area.

The structure of the snapback PMOS model is essentially the same as that of the NMOS model above, with the changes from NMOS to PMOS and NPN BJT to PNP BJT.

The similar snapback NMOS module can be further reused in the compact model for the LVTSCR ESD device. To provide SCR-type positive feedback on the device level additional p-n-p BJT device is added in the sub-circuit (Fig. 1.32). Together with already included in the circuit parasitic NPN, this circuit creates a typical Ebers equivalent of an SCR. Additional components are the resistors: *Rbpnp* to control the base resistance, *Repnp* to control the emitter resistance, and *Rcpnp* to control the collector resistance of the effective PNP device. The remaining components represent the part of the Pwell current (and associated potential) that does not contribute to the operation of the NPN device and has a rather large value.

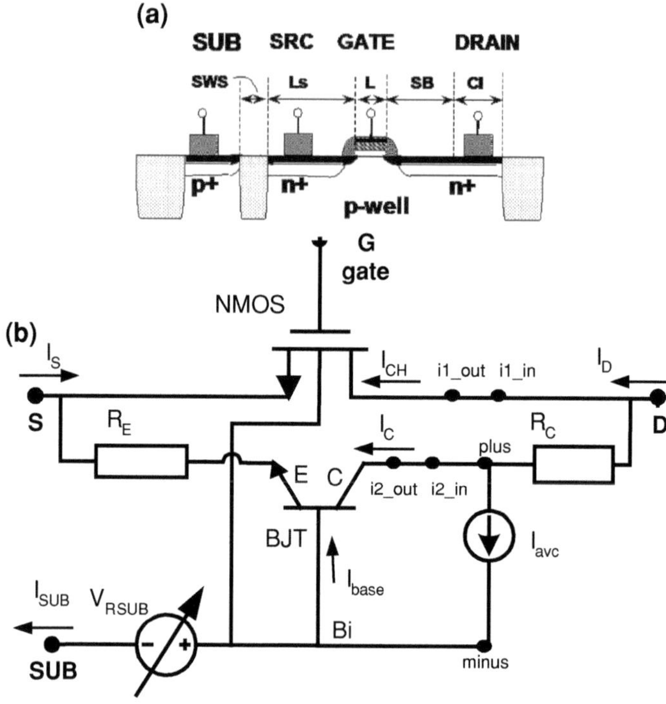

Fig. 1.31 Cross-section and ESD layout parameters (**a**) and the equivalent circuit (**b**) for the compact model with the avalanche current source AHDL of the snapback NMOS device

Fig. 1.32 Equivalent circuit for an LVTSCR compact model

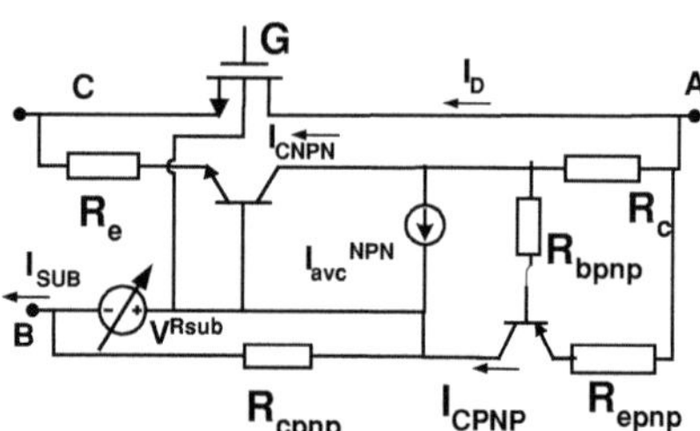

1.4.2 ESD Compact Model for HV Devices

ESD compact models for extended drain high voltage devices can be composed using the same sub-circuit approach. In case of NLDMOS (Fig. 1.33a) and NLDMOS-SCR (Fig. 1.33b) devices the extraction of several important parameters can be made using TLP I-V characteristics at different gate bias (Fig. 1.33c, d) [16]. The equivalent circuits of the snapback compact models for the NLDMOS

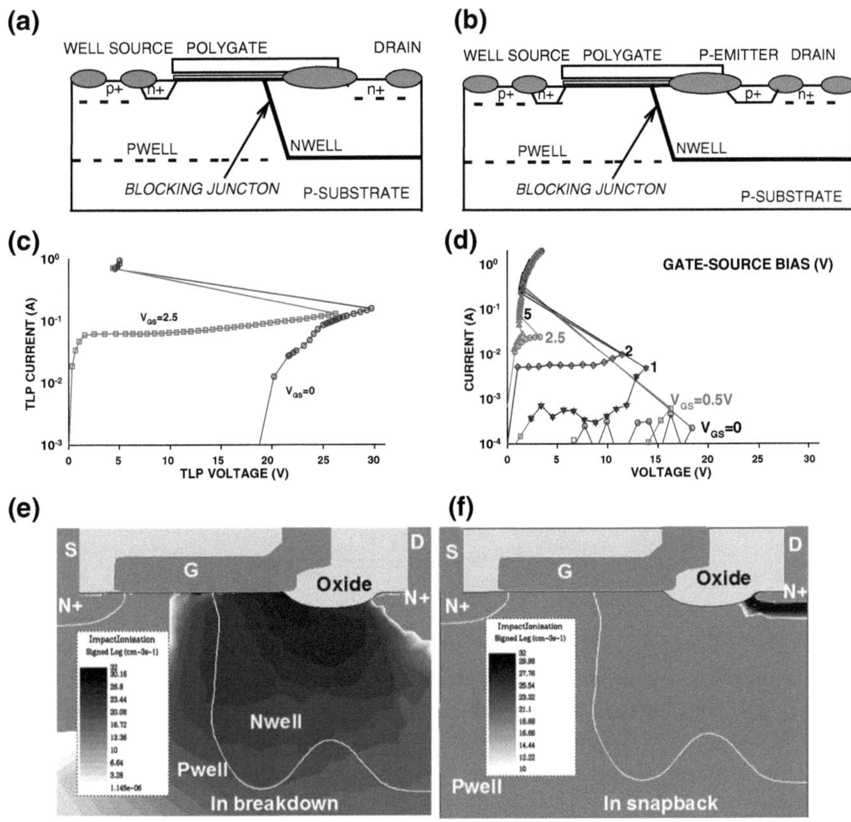

Fig. 1.33 Simplified cross-section of 12 V NLDNMOS (**a**) and 12 V/20 V NLDMOS-SCR (**b**) devices and corresponding measured TLP characteristics ($\tau_P = 100$ ns, $\tau_R = 10$ ns) at different gate bias for W = 200 μm (**c**, **d**). Simulation of the Kirk effect inside the NLDMOS device as a change of impact ionization region localization for the condition of the avalanche breakdown (**e**) and the high current in snapback mode (**f**)

and NLDMOS-SCR are similar to low voltage counterpart NMOS and NLVTSCR devices with one major exception.

For the NLDMOS device at high drain-source bias, the breakdown is initiated in the depletion zone of the N_{WELL}-P_{WELL} junction, as illustrated in the TCAD simulation shown in Fig. 1.33e, f. Increasing the drain bias causes a significant voltage drop across the low-doped N_{WELL} drain region, since it is depleted of carriers. This is seen in the I–V characteristics after junction breakdown. When the high-field depletion region expands with increasing drain bias, it touches the highly doped $N+$ drain region. The conductivity of the N_{WELL} region becomes completely over-modulated by the generated electrons and holes. This results in a shift of the maximum electric field (and impact ionization generation) toward the $N+/N_{WELL}$ interface (Fig. 1.33f). Often, this effect is attributed to the Kirk or base-push out effect [9]. In this regime, the parasitic bipolar device is activated, triggering the structure into snapback.

This behavior of the device has been used to develop circuit models for NLDMOS and NLDMOS-SCR structures operating in ESD conditions [16]. Initially, at low drain bias, the equivalent bipolar device and avalanche current source I_{avc}^{BJT}, representing the impact ionization at the $N+/N_{WELL}$ junction, are not active. The I_{avc}^{Nw} avalanche current source describes the breakdown of the $N_{WELL}- P_{WELL}$ junction and is expressed as $I_{avc}^{Nw}= M_{ch}I_S$, where M_{ch} is the multiplication factor for the channel current. It is given by the standard model

$$M_{ch} = \frac{1}{1 - \left(\frac{V_D}{V_{BR}NW}\right)^n} - 1, \tag{1.3}$$

where V^{Nw} represents the voltage drop across the carrier-modulated N_{WELL} region. It accounts for the modulation of N_{WELL} resistance with the increase of injected carriers and is modeled as

$$V^{NW} = \frac{I_{avc}NWl_w}{A_{eff}q\mu_n\left(N_d + \frac{I_{avc}NW}{A_{eff}q\upsilon_{sat}}\right)}. \tag{1.4}$$

In (1.4), l_w and A_{eff} represent the effective length and cross-section of current flow in the N_{WELL} region and are treated as fitting (extracted) parameters; q, μ_n, and v_{sat} are the electron charge, mobility, and saturation velocity, respectively while N_d is doping. A similar equation is used for modeling the voltage source V^{Rsub} that represents the increase of substrate potential in junction breakdown conditions.

The avalanche current source I_{avc}^{BJT} represents the shift of the avalanche region from the N_{WELL}-P_{WELL} to the $N+/N_{WELL}$ region, which leads to the activation of the parasitic bipolar structure. I_{avc}^{BJT} is described as $I_{avc}^{BJT}= M_{BJT}(kI_S+ I_C)$. M_{BJT} is described using a similar equation, as M_{CH} (with different breakdown parameters) and k are parameters used to control the gate coupling effect in snapback operation. Note that when I_{avc}^{BJT} is activated, I_{avc}^{Nw} consistently self-deactivates due to the bias V^{Nw}. This voltage drop reduces the effective bias across the $N_{WELL}-P_{WELL}$ junction and, correspondingly, the avalanche multiplication that generates $I_{avc}^{Nw.}$ Such model behavior is equivalent to the above-mentioned shift of the avalanche region inside the N_{WELL}.

A properly calibrated compact model can provide rather good correlation of both the TCAD I–V characteristics calibrated to the experimental data and the circuit simulation characteristics generated by the model [16]. The model also accurately represents the breakdown and trigger behavior for different structure dimensions.

Similarly to LVTSCR, the NLDMOS-SCR model can be constructed by attaching a PNP bipolar device to the NLDMOS snapback model, as shown in Fig. 1.34b. The equivalent emitter and base resistance R_{epnp} and R_{bpnp} are used to control the activation and the high-current operation of the equivalent PNP

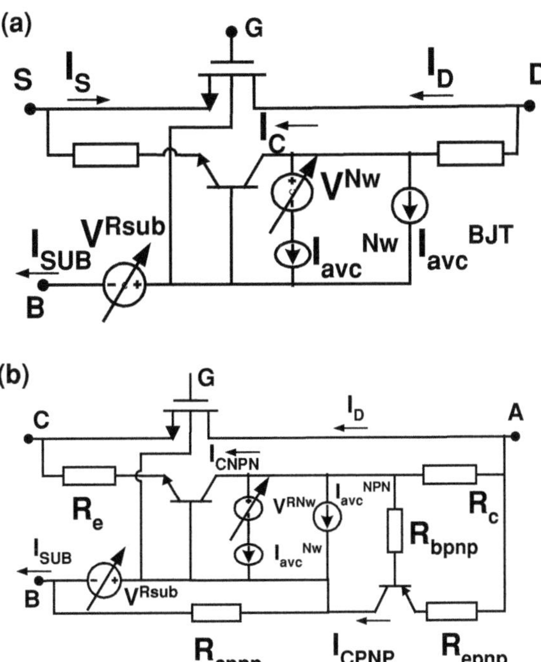

Fig. 1.34 Equivalent circuit of the snapback model circuits for NLDMOS (**a**) and NLDMOS-SCR (**b**)

transistor. R_{cpnp} represents the P_{WELL} resistance and substrate current flow that does not contribute to the operation of the NPN device.

An example of simulation with ESD compact models is represented by a simplified schematic (Fig. 1.35a) of an open drain driver circuit, which combines HV NLDMOS and NLDMOS-SCR structures. The NLDMOS device has an attached 30 kΩ gate resistor to mimic the dynamic coupling load from the driver circuit. In ESD conditions, depending on the circuit and structure parameters, the ESD current can discharge either through the NLDMOS-SCR ESD protection clamp or through the NLDMOS device. As can be seen from the comparative analysis of the waveforms (Fig. 1.35b, c), two different scenarios are realized. When R_{GATE} = 10 kΩ, under given circuit parameters for a 2 kV HBM, the gate coupling on the NLDMOS-SCR gate is sufficient to provide the early turn-on. As a result, the SCR clamp takes over the ESD current (Fig. 1.35b). On the contrary, in the case of insufficient gate coupling where R_{GATE} = 1 kΩ, the clamp does not turn on and the current path is formed through the 5 mm NLDMOS device (Fig. 1.35c).

Thus, depending on the circuit parameters and the HBM pulse amplitude, the critical regime can be determined as a condition of where the current changes direction in the turning-on circuit from the SCR clamp path to the NLDMOS path. If the current through NLDMOS is uniform, the high-current operation will be non-destructive. However, in a real 3D situation, the local snapback turn-on might result in irreversible failure [17].

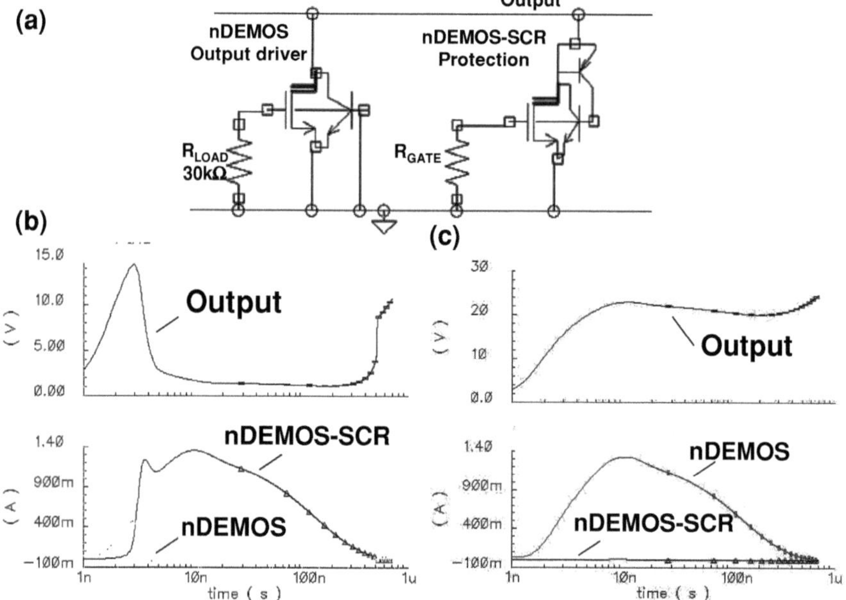

Fig. 1.35 Simplified schematic of ESD protection for open drain output driver (**a**) voltage current waveforms for two values of R_{GATE}—10 kΩ (**b**) and 1 kΩ (**c**), representing two possible circuit designs

A similar scenario of competition between the snapback clamp and array turn-on can result from the substrate potential effect. The substrate coupling technique is also one of the most useful methods in controlling ESD devices [18].

1.5 On-Chip ESD Design with Mixed-Mode Circuit Simulation

This section presents the key simulation-based methodology for industrial on-chip and off-chip system level ESD design. The examples of this methodology application are widely presented across the book chapters.

1.5.1 Industrial ESD Development Workflow with TCAD

Industrial ESD IP development for a new mixed-signal process technology is usually initiated simultaneously with the process specification definition. Typically, several learning cycles toward the physical development of ESD devices and

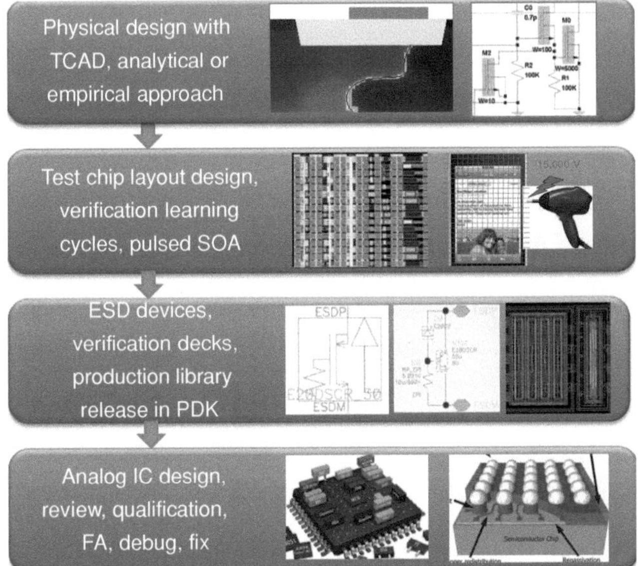

Fig. 1.36 Illustration of the industrial ESD development workflow for new technology process development

clamps are required and combined with an extensive test chip experimentation to target the desired parameters. When all the development objective choices are finalized, the ESD design and collateral are released together with the process development kit. The ESD IP typically combines layout, schematic and symbol views supported by the ESD library datasheet and design guidelines. In addition latch-up, IC-level ESD scheme and layout verification decks are often provided. At the process development and qualification states design and verification of the lead analog products are conducted. After this development period a number of custom ESD solutions are often added across the entire life span of the process technology, which targets additional optimization and specific product cases.

With each next step of this ESD process development workflow (Fig. 1.36), redesign involves a significantly elevated cost. The design of experiment choices and the changes in layout design of test chip structure are of a relatively low cost, the changes to the PDK library require substantially more resources, while a redesign and re-qualification of the IC product result in significant cost penalties. The incurred penalties are not just directly related to the resources consumed by the redesign. Often, the delay can impact time-to-market for the product, thus subtracting a significant fraction of the planned revenue. The delay in production, downward trends in overall sales price or sometimes even loss of the product to competitors may follow.

Therefore any proactive measures to improve the efficiency, verification and validation of the physical design solution become an important part of the

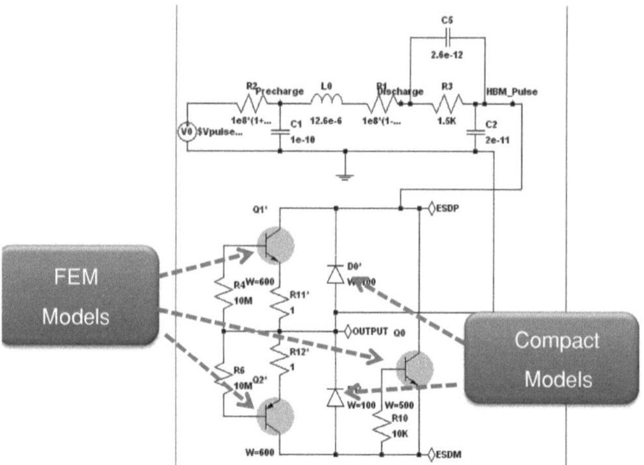

Fig. 1.37 Illustration of the mixed mode simulation approach with FEM devices

development approach. Among these measures a special role is dedicated to the numerical simulation due to virtually real time predictability and response on the development needs in comparison with experimental approach. The properly set numerical experiments allow the developer to target the design goals at reduced cost at all workflow stages (Fig. 1.36).

Successful design of ESD solutions requires a combination of conventional circuit operation with rather specific device operation transient conditions that include avalanche breakdown, strong injection and conductivity modulation.

To address these needs the mixed-mode simulation becomes the TCAD approach of choice. To realize this approach, at each time step, the simulator simultaneously solves a single matrix that represents both the circuit equations with compact models similarly to SPICE simulators and the carrier transport equations in the finite element model (FEM) semiconductor structures introduced into the circuit (Fig. 1.37). For active devices the non-ESD compact models devices can be used in case if they are operating in a small signal non-breakdown conditions. Simulation analysis with FEM devices that support accurate physical solutions for breakdown, injection and conductivity modulation is the main methodological breakthrough for ESD design problems. It eliminates the limitations or even the need of the ESD compact models development based upon empirical characteristics matching as described in the previous section.

After many years of industrial use the mixed-mode simulation approach appears to be logical and simple. However it presents several challenges. The primary challenge is in the generation of FEM devices for the mixed-mode simulation framework. A secondary objective is the simulation setup itself, which should be transparent and user-friendly for an involved ESD practitioner or analog circuit designer that cannot afford to dedicate a full time effort for TCAD analysis.

There are two known ways to generate FEM devices for mixed-mode TCAD numerical experiments. The first is based upon accurate physical process simulation steps with calibrated process flow if it is originally available as part of the process technology integration. The main process simulation steps involve deposition, implantation, diffusion, epitaxy and etch. The second is based upon an approximate analytical definition of the regions and diffused doping profiles of semiconductor device structure using an input file script.

1.5.2 New Approach with Parameterized Device and Process

The physical process simulation flow is usually supported by companies that develop new process technology as a standard industrial development procedure. This is done both by foundries and the companies with in-house fabrication. Unfortunately, overtime the level of calibration typically represents the process at the initial stage of development only. As soon as the process integration team receives more-or-less on-target silicon results, TCAD-based numerical experimentation is usually substituted by experimental data. As a result, the calibration of TCAD process flow is often not up to date with the most recently recorded variations. Another typical problem is a lack of consolidated process flow that has a flexibility to support all the standard devices for the process. Instead, often there are parallel TCAD process flows optimized and calibrated separately for different devices.

A more critical situation with calibrated process flow is in case of relatively small fabless IC development companies. It is practically impossible to obtain access to a calibrated TCAD process flow due to extreme confidential policies enforced by major foundries to protect their IP. At the same time, most foundries provide almost no relevant analog ESD IP. When needed for analog design such IP has to be developed by either fabless companies themselves using expensive TCAD-less experimentation on the shuttle test chip runs or acquired from external IP development parties. Thus, the traditional TCAD approach of generating FEM devices based on physical simulation is not directly applicable in this case. Ultimately it can be used only by combining the best guess approach for the process steps with SRP, SIMS profiles and TEM pictures obtained using external services.

The next challenge is the setup of and running the simulation. It is critical that the tool supports a user-friendly interactive GUI for circuits, parameterized FEM devices and data analysis. These features are not commercially available in traditional TCAD tools. Historically they are based on a few-decades-old university-produced code.

With traditional commercial TCAD tools, assuming a feasible use of the traditional TCAD approach, the simulation workflow can be visualized according to Fig. 1.38.

With physical process simulation approach to obtain a new version of FEM device after even one parameter change the process steps must be completely redone. Any new variations of the mask parameters require re-run of the entire

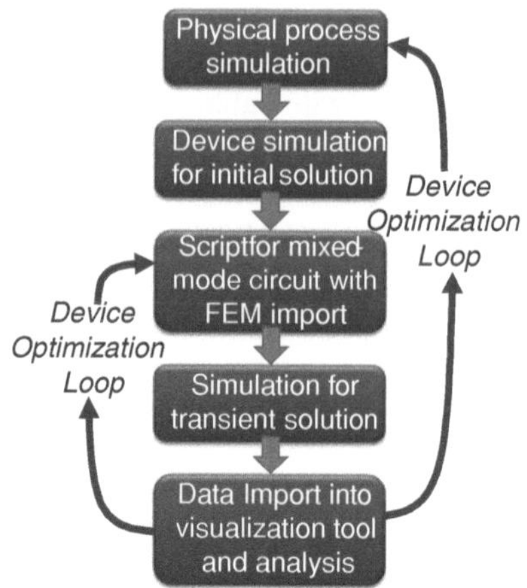

Fig. 1.38 Mixed-mode simulation workflow with traditional industrial TCAD tools

physical process deck again for deposition, etching, implantation, diffusion and other steps. If the change occur in the process recipe itself, then the calibration for the physical simulation steps is likely required too. This rather challenging task requires a substantial TCAD engineering effort to adjust the flow and introduce mask and mesh parameters especially for large HV analog devices.

After it is generated by the process simulator, the FEM device file is imported into the simulation (with corresponding conversion of the activated species to donors and acceptors) to be solved in a static regime to obtain the initial condition. The electrode conditions of this regime correspond to the initial conditions of this particular device in the circuit. At this step, the mesh usually needs to be redefined to eliminate unnecessary multiple nodes generated by the physical process simulation and to bring down the total mesh point count of the mixed mode circuit to a reasonable level. At the same time, the initial static solution for the mixed circuit is not always known and may require several iterations or even limit the type of problems that can be solved. Similar FEM files for the mesh and initial solution must be obtained for every different device type included in the mixed mode circuit, through tediously rerunning the entire physical process simulation flow (Fig. 1.38).

Finally the mixed-mode simulation input file script is rewritten to import the FEM files and DC solutions. This concludes the mixed-mode simulation setup for the currently popular old generation TCAD tools implementing the traditional simulation approach.

During the mixed-mode simulation, the transient solution for a mixed-mode circuit generates output files that need to be imported into another program for visualization and data analysis. As well, new device/circuit optimization loops require repeating the entire tedious workflow (Fig. 1.38).

Apparently, such a cumbersome and time-consuming approach is rather challenging even for full-time dedicated TCAD professionals. So, it is no wonder why TCAD simulation is not so popular among the majority of ESD engineers and

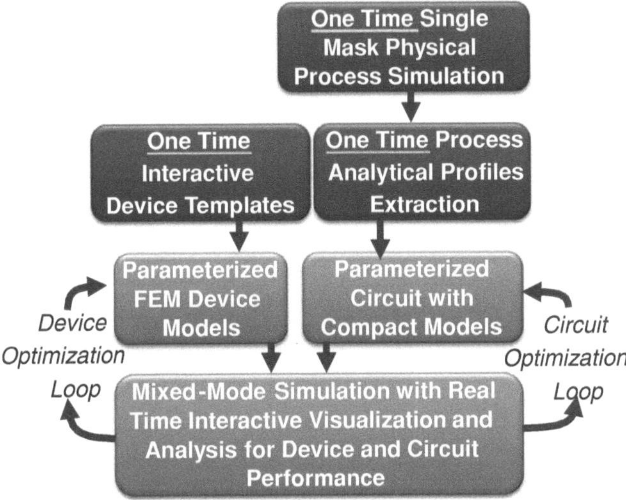

Fig. 1.39 Mixed-mode simulation workflow with DECIMM^{TM} [19]

analog IC designers, who desire rapid outcome and can afford to run TCAD only on part-time a casual basis. Thus, in spite of good accuracy of results obtained with a well-calibrated physical process simulation the methodology in general is impractical.

Fortunately for ESD engineers, a new tool DECIMM [19] has emerged over last few years. This tool completely replaces the physical process simulation by generating parameterized FEM devices in real-time during the mixed-mode simulation, using as input the diffused profiles and device templates. Similarly, fully parameterized circuits support mixed-mode simulation analysis, while real-time data extraction, visualization, and analysis are integrated in this single GUI simulator.

This simulation tool enables a revolutionary mixed-simulation workflow (Fig. 1.39) that combines procedures which must be accomplished for the entire process technology only once for each semiconductor device type. Thus, the mixed-mode analysis can be done for any circuit and FEM device parameters through a one-time input.

The first step of this approach is to obtain the diffused profiles for the given process. This can be done in several ways. For in-house processes, a single mask process simulation can be followed interactive extraction of the profile parameters supported by DECIMM tool. This is described below in more details. For foundries or in-house processes without physical process simulation flow data, alternative approaches consist in (i) receiving the profile information from the foundry; (ii) obtaining profiles from SIMS and SRP data by using external services and one-time test chips (iii) defining profiles manually using adjustments based on electrical results.

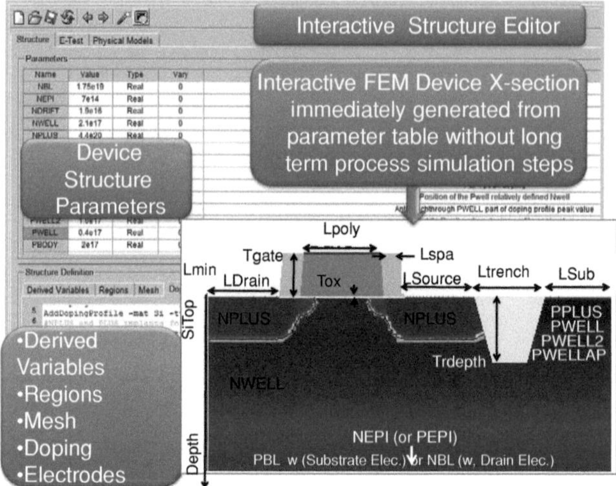

Fig. 1.40 Interactive device parameterization and real-time FEM generation in the DECIMM tool

This IP "investment" enables developers to simulate the entire process technology. Creation of interactive FEM parameterized templates and circuit is supported by the DECIMM tool. The transient mixed-mode solution combined with real-time data visualization and analysis enable optimization loops at both circuit and device level (Fig. 1.39).

A few more details of this approach are provided below. The parameterized device generation is interactively defined in the tool by providing a set of physical device parameters supported by the interface and script-derived variables for related template regions and doping profiles (Fig. 1.40). Device regions and electrodes are set using an interactive device editor. All device regions, doping, and even semiconductor material parameters can be changed in the mixed-mode simulation since FEM generation is done in real-time, rather than by imported mesh and solution files. The parameterized device is equivalent to a device generated by the physical process simulation. The procedure of interactive automatic fitting and extraction from a single mask process simulation cross-section is illustrated in Fig. 1.41.

In this example, a two-component Gaussian profile for the n-Well donor implant is obtained from the cutline after importing the single mask process simulation profile into the DECIMM tool. The two points on the cutline profile (Fig. 1.41b) define the fitting region. After running fitting iterations, the extraction results are shown in a fitted plot together with the set of extracted parameters for a two-peak analytical Gaussian profile (Fig. 1.41c). A table of diffused profiles is compiled with one-by-one cutline and extraction of the lateral and vertical profile parameters for each implant (Fig. 1.41d). This table shows a complete analytical representation of the process technology, combining the vertical Gaussian profiles

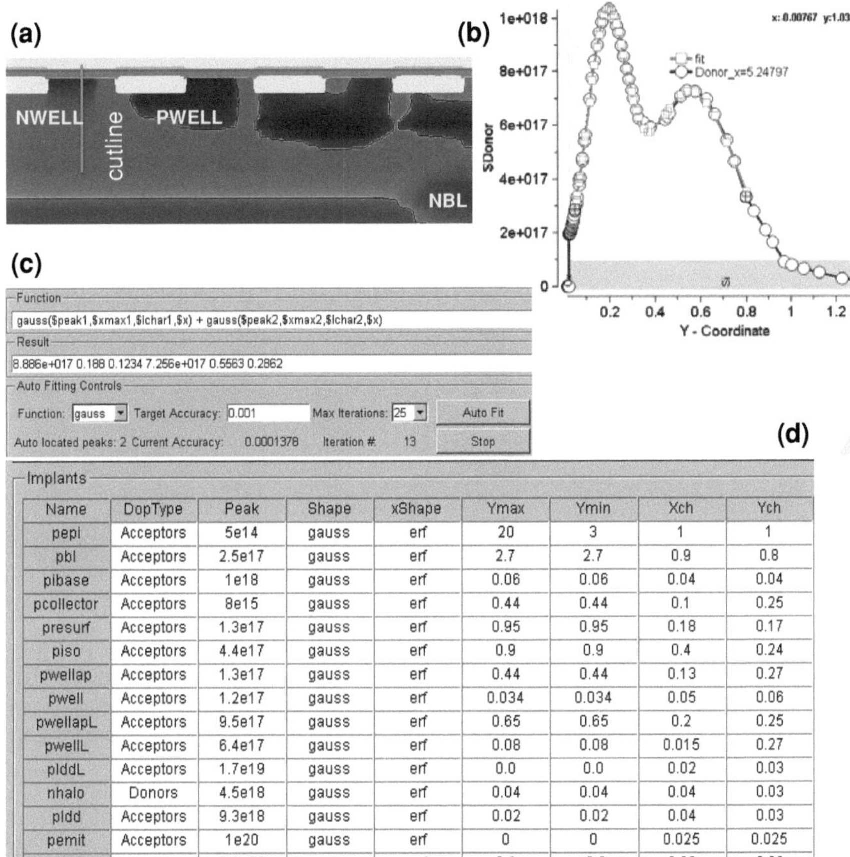

| Implants | | | | | | | | |
Name	DopType	Peak	Shape	xShape	Ymax	Ymin	Xch	Ych
pepi	Acceptors	5e14	gauss	erf	20	3	1	1
pbl	Acceptors	2.5e17	gauss	erf	2.7	2.7	0.9	0.8
pibase	Acceptors	1e18	gauss	erf	0.06	0.06	0.04	0.04
pcollector	Acceptors	8e15	gauss	erf	0.44	0.44	0.1	0.25
presurf	Acceptors	1.3e17	gauss	erf	0.95	0.95	0.18	0.17
piso	Acceptors	4.4e17	gauss	erf	0.9	0.9	0.4	0.24
pwellap	Acceptors	1.3e17	gauss	erf	0.44	0.44	0.13	0.27
pwell	Acceptors	1.2e17	gauss	erf	0.034	0.034	0.05	0.06
pwellapL	Acceptors	9.5e17	gauss	erf	0.65	0.65	0.2	0.25
pwellL	Acceptors	6.4e17	gauss	erf	0.08	0.08	0.015	0.27
plddL	Acceptors	1.7e19	gauss	erf	0.0	0.0	0.02	0.03
nhalo	Donors	4.5e18	gauss	erf	0.04	0.04	0.04	0.03
pldd	Acceptors	9.3e18	gauss	erf	0.02	0.02	0.04	0.03
pemit	Acceptors	1e20	gauss	erf	0	0	0.025	0.025

Fig. 1.41 Process simulation cross-section for single mask diffused profiles (**a**) and fitted analytical profiles (**b**) with extraction of implant parameters (**c**) and the process definition table (**d**)

for the implant and the error functions with the lateral profile parameters similarly extracted using horizontal cut lines (Fig. 1.41d). Unlike traditional process simulation, once the process parameterization implant table is obtained, any FEM device in the process can be generated in real-time using parameterized device templates.

Once the mixed-mode circuit is set, the simulation will run for any variation in physical devices or circuit parameters. An example of the TLP simulation for an NMOS device is presented in Fig. 1.42.

There are several experimentally verified theorems that support this new methodology [20]. A 2D FEM device with analytical profiles is equivalent to an FEM device from calibrated process flow within a desired accuracy of $\sim 5\ \%$. The results are simulator-independent when compared to traditional TCAD tools for the same imported process files and the same set of activated models. Appropriate

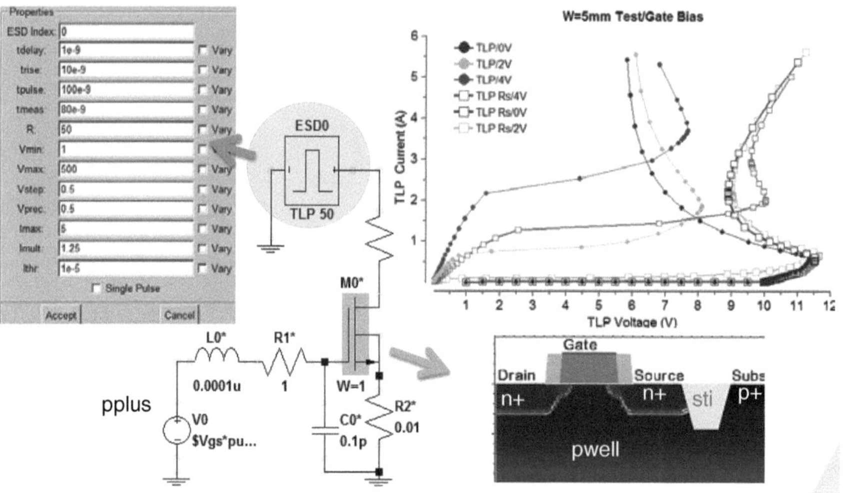

Fig. 1.42 Automated mixed-mode analysis: circuit with automated TLP pulse source demonstrates the tester parasitic effect for the source resistance

resultant implant profiles can be defined by a linear combination of diffused implant components using vertical Gaussian and lateral error functions. The parameterized FEM structure can be defined relative to real mask coordinates and device region boundaries—no mask bias is needed. Analytical diffused profiles extracted separately using the single mask approach can adequately generate FEM devices for the entire process.

So far, it has not been found that the user-friendly DECIMM tool [19] is indeed helpful in proceeding with simulation compared to the widely used empirical approach with its many costly re-design cycles. The material of the following chapters will include many examples of the DECIMM-based mixed analysis involved to provide an in-depth understanding of the subject.

1.6 Summary

There are many design aspects that should be taken into account during the challenge of a functional and reliable product design. The trends towards portable system design, higher data rates, faster signal speed, lower power consumption and lower operation voltages combined with high level SoC and SoP integration trigger an elevating complexity of the required design solutions, substantially based on innovation and novel approaches.

The demand on new age electronics creates a significant design paradigm shift both in on-chip system-level design with integrated system-level ESD protection devices and off-chip PCB design with new Si TVS solutions that deliver both low parasitic capacitance and more accurate clamping voltage waveforms in comparison with outdated polymer or varistor type TVS components. The transient voltage suppression components are now designed to combine a more precise transient voltage waveform, appropriate dynamic characteristics with a record low parasitic capacitance of ~ 0.1 pF to support the board design.

At the same time the IC product pins interfacing with the system-level ports experience the secondary ESD current stress partly dissipated by TVS and PCB network. Often those pins are required to pass certain levels of system ESD test under power-on and power-off conditions in order to provide a predictable second stage current conduction. Thus those pins need to be protected for an order of magnitude higher current level than standard component specifications CDM, MM, and HBM.

In contrary to the trend of lowering component-level ESD target levels from 2 kV HBM below 500 V HBM, the severity of ESD stress in the real-user environment is much higher. It impacts the reliability of the consumer products directly unless a system-level protection is implemented.

ESD "solutions" for the system are no longer a simple choice of a suppressor component to place at the system port. An effective solution requires application of a new design methodology that takes into account the layout of the circuit board, the pulsed electrical characteristics of the suppressor and ESD characteristics of the IC itself.

System-level protection strategies for the ESD protection network design were already preliminary discussed in Chap. 1. They will be further elaborated across the remaining chapters of this book. They need to involve an understanding of the test methods and the correlation factors for the on-chip components (Chap. 2). Engineering of the on-chip ESD protection devices (Chap. 3) and overall on-chip design including latch-up (Chap. 4) is an important part for the IC manufacturing and the IC specification. The off-chip and on-chip system-level co-design methodology (Chap. 5) becomes a logical step to address the expectation from the new era consumer, medical, automotive, industrial and other electronic applications.

Chapter 2
System Level Test Methods

In Chap. 1 the major trend toward SoC and SoP integration with system level pins was emphasized. This trend results in the design paradigm shift toward integration of the system level ESD protection capability on-chip. By providing the second stage ESD current capability the on-chip ESD protection can be both used for the IC-system co-design with the PCB components (Chap. 5) or provide a complete system level compliant pin protection.

To support this trend a significant gap between component and system level test methods and standards is bridged. The system level standards, for example IEC 61000-4-2, have been developed to support ESD and EMI compliance qualification of the systems rather than to validate the passing level of IC components. The accomplished system design in general significantly impacts the ESD test results and pulse waveforms. However not only the system blocks, but also the PCB design are usually not finalized or communicated to the IC developers and included in the initial specification of the component product.

Needless to point out that the ESD clamp solution development prior to the IC design itself creates a challenge on the predictive experimental validation. Thus both from technical realization perspective and from overall methodological point of view, the on-chip system level ESD design requires at least a good understanding of the test standards, procedures and their adaptation for the component level and on-wafer verification. The aspects of such understanding include the measurements of the quasi-static I-V characteristics, transient ESD pulse current and voltage waveforms, and the establishing of the correlation factors between particular device types, ESD protection capability and different pulse types.

This Chapter describes the physical aspects of the key test methodologies and their applications for the on-chip ESD system level design at each development stage. The focus is made on the explanation of the ESD gun test on board level, followed by the package and wafer level test methods towards an approach for a more effective on-chip design rather than just make a citation of the standard documents or to provide a reference guide.

The first section focuses on system level tests like the commonly used IEC 61000-4-2 and ISO 10605 standards. This material is followed by the key methodological approach of Human Metal Model (HMM) testing as the first

V. Vashchenko and M. Scholz, *System Level ESD Protection*,
DOI: 10.1007/978-3-319-03221-4_2, © Springer International Publishing Switzerland 2014

component-level emulation of system level ESD stress. The following section presents the transmission line pulse methodology as a very common tool for the on-chip design. The next section introduces the ESD waveform capturing as an essential approach for analysis and verification of the transient device and circuit characteristic. A discussion of the correlation factors for different pulses, devices and test conditions concludes this chapter.

2.1 Board Level Test Methodology

The purpose of ESD testing in the laboratory environment is both to emulate the real life ESD event and to verify the protection capability of a component or a system in order to comply with the corresponding IEC and ISO standards. In case of on-chip system level design the differentiation can be made between the printed circuit board (PCB) level test and the stand-alone component-level IC tests. The further expansion of the methodology is the on-wafer validation of either an IC or even standalone ESD structures from a test chip. As discussed in Chap. 1, the application of the component level standard pulses (HBM, MM and CDM) even with elevated pulse amplitude is not an adequate experimental and qualification approach to verify the system level ESD robustness. This is mainly due to the different ESD pulse waveforms, absence of power-on conditions and in general different ESD current paths through the IC. The component-level HBM, MM and CDM tests are performed to ensure the robustness of an integrated circuit or discrete component during IC and system manufacturing in the ESD protected environment. This is done under expectation that future designed systems or system blocks reliability will not be impacted during assembly and manufacturing rather than to add an ability to pass the system level ESD tests. Therefore system-level ESD qualification to obtain the certification is carried out at least for an equivalent of a system to ensure the functionality during operation, handling and maintenance.

2.1.1 General Electrical Equipment IEC 61000-4-2 Standard and Test Methodology

IEC 61000-4-2 [21] is the most commonly used standard released by the International Electrotechnical Commission (IEC). It defines electromagnetic compatibility (EMC) for test and measurement techniques for the electrostatic discharge immunity test. The document defines the corresponding system level ESD pulse waveform parameters, that should be delivered by ESD tester, and outlines the corresponding test methodologies. Since the standard is originally defined for system tests, it brings no straight forward understanding on how to apply it for the

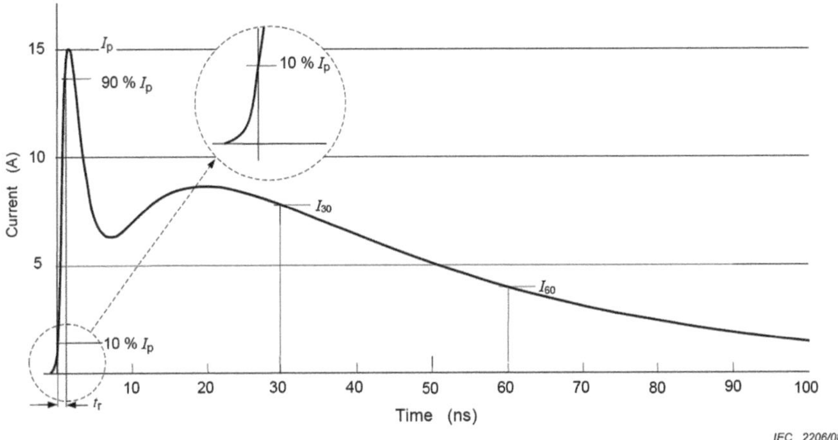

IEC 61000-4-2 ed.2.0 "Copyright © 2008 IEC Geneva, Switzerland. www.iec.ch"

Fig. 2.1 IEC 61000-4-2 Ideal contact discharge current waveform at 4 kV with defined parameters: I_p 3.75 A/kV ± 15 %, I_{30ns} 2 A/kV ± 30 %, I_{60ns} 1 A/kV ± 30 %, t_r rise time of 0.7 ns ± 25 %

verification of the protection capability of IC pins with system level ESD requirements. It is even more undefined how to apply the test for standalone on-chip ESD protection clamps.

The major difference between the component and system ESD pulse waveforms and the energy of the pulses was already compared in Chap. 1. The waveform of the ESD current supported by this standard (Fig. 2.1) physically represents an event of a discharge of a conductive object through a system port. The double-peak waveform (Fig. 2.1) represents the physical discharge produced by a conductive distributed object suddenly connected or approached to the discharge point. The first short peak corresponds to the discharge of the conductive peripheral region of the object in the immediate vicinity connected to the port. The second peak represents the discharge of the remaining body of the object. Assuming a constant object resistance, the periphery is quickly discharged with a short rise time and low resistance forming the first high amplitude peak. The remaining discharge of the object body has a bigger pulse propagation delay and higher current path impedance thus forming a longer, but smaller amplitude second peak.

The ratio between peak amplitudes and duration is defined by the standard itself. In the IEC 61000-4-2 short circuit contact discharge current waveform the amplitude of the first peak *Ip* is defined at $\sim$3.75 A/kV ± 15 % with rise time of 0.7 ns ± 25 %, while the current levels after 30 ns and 60 ns from the beginning of the pulse $I_{30ns} \sim$ 2 A/kV ± 30 % and $I_{60ns} \sim$ 1 A/kV ± 30 %, respectively. Thus in general according to the standard the pulse waveform has significant margins to be varied in different test setups.

The real life system level stress in uncontrolled environments has much higher energy than an ESD event in an ESD protected areas (EPA). However, the overall difference between the system and the component level standards are not only based in the much larger stress current level and the corresponding higher energy (Fig. 1.5). Another major differentiating factor for the system standard in comparison with the component level standards is that the tests are accomplished both under power-on and power-off conditions. This is done to reflect the real life applications where the system can experience an ESD event both under power-on and power-off conditions. Thus when a component is subjected to ESD stress the stress may result in a failure which causes irreversible changes in the device and interconnects.

Four possible system states are classified by IEC 61000-4-2 standard as a result of system-level ESD stress:

A   normal performance within given limits;
B   temporary loss of function or degradation of performance which ceases after the disturbance ceases, and from which the equipment under test recovers its normal performance, without operator intervention;
C   temporary loss of function or degradation of performance, the correction of which requires operator intervention;
D   loss of function or degradation of performance, no recovery possible.

The goal for most of the system designs is to pass the qualification test with the system state class A. Classes B and C represent so-called soft failures that are either self-restored or require a system reset. The classes B and C bring another major differentiation to component-level testing which correlates only the failure classes A and D.

Since the waveform is defined by the Standard, a number of testers have been manufactured by ESD equipment vendors to enable the test in laboratory or even in a field environment. In the standard the simplified discharge circuit uses a 150 pF capacitor and a resistance of 330 Ω to create the system-level ESD stress pulse (Fig. 2.2). Although the standard proposes the equivalent circuit (Fig. 2.2a) for the test setup, it is obvious that this simplified circuit cannot support double-peak standard waveform. The exact tester schematic is usually more complex to fit to the pulse waveforms. However the circuit (Fig. 2.2b) provides a relatively good waveform to meet the standard with the component parameters for example: $L_1 = 4.5\ \mu H$; $C_1 = pF$; $C_B = 20$ pF and $L_1 = 200$ nH.

The results of stand-alone IC test or packaged ESD component verification unlikely can be extrapolated to predict an arbitrary system passing level. Perhaps, a more realistic goal can be set to understand the aspects of the "worst case scenario", the impact of the test conditions and the test methodology. This is an important part of the IC-System co-design approach presented in Chap. 5. The critical step toward this goal is to build an understanding of the contact and air discharge methods in case of the bench test setup specific for the IEC 61000-4-2 standard with the test methodology.

Fig. 2.2 Simplified (**a**) and real tester (**b**) schematic of an IEC 61000-4-2 compatible discharge circuit, the L-C parasitic may vary in a real tester discharge circuit

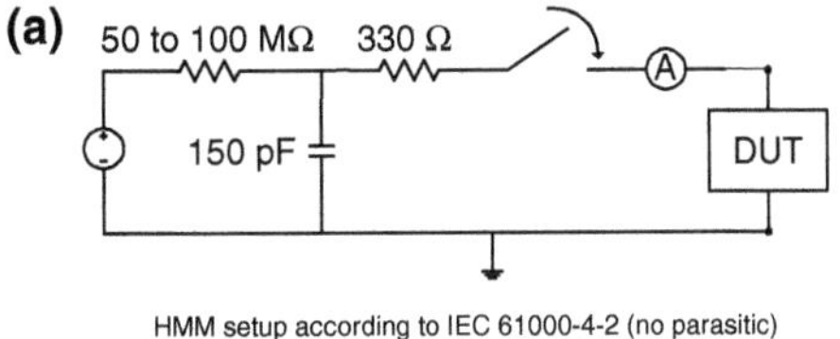

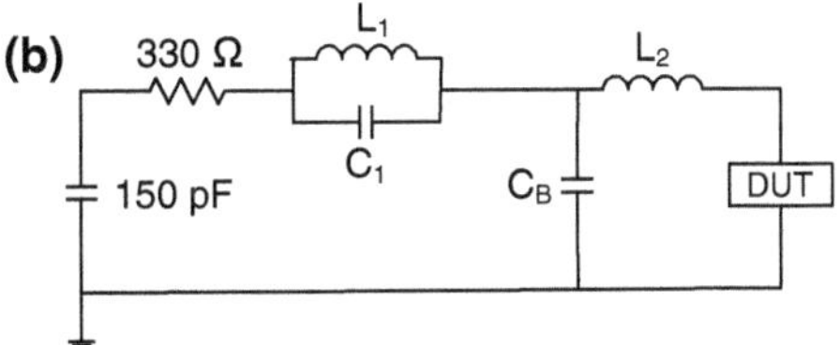

The ESD pulse current waveform as defined in the standard (Fig. 2.1) is mostly reproduced only during tester calibration with a dedicated target tool. In the majority of other test conditions the waveform shape is dependent on the ESD protection components used in the object under test, system or PCB design.

A significant deviation of the ideal standard waveform is observed for the air-gap test. Originally, both contact and air gap discharges are defined as a requirement when testing for IEC 61000-4-2 specifications with the preference of the contact discharge. However, typically by manufacturers the requirement for IC pins to pass a given level during both the contact and air discharges are specified. Air discharge is usually defined at a higher level than contact discharge. A common requirement for the level 4 of IEC 61000-4-2 specification is the passing of 8 kV contact and 15 kV air gap discharge. Thus to avoid "surprises" with inadequate performance at the system-level an IC with system level pins must be verified by a methodology that would adequately predict this performance or at least cover the worst case scenario.

Contact discharge is applied to conductive surfaces (e.g. connectors) and air discharge to insulating surfaces (e.g. housing) of system blocks. In case of air discharge, the current level and rise time are less reproducible and more related to environmental conditions (humidity, speed of the tip approach, etc.). For the air discharge test the round ESD gun tip is used (Fig. 2.3a). When the test voltage is set the ESD gun is moved towards the discharge point until a spark appears and then further until the ESD gun tip touches the discharge point on the system surface. This action is usually repeated with positive and negative polarities depending on the applied test procedures, for example, ∼ 10 times. Touching the systems surface is important for the removal of the residual induced charges from the system since only some guns have an integrated charge remover. To remove the residual charge the system must be grounded, brushed, etc. The gun must touch the object because it is intended to model the event when the user, after the first air-gap discharge, will or can eventually touch the apparatus.

Fig. 2.3 Examples of the contact (**a**) and air-gap (**b**) discharge gun testing on PCB

At the contact discharge test the sharp ESD gun tip (Fig 2.3b) is used to obtain a good electrical contact with the conductive system surface. Similarly to the air-gap test about 10 discharges are typically applied with positive and negative polarities.

The test setup for the system level ESD verification is recommended in the standard (Fig 2.4a). These main components critically impact the current pulse waveforms. If inappropriately used, they may significantly change the outcome of the test. In the standard, two different types of tests are distinguished: tests performed in laboratories and post installation tests performed on equipment in its final installed conditions.

When the system is not grounded, for example in case of battery powered systems, there is no self-discharge like in grounded equipment. If the charge is not removed before the next ESD pulse is applied, it is possible that the EUT or part(s) of the EUT are stressed significantly lower than the intended test voltage level or higher if the polarity of the stress pulse is reversed. To avoid the charging of EUT to an unrealistically high charge the test setup is modified (Fig. 2.4b). The possible additional charges are removed prior to each applied ESD test pulse by increasing the time interval between successive discharges or by sweeping away the charges from the EUT through a grounded carbon fiber brush, or with bleeder resistors (for example, 2×470 kΩ) in the grounding cable.

In general the components form a very large scale capacitor $C_T \sim 56$ pF between the ground reference metal plate and the upper metal plate, mounted on the table with a specified height. Also there is a capacitance C_B between the upper metal plate and the board. The distance between the corresponding upper plate and board is determined by the dielectric mat defined by the standard.

The value of capacitance C_T is practically fixed by the design of the setup. The value of the board capacitor C_B is limited by the mat parameters, but otherwise variable depending on the board size and coupling with the upper metal plate.

The pulse waveforms are sensitive to the ground connection of the plates, tested boards and the gun. In the IEC 61000-4-2 standard the ground connection has very

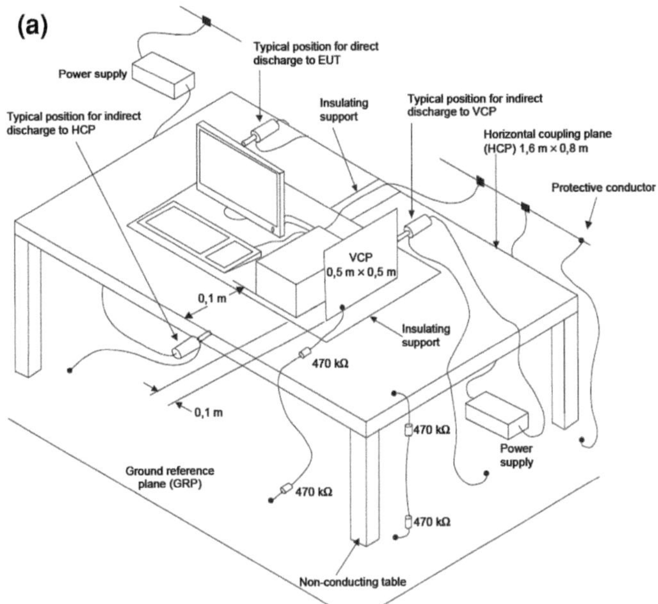

IEC 61000-4-2 ed.2.0 "Copyright © 2008 IEC Geneva, Switzerland. www.iec.ch"

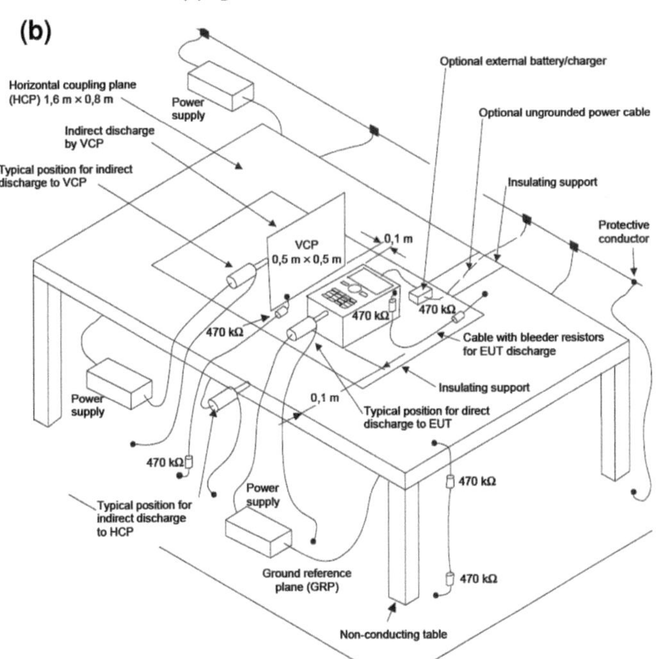

IEC 61000-4-2 ed.2.0 "Copyright © 2008 IEC Geneva, Switzerland. www.iec.ch"

Fig. 2.4 Example of IEC 61000-4-2 standard test setup for floor-standing equipment laboratory tests (**a**) and for ungrounded table-top equipment (**b**)

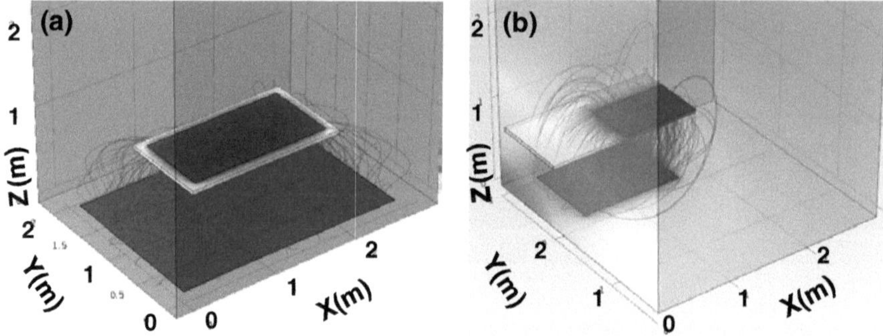

Fig. 2.5 3D FEM electrostatic simulations for electric potential carried out for IEC standard setup with the upper plate positioned in the middle of the table (**a**) resulting in with calculated capacitance ~ 56 pF (**b**) and an incorrect custom setup with upper plate shifted to the corner of the table calculated capacitance ~ 25 pF (courtesy of Augusto Tazzoli)

high impedance through two 470 kΩ charge bleeding resistors and additional inductive wire impedance. These bleeding resistors are only capable to remove the residual charge from the system during the long time between the stress events. Therefore the major ESD current path is provided by low impedance of the capacitors formed by the test setup. This impedance is only a few Ohms for the first current peak. The major difference in the ground connection for the power-on test is translated from the power supply source implementation that can be battery or power grid based.

The stress waveform depends on the current path during the discharge. In case of no direct ground connection to the upper plate and to the board, the bleeding resistors cannot conduct any large system level current. Thus the discharge path is realized through the two serial capacitors C_T and C_B. In this case, the position of the metal plates and the dimension is directly impacting the discharge capacitance which changes the waveform of the ESD current pulse. A simple electro-magnetic simulation of the equivalent capacitance of the table and the ground reference plate demonstrates a significant variation of the test setup capacitance depending on the upper plate—Horizontal Coupling Plate (HCP) size and position (Fig. 2.5).

An important part of the system level test is the test board design. The devices for customer systems that require either IEC 61000-4-2 or ISO 10605 are expected to be tested in conditions that more or less accurately emulate the system in the final application. Thus at least the board level stress is an approach to meet a more ideal expectation for robust ESD design without redesigning the final system.

Initially the test boards can be either demo boards or actual customer boards (Fig. 2.6). To enable the gun test in case of present standard connectors, a port pin extenders is introduced. However it is ideal to understand the customer board environment and to clarify how the case/housing is grounded. The following question should be answered before the design:

Fig. 2.6 Example of the industrial evaluation board design with the ground plane and extended connectors

(i)   *Does demo board need a ground plane?*
(ii) *Is the case floating?*
(iii) *Are cables being connected?*
(iv) *Are there any other specific requirements?*

Through evaluation of the current waveforms the capacitor C_B, which is formed by the board ground plane with the upper metal plate, is now well understood in the industry. This understanding has been translated into the supported practice to mount all custom boards on an additional ground metal plane. In addition, metal pins extended from the board surface are spaced apart about ~ 10 mm. Additional pins of ~ 5 mm are introduced to guarantee the discharge of the current pulse through the tested pins rather than through non-dedicated PCB components.

Gun verification techniques, practical aspects and recommendations are described by the gun manufacturers [22]. They cover the parameters which must be measured. Those parameters are the tip voltage, the contact discharge current waveform which is defined for its peak, rise time and the currents after 30 and 60 ns, and in case of air discharge the rise time and time constant.

For stress waveform verification a low impedance shunt ($<2.1\ \Omega$) is used. It represents a discharge into a large metallic object rather than just a piece of wire (Fig. 2.7). A new calibration target design with a frequency response flatness of up to 4 and a 2 GHz oscilloscope was introduced in IEC 61000-4-2:2008. Environmental factors also affect the calibration results. Therefore no air discharge verification procedure was included due to too high variability of the approaching speed, humidity, and the arc and ionization length.

A few simple experiments can be used to demonstrate the impact of the test setup grounding on the air gap test current waveforms. In the setup for the packaged component test on PCB with no external ground (Fig. 2.8) the current often can be a source of miscorrelation between the contact and the air discharge. It largely depends on both the PCB ground plate dimension (Fig. 2.9a) and the bleeder resistor value (Fig. 2.9b).

Certainly the presence of the high value bleeder resistor is negligible for the alternative current path. A significant change in the pulse waveform results in formation of the negative current peaks that may result in a damage of the device

Fig. 2.7 ESD simulator verification setup (**a**) and the verification target (**b**)

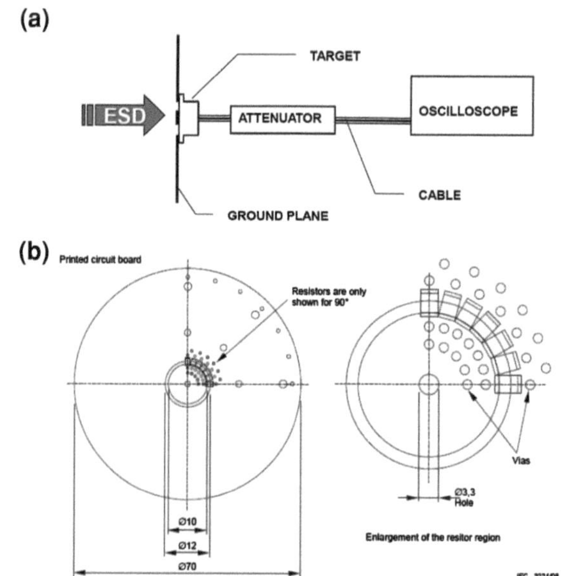

IEC 61000-4-2 ed.2.0 "Copyright © 2008 IEC Geneva, Switzerland. www.iec.ch"

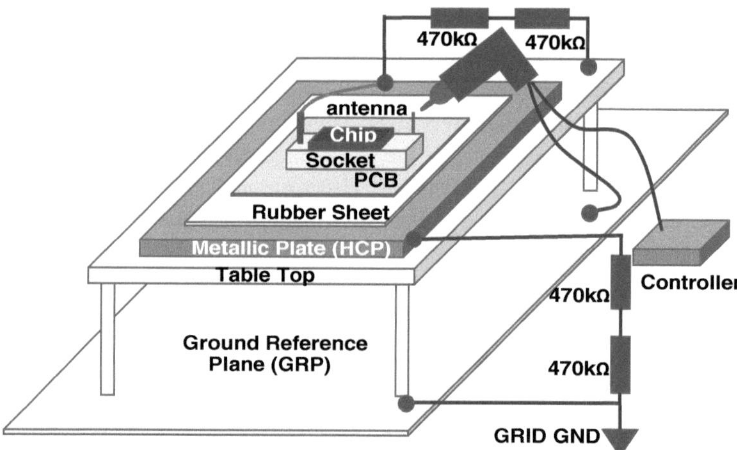

Fig. 2.8 Setup for system level test with the bleeding resistors connected to the upper metal plate only

in rather unexpected conditions. For example a blocking junction of a dual-direction protection design can be destroyed during a positive air discharge rather than negative, as it would be originally expected.

An ideal air discharge waveform for different load conditions is not defined even by the standard. It is realized only when the gun ground is connected directly

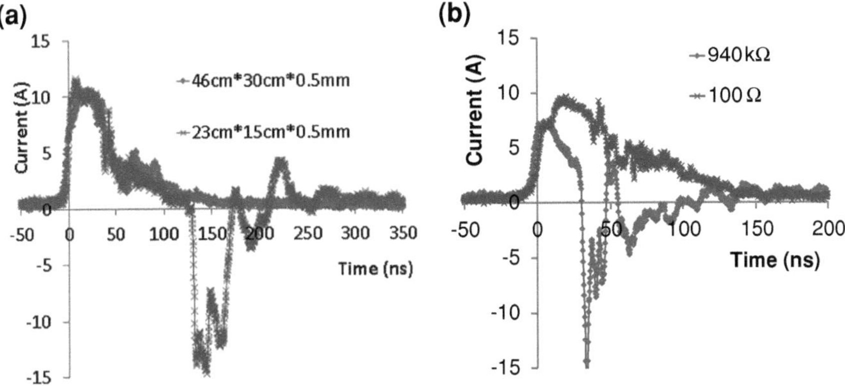

Fig. 2.9 Current waveforms for 8 kV air gap discharge realized in the set up (Fig. 2.8) for the conditions of different PCB plate size (**a**) and the resistors values (**b**) (courtesy of Yunfeng Xi)

to horizontal coupling plane (HCP, Fig. 2.10). In this case the waveform with a good accuracy can be assumed similar to the standard pulse (Fig. 2.1) with simply eliminated first peak. In this conditions according to the experimental waveforms the only peak has the slow rise time until the maximum current of ~ 20–40 ns with the same current levels after 30 and 60 ns from the beginning of the pulse $I_{30ns} \sim 2$ A/kV $\pm$ 30 % and $I_{60ns} \sim 1$ A/kV $\pm$ 30 %, respectively. To reproduce this waveform in simulation namely the circuit (Fig. 2.2a) can be used, rather than (Fig. 2.2b). The additional parasitic components and circuit parameters can be adjusted to form the single peak pulse.

2.1.2 Automotive Standard ISO 10605

Another widely used common standard for system-level ESD test was released by ISO (the International Organization for Standardization) as a major guidance in the automotive industry. This ISO 10605 standard [23] defines the road vehicles test methods for electrical disturbances from electrostatic discharge.

While IEC 61000-4-2 is created to establish a common and reproducible basis for evaluating the performance of electrical equipment subjected to ESD, the ISO 10605 Standard is created to specify ESD test methods necessary to evaluate electronic modules intended for vehicle use based on IEC 61000-4-2.

This Standard has many similarities with IEC 61000-4-2 especially if it is compared to the packaged level component tests (HBM, MM, CDM). However there are a number of differentiating aspects. First of all the ESD stress level that is usually required to be passed in the automotive industry is much higher. A passing level of 30 kV is often requested.

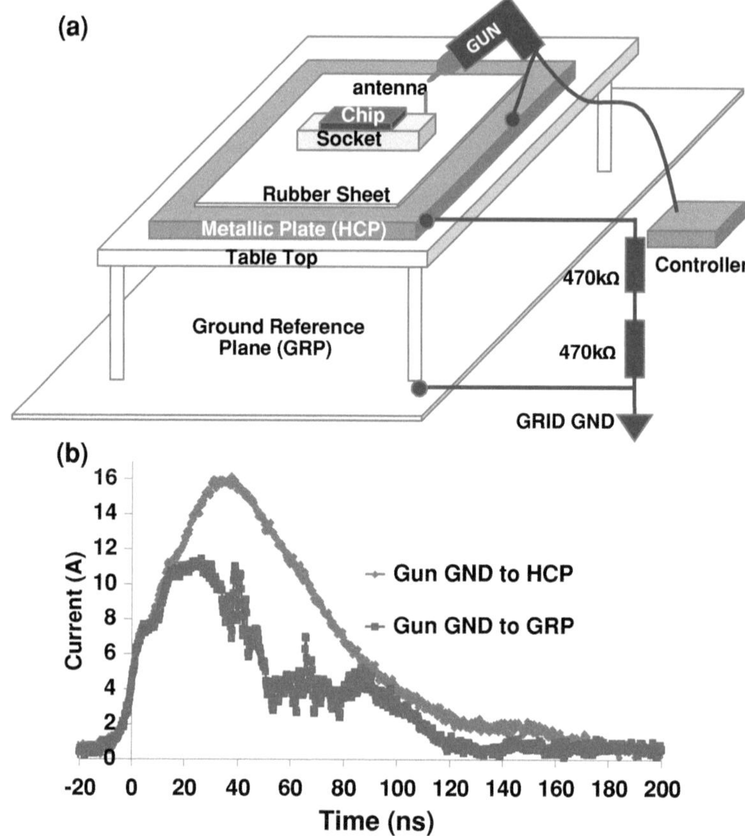

Fig. 2.10 The non-standard setup with the IEC gun ground connected directly to the HCP (**a**) and the 8 kV air gap discharge current waveforms if the gun ground is connected to HCP and GRP (**b**) (courtesy of Yunfeng Xi)

The equivalent circuits for this test setup (Fig. 2.11a) include four discharge networks composed of a 150 or 330 pF charging capacitor and a 330 or a 2000 Ω discharge resistor. These are the only two combinations required by the standard in spite of that some gun manufacturers offer the option of 330 pF charging capacitor with 330 Ω resistor. Physically 150 pF–330 Ω combination represents the discharge of a human body through a metallic part to the system port, while 330 pF–2 kΩ combination represents a discharge of a human body directly through the skin.

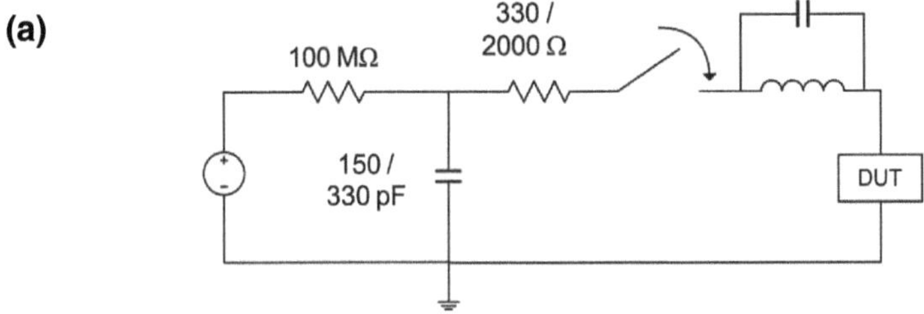

Fig. 2.11 Principle schematic of an ISO 10605 discharge circuit, (a) the L-C parasitic may vary in a real discharge circuit, and (b) defined stress level [23]

Unlike in the IEC 61000-4-2, in the ISO 10605 the upper table plane (HCP) is directly connected to the gun ground. Another differentiation aspect of this standard from IEC 61000-4-2 is the preferred test levels. Although in addition to the 4 main levels IEC 61000-4-2 defines any arbitrary level x, many IEC test guns and ESD stress generators had 16 kV pre-charge limit. ISO 10605 clearly states test severity levels (Fig. 2.11b). The test severity levels are differentiated for the direct and indirect discharge.

Among more detailed *vs.* IEC 61000-4-2 definitions in ISO 10605 the speed of the gun tip approach speed should be between 0.1–0.5 m/s for any test. Because the approach speed is not trivial to measure, in practice the ESD generator should approach the DUT as quickly as possible until the discharge occurs or the discharge tip touches the discharge point without causing damage to the DUT or generator.

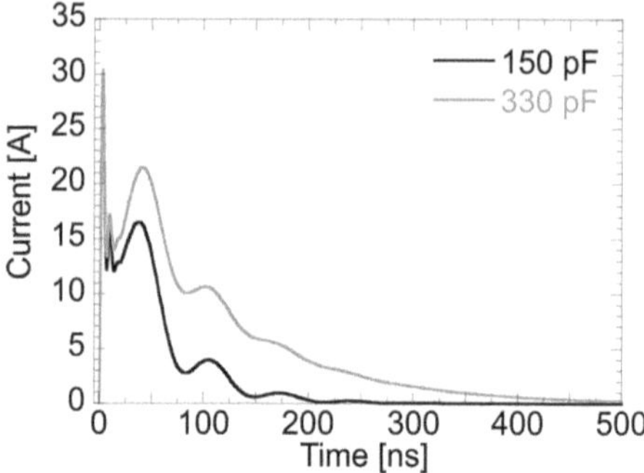

Fig. 2.12 ISO 10605 current waveforms into short load, discharge network: 150/ 330 pF and 150 Ω, stress level: 8 kV

Similarly the charge removal measures are defined as the charge build-up can be eliminated by briefly connecting a bleeder wire with high resistance (1 MΩ) in the following sequence: (i) between the discharge location and ground, and (ii) between the ground point of the DUT and ground. If there is evidence that the wire does not have any impact on the test result, it can remain connected to the DUT.

The evaluation test results are similar to IEC 61000-4-2, but according to Section C of the standard reports in detail on "function performance status classification (FPSC)".

Two waveforms into a short circuit load (Fig. 2.12) have much longer discharge duration in comparison with the IEC 61000-4-2 stress when using the larger precharged capacitor. The ISO 10605 standard is not directly intended to be used in component-level testing, but it is a requirement of many automotive manufacturers to their suppliers.

The longer stress durations of some of the ISO 10605 discharge currents directly impacts the design of on-chip ESD protection structures. Figure 2.13 shows the current and the maximum temperature in an ESD diode during IEC 61000-4-2 and ISO 10605 discharge. The latter is created with a 330 pF/330 Ω network. Because of the longer stress duration the device self-heating is significant higher for ISO 10605 discharge stress conditions and needs to be taken into account during the ESD protection design to prevent unexpected failure during ESD qualification. One approach for the on-chip ESD protection design has been used by [24] where long duration TLP testing emulates the longer duration of an ISO 10605 stress.

To summarize the above two sections the comparison of the major specifications of the IEC 61000-4-2 and ISO 10605 standards is compiled in Table 2.1. The specific characteristic of IEC 61000-4-2 is that the ungrounded DUT cannot discharge itself like grounded equipment. Therefore the charge shall be removed prior to each applied ESD test pulse either by waiting a sufficient time between zaps through 2 × 470 kΩ bleeding resistors or by a carbon fiber brush. During Air-discharge test methods, the gun should approach the DUT as fast as possible and touch it after the discharge occurred.

In ISO 10605 the speed of approach should be between 0.1–0.5 m/s for any test. Charge build-up can be eliminated by briefly connecting a bleeder wire with high

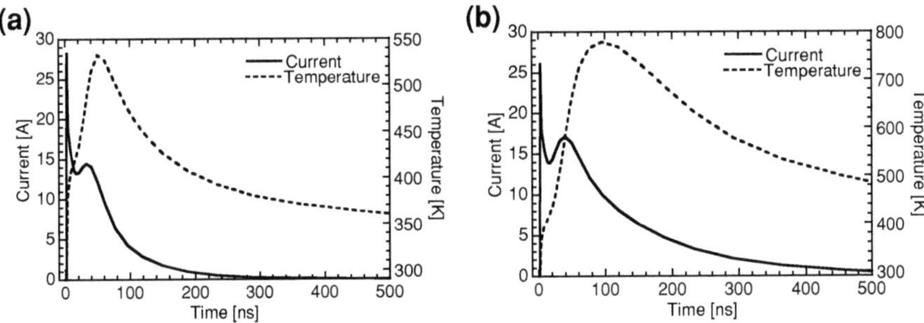

Fig. 2.13 Current and temperature in an ESD clamp during system-level ESD stress: **a** IEC 61000-4-2 and **b** ISO 10605 with 330 pF/330 Ω discharge network, stress level: 8 kV

resistance (1 MΩ) in the following sequence: (1) between the discharge location and ground, and (2) between the ground point of the DUT and ground. If there is evidence that the wire does not have any impact on the test result, it can remain connected to the DUT.

2.1.3 The Surge Standard IEC 61000-4-5

Similar to the ESD pulses the trend to propagate system level requirements for on-chip solutions has involved the specification of the surge requirements. The surge pulse specification is captured in the standard Electromagnetic compatibility (EMC) standard IEC 61000-4-5 *Part* 4–5 [25]. The standard formalizes the testing and measurement techniques for the surge immunity test.

In real life the surge events are represented by systems switching transients or lightning strike events. The power system switching transients are associated with major power system switching disturbances, such as capacitor bank switching; minor local switching activity or load changes in the power distribution system; resonating circuits associated with switching devices, such as thyristors; various system faults, such as short circuits and arcing faults to the grounding system of the installation [25]. The major mechanisms of lightning surge include direct lightning strikes to an external (outdoor) circuit injecting high currents producing voltages by either flowing through the ground resistance or the impedance of the external circuitry; an indirect remote lightning strike which produces electromagnetic impulses that induce voltages/currents on the conductors outside and/or inside a building; lightning ground current flow resulting from nearby direct-to-earth discharges coupling into the common ground paths of the grounding system of an installation. The rapid change of voltage and flow of current which can also occur as a result of the operation of a lightning protection device can induce electromagnetic disturbances into adjacent equipment [25].

Table 2.1 Comparison of the essential features of IEC 61000-4-2 and ISO 10605 standards

Standard	IEC 61000-4-2	ISO 10605
Target	General electrical equipment	Electronic modules for vehicle
Preferred test	Contact discharge method	Air discharge method Direct ESD, powered DUT
RC network	150 pF 330 Ω	150/330 pF 330 Ω 150/330 pF 2 kΩ
Connection of ESD gun ground	To ground reference plane (GRP) through 2 × 470 kΩ charge bleeding resistors	*Direct* powered DUT: HCP and DUT GND *Indirect* powered DUT HCP or GRP
Preferred stress levels (kV)	Contact: 2, 4, 6, 8 Air: 2, 4, 8, 15	Contact: 2-8; 4-15 Air: 2-15; 4-15; 6-25 Indirect: 2-8; 2-15; 4-20
Number of discharges	At least 10 single discharges in the most sensitive polarity	*Direct, unpowered or Vehicle test method* at least 3 discharges are applied to all direct discharge test points for each specified test voltage and polarity *Indirect* 50 discharges are applied to all indirect discharge test points for each specified test voltage and polarity
Short circuit contact discharge waveform	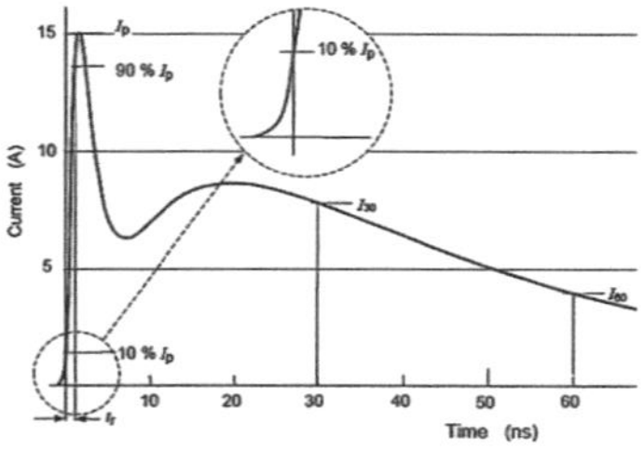	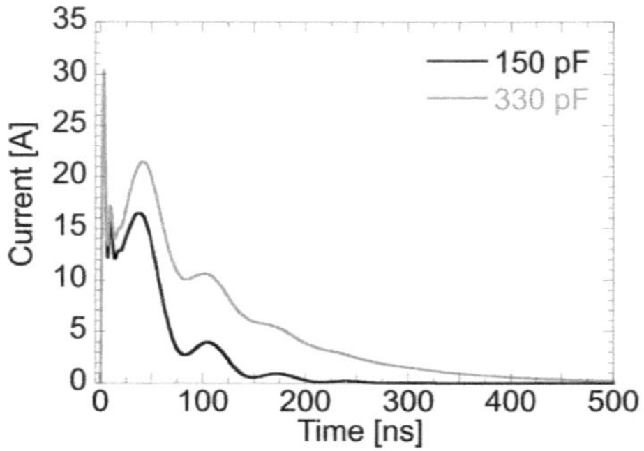

Similar to other standards the simulation of the transients in the laboratory environment is defined in the standard for the test generator to simulate the above-mentioned phenomena as close as possible to real physical conditions. This requires a corresponding surge generator in comparison with ESD pulse generators.

Respectively to the direct and indirect physical events, mentioned above, the surge testers simulate surge events to target both the direct and the indirect coupling conditions. During direct coupling the source of interference is in the same circuit, for example in the power supply network. Inductive spikes or load dumps can be another source. In those cases the generator simulates a low impedance source at the ports of the equipment under test. In case of an indirect coupling the source of interference is not in the same circuit as the victim equipment. In such case the generator simulates a higher impedance source.

Unlike system and component level ESD pulses, the surge tests standard waveforms are specified both as open-circuit voltage and short-circuit current. Similarly to the system level ESD guns the waveforms for surge tester are verified without the equipment under test (EUT) connected. The output can be also specified for the cases of AC or DC powered products.

The surge generator is intended to generate a surge having an open-circuit voltage front time of 1.2 µs; an open-circuit voltage time to half value of 50 µs (Fig. 2.14a); short-circuit current front time of 8 µs; and a short-circuit current time to half value of 20 µs (Fig. 2.14b).

A simplified circuit diagram of a generator as provided by the standard is presented in (Fig. 2.15) with the high voltage source U, the charging resistor R_C, the energy storage capacitor C_C, the pulse duration shaping resistors $R_{S\#}$, the impedance matching resistor Rm, and the rise time shaping inductor L_r. The values for the generator components are selected so that the generator delivers the corresponding standard pulses for the 1.2/50 µs voltage surge at the open-circuit and a 8/20 µs current surge into the short circuit conditions. Most surge pulse generators produce pulses with peak currents in the range from 250 A to 2 kA.

For convenience, the ratio of peak open-circuit output voltage to peak short-circuit current of a combination wave generator may be considered as the effective output impedance. For this generator, the ratio defines an effective output impedance of 2 Ω. The resulting waveform of the voltage and current is a function of the EUT input impedance. This impedance may change during surges to equipment due to either proper operation of the installed protection devices, or due to flash over or component breakdown if the protection devices are absent or inoperative. Therefore, the 1.2/50 µs voltage and the 8/20 µs current waves have to be available from the same generator output as required by the load.

The circuit simulation for the surge tester in a Spectre simulator environment has been done in [25] for a circuit (Fig. 2.16a) with the components value $C_C = 6.038$ µF, $L_T = 10.37$ µH, $R_{s1} = 25.105$ Ω, $R_{s2} = 19.8$ Ω, $R_m = 0.941$ Ω and $U = 1082$ V, demonstrating adequate short and open circuit waveforms (Fig. 2.16a).

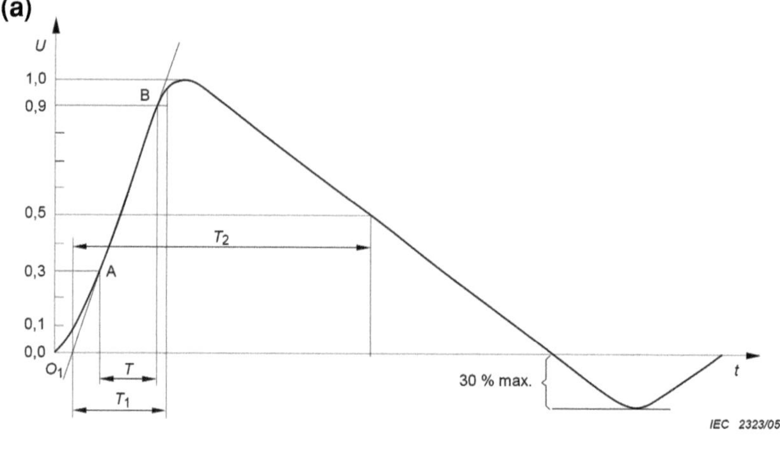

Front time: $T_1 = 1{,}67 \times T = 1{,}2\ \mu s \pm 30\ \%$
Time to half-value: $T_2 = 50\ \mu s \pm 20\ \%$.

IEC 61000-5-4 ed.2.0 "Copyright © 2005 IEC Geneva, Switzerland. www.iec.ch"

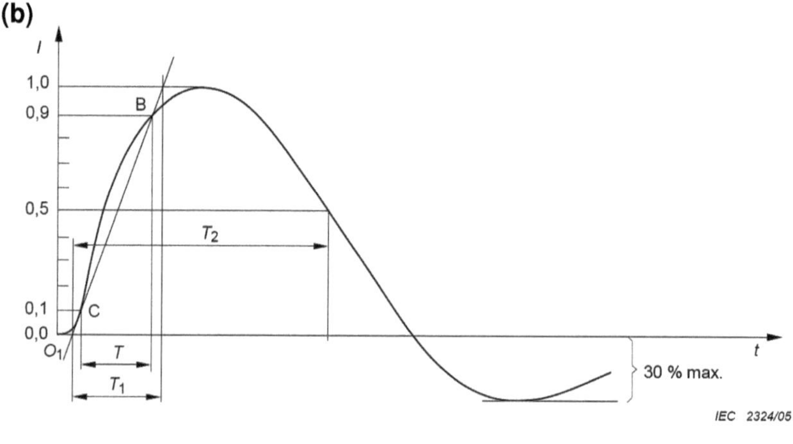

Front time: $T_1 = 1{,}25 \times T = 8\ \mu s \pm 20\ \%$
Time to half-value: $T_2 = 20\ \mu s \pm 20\ \%$

IEC 61000-5-4 ed.2.0 "Copyright © 2005 IEC Geneva, Switzerland. www.iec.ch"

Fig. 2.14 Waveform of open-circuit voltage (1.2/50 µs) at the output of the combination wave generator with no Coupling-Decoupling Network (CDN) connected (waveform definition according to IEC 60060-1) (**a**), waveform of short-circuit current (8/20 µs) at the output of the generator with no CDN connected (waveform definition according to IEC 60060-1) (**b**)

The waveforms into a short circuit load (Fig. 2.16) have much longer discharge duration in comparison to the IEC 61000-4-2 stress. ESD pulses surge tests create the stress in a different time domain where not only adiabatic electrical phenomena

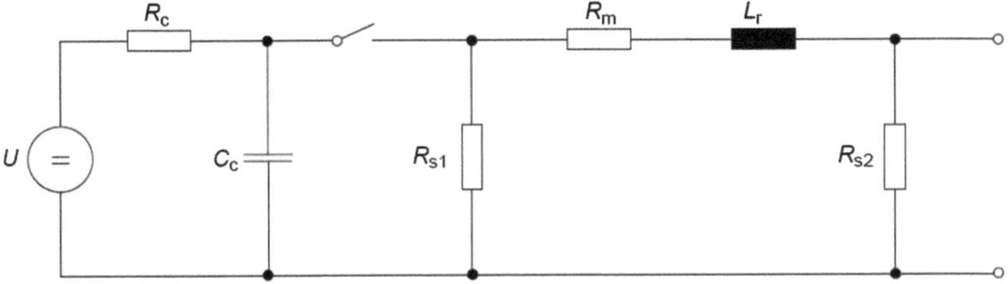

IEC 61000-5-4 ed.2.0 "Copyright © 2005 IEC Geneva, Switzerland. www.iec.ch"

Fig. 2.15 Simplified circuit diagram of the combination wave surge generator

are responsible for the passing level but also the electro thermal phenomena are dominating. In general, ESD protection devices originally designed to withstand the electrical, rather than thermo-electrical current can only provide the ESD protection. In experimental results the typical correlation for SCR type of ESD protection devices results in a 10 times lower passing current.

According to [25] the selection of the source impedance of the combination wave generator depends on the type of cable, conductor or line. The differentiation is done for a.c or d.c. power supply networks, interconnections, the length of the cable lines, indoor/outdoor conditions, and the application of the test voltage for either line-to-line or lines-to-ground.

As for the Coupling Decoupling Network (CDN), the impedance of 2 Ω represents the source impedance of the low voltage power supply network. Therefore in the equivalent cases the generator with its initial internal effective output impedance of 2 Ω is used directly. The impedance of 12 Ω with additional 10 Ω serial resistor represents the entire low voltage power supply and ground network. The effective impedance of 42 Ω is provided by an additional 40 Ω resistor that represents the source impedance between all other lines and ground (Fig. 2.17).

Although the current of a surge can be lower, the total energy is much higher. Highly specialized ESD protection schemes can be ineffective against such slow transient, low voltage, but high current stresses. Stressing devices using relatively low level (up to 10–20 A) surge stresses can be also based on the IEC 61000-4-5 specification. Specialized surge protection may need to be added. A number of Transient Voltage Suppressors (TVS) have been characterized using the IEC surge stress for years. Their performance during an IEC surge stress is frequently found in TVS datasheets.

For the on-chip standalone devices surge-IEC correlation study TESEQ surge tester (NSG3040) was used with 1 kΩ resistor in series with tested device to limit the current down to reasonable level of indirect surge scenario. The majority of SCR devices passed at least 3 kV stress with corresponding current ~3 A demonstrating ~10 times pulsed current reduction in comparison with the ESD current level. Failures from surge were open, while the typical failure signature from ESD

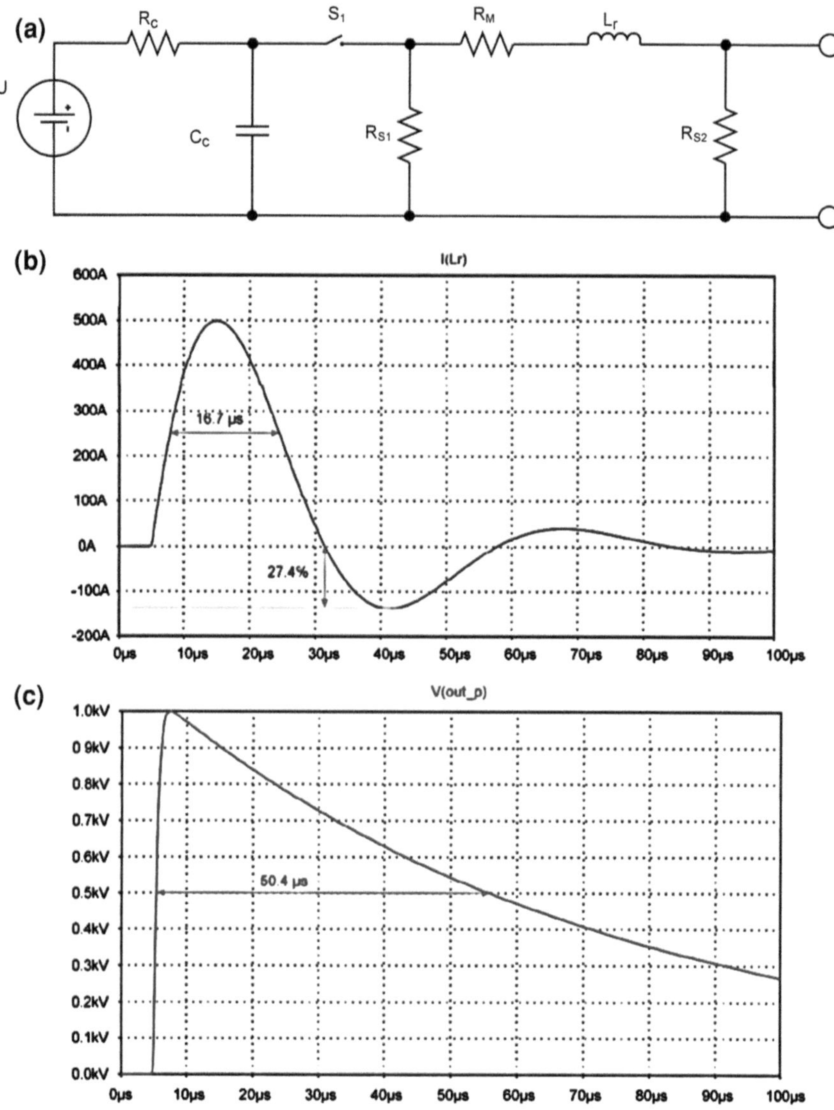

Fig. 2.16 Spectre simulation circuit for the combination wave generator (**a**) and the simulated waveforms for short circuit current (**b**) and open circuit voltage (**c**) [26]

stress is short circuit or elevated leakage (Table 2.2). In snapback mode the correlation factor between the surge current and standard component level test currents was ~ 0.1. In forward mode the backend limitations were dominant during the surge test.

A similar ~ 10 times correlation factor between ESD and surge pulse peak current was measured for the snapback NMOS (SNMOS). The increasing of the

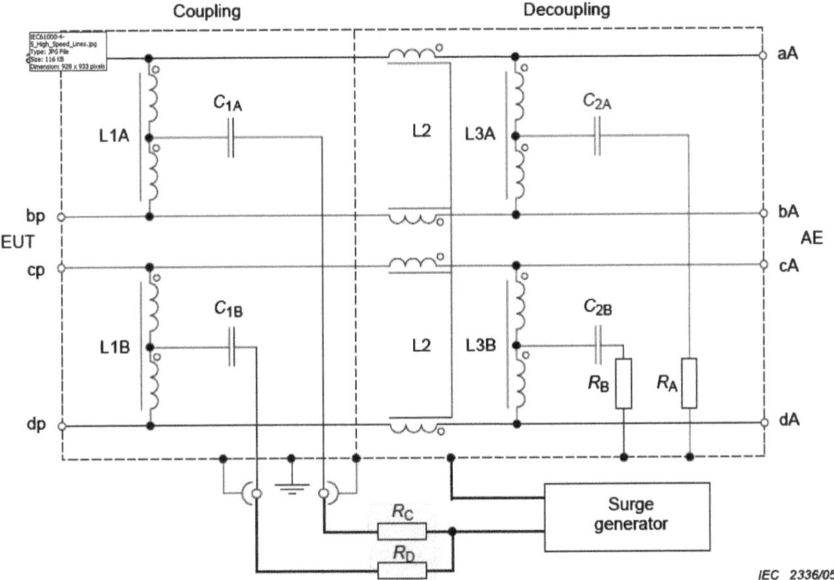

IEC 61000-5-4 ed.2.0 "Copyright © 2005 IEC Geneva, Switzerland. www.iec.ch"

Fig. 2.17 Example of a coupling/decoupling network for symmetrical high speed communication lines using 1.2/50 μs surge [25] with $R_C = R_D = 80$ Ohm and the decoupling capacitors C_{IA} and C_{IB}

Table 2.2 Comparison of the ESD passing level upon for different HV SCR device types

Device type	Breakdown voltage @ 1 μA	Positive TLP IT2	Positive surge kV/1 kΩ	Negative surge kV/1 kΩ
Dual-direction SCR (DIAC)	60 V	>15 A	Pass 3 kV($\sim$3 A) Fail 3.5 kV	Pass 3.5 kV ($\sim$3.5 A) Fail 4 kV
HV NLDMOS-SCR	40 V	>15 A	Pass: >4 kV (>4 A)	Pass: >4 kV (>4 A)

Table 2.3 Comparison of the ESD pulse and surge pulse performance for the snapback NMOS with width 800 μm for two drain silicide blocking regions

	TLP current IT2	HBM pulse	MM pulse	HMM pulse	Surge voltage over 1 kΩ
Positive stress (snapback mode)					
SB = 2.9 μm	4.6 A	7 kV 4.6 A	450 V $\sim$ 6.7 A	2 kV/ $\sim$ 6.1A	450 V
SB = 1.6 μm	2.8 A	5 kV 3.3 A	400 V $\sim$ 6.7 A	0.8 kV $\sim$ 2.4 A	400 V
Negatives stress (body diode)					
SB = 2.9 μm	11 A	>8 kV	600 V $\sim$ 9 A	5.6 kV $\sim$ >17 A	2.5 kV
SB = 1.6 μm	11 A	>8 kV	550 V $\sim$ 8.2 A	4.4 kV $\sim$ 13.3 A	2.5 kV

silicide block (SB) length in the drain ballasting region provided practically no improvement of the surge pulse passing level in spite of the significant effect on standard component and even HMM test results (Table 2.3). The description of the SNMOS and SCR ESD devices is presented in Chap. 3.

2.2 HMM Testing

A significant progress for the on-chip system level IC protection design was achieved by implementing the component and on-wafer test methods. These methods simulate the system level ESD discharge pulse waveforms with a margin which is acceptable for initial design steps. In particular, an understanding of the IEC 61000-4-2 standard requirements led to the development of a component-level test method using the same stress waveform.

The system-level ESD test standards do not guide directly how to apply system-level ESD stress on the IC component level. Nevertheless, there was a critical need of characterizing components and on-chip ESD clamps with a system-level equivalent stress. Thus, a new measurement method—the Human Metal Model (HMM), has been proposed. While the system level IEC and ISO standards for the contact discharge define the tests for the system ports, the HMM methodology is on the contrary primarily targeting the evaluation of the IC pin robustness. It was also successfully used for the evaluation of the standalone ESD solutions placed on test chips.

The methodology is done under the assumption that, if the HMM pulse waveform repeats the system level waveform, a certain passing level correlation can be expected. Several studies demonstrated that in many cases there is a correlation between standard system level gun test and HMM stress [51]. In spite of a number of reported miscorrelations, today HMM represents the most useful approach for the on-chip system-level design. This triggered the release of many industrial HMM laboratory tools [28, 29] now offered by variety of vendors. The details of the setup and associated issues are described in the following sections.

A complementary angle of view on the HMM practice is related to the propagation of the system level stress across the PCB. This results in a possible current overstress at the IC pins directly interfacing with the system port under stress. It is logical to expect that the ability to withstand this system level test is not automatically guaranteed just by the passing the standard component level stress (HBM, MM, CDM). Instead, such capability should be added by the dedicated on-chip design, discussed in Chaps. 3 and 4.

HMM tester applies stress waveforms with similar characteristics to the IEC 61000-4-2 standard waveforms. Perhaps the only parallel in physical understanding between the HMM test and the standard component level tests can be roughly drawn by a physical representation of HMM current pulse waveform as a superposition of the component-level CDM pulse, representing the first peak, and HBM pulse, representing the second peak, after both scaled in a right proportion. Although certain attempts can be made to find the correlation factors between

HMM pulse and these two component pulses unlikely this approach can provide a replacement for HMM. This is mainly because the current paths of the single pin stress CDM are not in general the same to the path during HMM stress. Also the duration of an HBM event is much larger than the duration of *e.g.* the second peak current during an HMM event.

2.2.1 HMM Setups with ESD Gun

The HMM standard practice document [30] describes three different measurement setups to apply the stress waveform to a component. The setups are using both gun type and 50 Ω pulse generators. The gun based setups include an IEC 61000-4-2 compliant ESD gun as a stress source. The packaged DUT is placed on a PCB which is mounted on a larger ground plane. The ground plane of the PCB and ground plane of the test setup thereby form a continuing ground plane (Fig. 2.18a). When stress is applied to the DUT, the ground wire of the ESD gun is connected to the ground plane of the measurement setup. A variation of the setup uses a vertical ground plane (Fig. 2.18b) which allows the shielding of the measurement equipment from the electromagnetic fields send out by the ESD gun during discharge.

Both setups allow the application of HMM stress when the DUT is powered up. In this case the supply pins of the DUT should be equipped with a by-pass capacitor to ground to decouple the supply line.

Several miscorrelation cases between different ESD gun models have been reported [31–33]. A common reason for most of the miscorrelation is related to the rather flexible tolerance range of the waveform parameters defined in the standard. For example the amplitude and rise time of the IEC 61000-4-2 stress current are defined with rather big tolerance of 30 % for the currents after 30 and 60 ns. Combined with the 25 % acceptable range for the rise time of the initial current peak it is logical to assume that different model ESD gun pulses might impact the passing level at the same equal conditions [33, 34]. In addition the electro-magnetic field around the discharge point depends strongly on the shape of the ESD gun discharge tip. Thus the gun based HMM measurement setups partially include some of the limitations related to the flexibility of the IEC 61000-4-2 compliant ESD guns.

2.2.2 50 Ohm HMM Setup

The alternative third HMM setup uses a 50 Ω ESD pulse source (Fig. 2.19). The stress is applied with coaxial lines to a test board with the DUT mounted on. This setup has been originally proposed in order to improve repeatability of the applied stress pulses [35], to remove the ESD gun issues and to enable a reliable measurement of voltage and current during the HMM stress. Similar to the other HMM setups a DUT can be tested when powered up. In this case the supply pins of the DUT are equipped with a by-pass capacitor to ground to decouple the supply line.

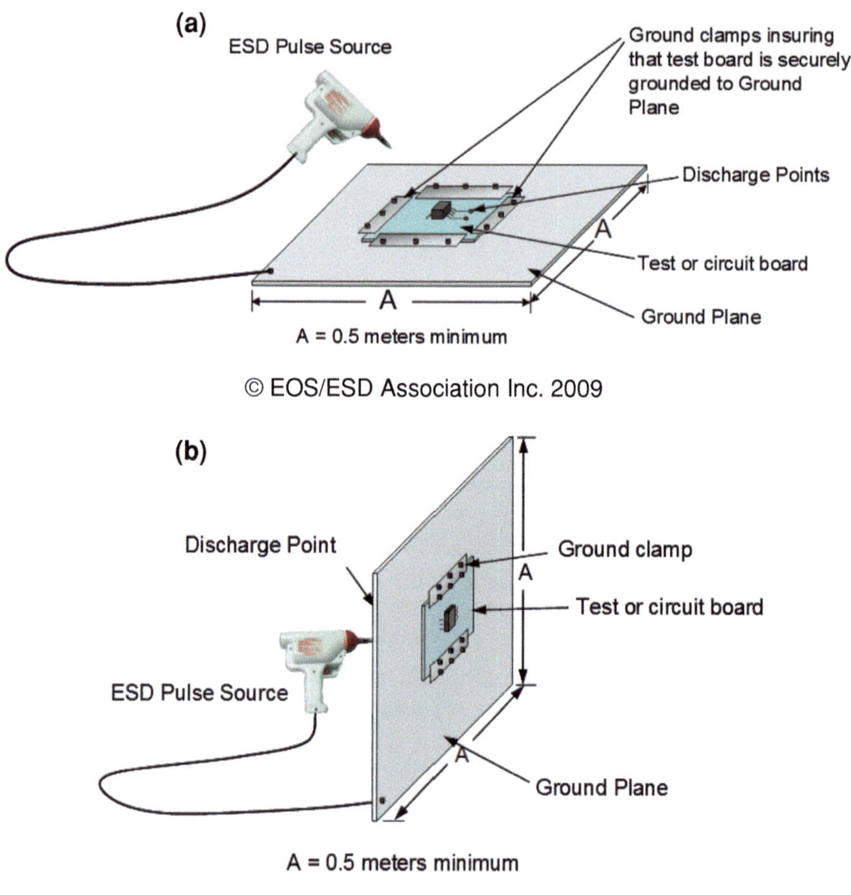

Fig. 2.18 HMM measurement setup with ESD gun and horizontal coupling plane (**a**) and with ESD gun and vertical coupling plane (**b**) [30]

The impact of reflections in 50 Ω HMM setups is an important issue. The source impedance of the 50 Ω HMM tester is much lower than for ESD guns. The difference in source impedance impacts the test results in 50 Ω HMM setups. The on-state of ESD protection devices typically results in a low impedance of a few Ohms. This causes an impedance mismatch between the turned-on device and the tester source impedance.

The HMM stress current is partially reflected back to the pulse generator which causes a disturbance of the pulse generator circuit (Fig. 2.20). This results in an appearance of a negative current after the decay of the HMM stress current. The current waveforms with and without matching DUT resistance become different (Fig. 2.21). Those reflections can impact the device failure level. Thus, there are miscorrelations in comparison to testing with an IEC 61000-4-2 discharge circuit

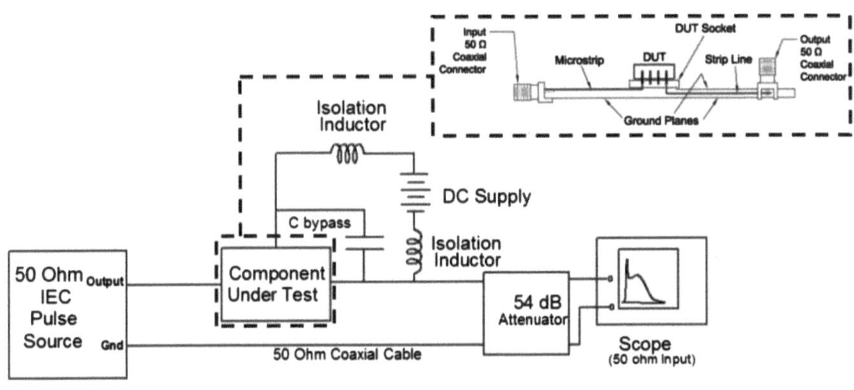

Fig. 2.19 50 Ω HMM setup [30]

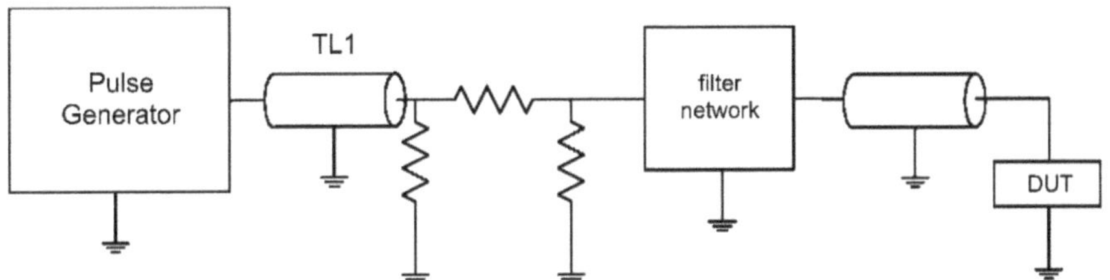

Fig. 2.20 Principle schematic of a 50 Ω HMM setup; *TL1* one transmission line for pulse shaping

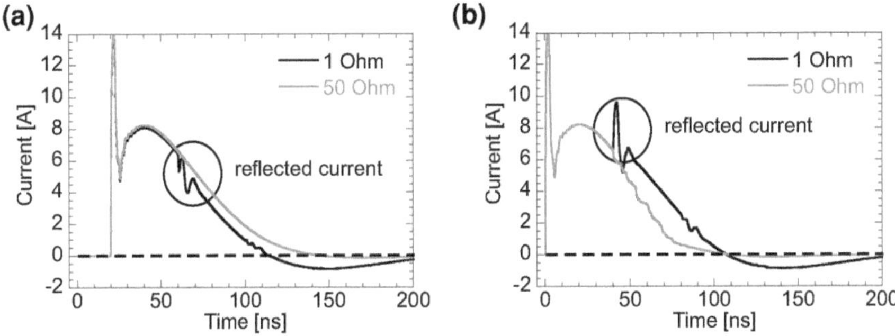

Fig. 2.21 Simulation of the impact of the DUT impedance on reflections in 50 Ω HMM testers: current through DUT (*left*) and current through TL1 (*right*); DUT represented with 1 and 50 Ω resistance

as HMM stress source. If a matched 50 Ω impedance is connected as DUT, then these disturbances do not occur. A detailed analysis of this phenomenon is provided later in this chapter.

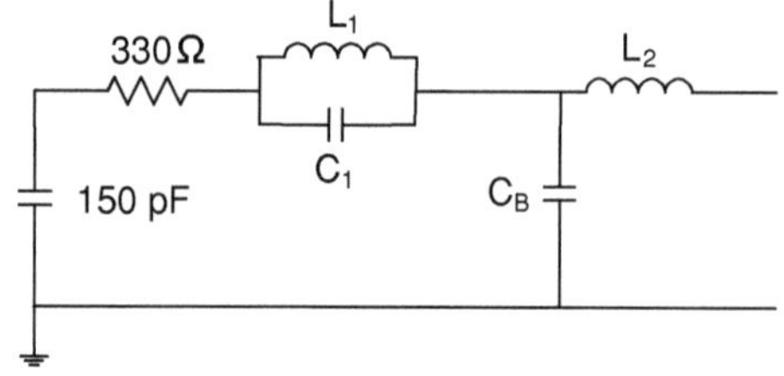

Fig. 2.22 Equivalent schematic of a 330 Ω HMM discharge circuit

The HMM setup with an IEC 61000-4-2 discharge circuit (Fig. 2.22) has an impedance of 330 Ω which is similar to ESD guns. Hence type of reflections as observed in 50 Ω HMM setups cannot occur. The main advantage of this HMM tester over any ESD gun is the compact form factor of its discharge module. It can be conveniently mounted into wafer-level measurement setups. This enables HMM characterization in a very early stage during the ESD protection design. Additionally the design of the module limits the radiation of electromagnetic fields during the HMM test. Alternatively, the discharge can be applied via connectors to application boards. The stress current complies with the IEC 61000-4-2 standard and the HMM standard practice.

2.3 Transmission Line Pulsed Characterization

Passing ESD qualification level for devices and systems under test is a primary goal of any ESD design. However it would be rather difficult to rely only on the passing level during the development of ESD solutions and ESD case studies. Neither capturing of the current and voltage waveforms is always simple nor an easy way to run comparative and conclusive evaluation of ESD solutions and the internal circuit response. Therefore Transmission Line Pulsing (TLP) I–V characterization becomes one of the major steps in the on-chip development process. After more than two decades of evolution the TLP methods have significantly improved. Many tools have been released by test equipment vendors and are now widely applied both on the component and system level evaluation. The TLP I–V characteristics measurement methodology including challenges and pitfalls related to the on-wafer measurement techniques are addressed in this section.

2.3.1 TLP Test Method

TLP testing has been introduced for the first time in [36]. Today, commercial TLP testers are available from several vendors and provide a variety of user friendly features and GUIs. The original motivation for introducing TLP was to have a measurement setup which allows the device characterization and the capturing of voltage and current in the HBM time domain. Since then, it became the most

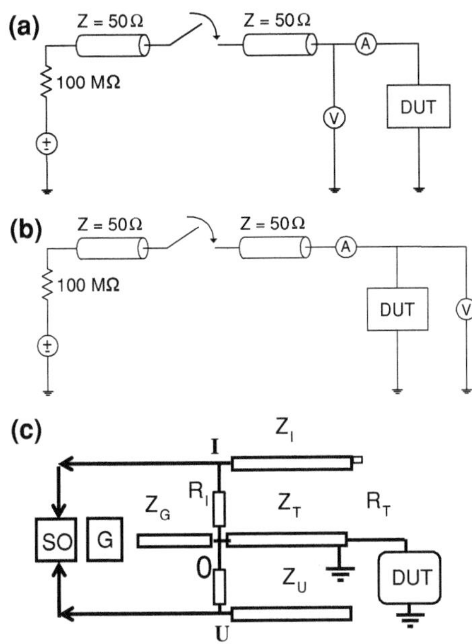

Fig. 2.23 Simplified diagram of two typical TLP experimental setups: Time Domain Reflected (TDR) TLP (**a**) and Time Domain Transmission (TDT) TLP (**b**) and simplified diagram for the voltage and current measurements by summation of the reflected pulses (**c**)

important component-level ESD characterization tool for obtaining the device parameters which are required for the design of on-chip and off-chip ESD protection circuits.

One typical TLP setup is the Time-Domain Reflected (TDR) TLP measurement setup (Fig. 2.23a). Certainly the voltage measurements include the probes with substantial attenuation. A transmission line of a certain length is charged by a high-voltage source. The length of the transmission line defines the width of TLP pulse. When the switch is closed the transmission line is discharged to the DUT. The incident and reflected voltage and current in time are measured with an oscilloscope. Typically the delay between incident and reflected waveform is not long enough. Therefore incident and reflected waveforms can been seen "overlaid" on the screen of the connected oscilloscope. The current and voltage at the DUT are obtained by adding incident and reflected data. This is done either in the control software of the TLP tester and/or during processing of the obtained waveform data.

The second typical setup is the Time-Domain Transmission (TDT) TLP measurement setup (Fig. 2.23b). Like before a transmission line is discharged to a DUT. The incident and reflected current in time are measured with an oscilloscope. The voltage, however, is measured directly at the device under test. The TDT setup can be also compared to a Kelvin type of measurement setup as the voltage is not measured through the same wire where the stress current is flowing.

The principle of the voltage and current measurement with only 2-pin DUT connection is illustrated in Fig. 2.23c. The main pulse from the generator G is propagating in the line Z_G. In the point "0" two small fractions are split out from the main pulse that is continuing to travel in the line Z_T toward DUT. These two small fractions are routed into the coaxial lines Z_I and Z_U. The attenuated pulses are obtained using 1/200 dividers formed by the corresponding resistors R_I and R_U into. All three coaxial lines Z_I, Z_U and Z_T are matched to the same length thus introducing the same propagation delay time. The line Z_I has with the opposite end shorted to the ground, while the coaxial delay line Z_U with the open circuit (Fig. 2.23c).

After corresponding propagation line delay time each split pulse reflected from opposite ends of the lines Z_I and Z_U arrive at the points I and U, respectively meeting with the new two fraction of the main pulse reflected from the DUT. Summation of the DUT pulse with the pulses in the reflected pulses in the lines with shorted and open circuit ends form the signal proportional to the current and voltage through the DUT, respectively. This signal are brought to the two channels of the oscilloscope SO.

In case of low voltage devices at high current the voltage signal at DUT at high current is the result of substraction of rather big values of the reflected signals. This significantly impacts the accuracy of voltage measurements. To improve the accuracy a direct voltage probe with corresponding transmission line delay can be used as an alternative (Fig. 2.23b).

The typical current and voltage waveforms of the TLP discharge pulse are rectangular with a user-defined rise time and pulse width. TLP pulses are usually 100 ns wide and rise within 200 ps to 10 ns which is partly equivalent to the rise times of the HBM current (Fig. 2.24).

To obtain the quasi-static DUT response the TLP pulses are analyzed for each stress level only in a selected time window that is usually selected between 70 and 90 % of the TLP pulse width. In this time window voltage and current are averaged (Fig. 2.25, left plots). By plotting each averaged voltage and current value the TLP I–V curve of a DUT is obtained after the corresponding multiple zap steps (Fig. 2.25, right).

TLP evaluation is used not only for the confirmation of the maximum current level provided by the clamp but also to deliver insights on the quasi-static characteristics of the ESD device. This is achieved by extracting several figures of merit from the TLP I–V characteristics. In industrial testers the TLP test is combined both with the functional leakage test and with the automatic bias conditions used for example for pulsed SOA measurements of the standard devices [37]. The leakage current is measured between each pulse (Fig. 2.26) under power down condition of the controlled pins prior to the measurement.

From a physical point of view, the ESD device in snapback mode operates similarly to a voltage-controlled switch with a resistive load. The first pair of parameters, the triggering voltage V_{T1} and triggering current I_{T1}, is used as figures of merit for the turn-on of the device into snapback. These parameters are important when defining the device turn-on within a so-called *ESD Protection Window* and to guarantee that during normal circuit operation the device does not accidently turn-on thereby having the risk for transient induced latch-up.

The next important figure of merit parameter is the holding voltage V_H. This parameter depends on the load impedance of the TLP tester and the snapback

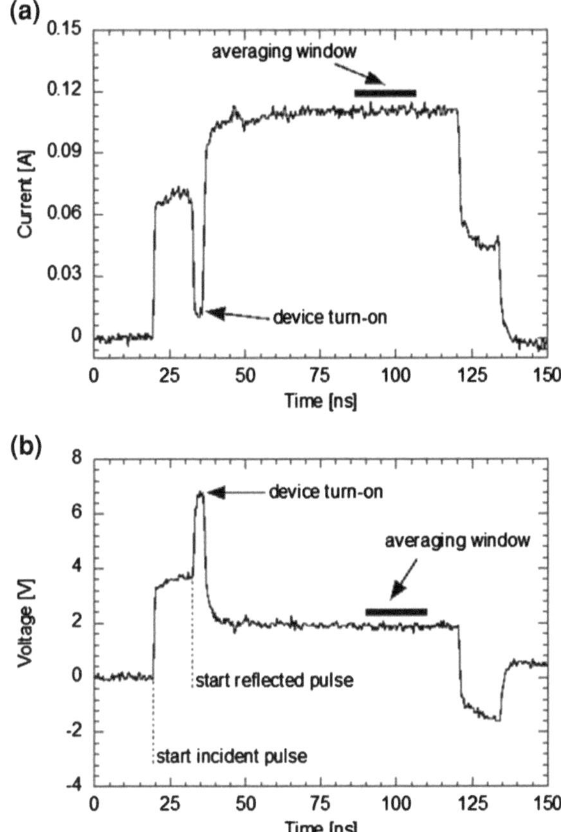

Fig. 2.24 100 ns TDR TLP waveforms: **a** voltage and **b** current; captured during TLP stress on a low-voltage-triggered SCR

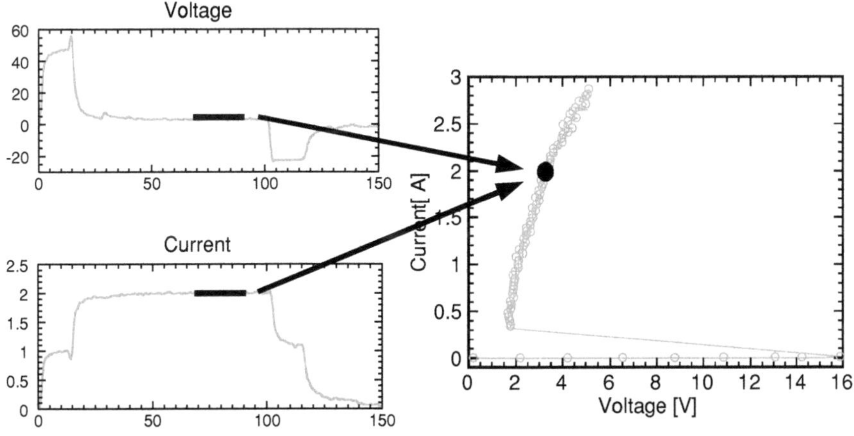

Fig. 2.25 Measuring TLP I-V curve: TLP waveforms (*left*) and extracted TLP I–V curve (*right*)

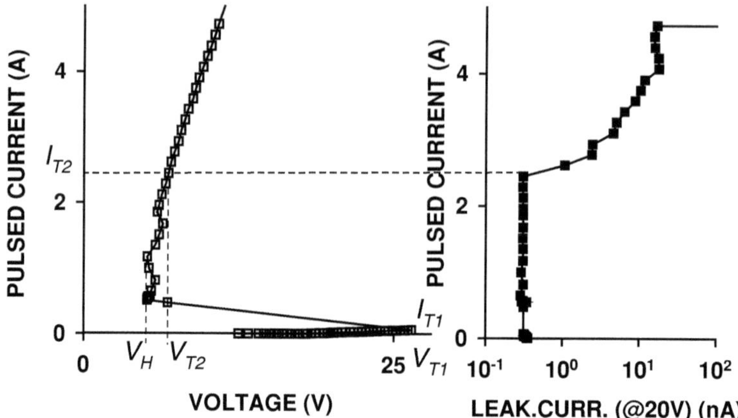

Fig. 2.26 Typical double plot for TLP snapback characteristics with major figures of merit indicated

voltage of the device. Thus, for example, when using a 50 Ω TLP system, the measured holding voltage generally is higher than the real device holding voltage which can be obtained by characterizing it with a waveform HBM system or a DC measurement.

After turn-on into the high current state, the device provides a certain on-state resistance due the internal positive feedback and a saturation region that determines the voltage waveform in the ESD pulse domain at a high current level. Usually, this parameter is an important practical criterion. For standard package level specifications it can be defined, for example for 1.33A (2 kV HBM). Finally, at a certain stress level, the physical limitation of the device results in irreversible changes to the devices structure. This equivalent TLP I–V curve point is usually referred to as I_{T2} and V_{T2} (Fig. 2.26)

In general the parameters I_{T2} and V_{T2} cannot be seen from the pulsed I–V characteristic alone. Even after irreversible changes the TLP I–V curve still may show the same trend. A separate functional test is usually required to establish the detection of device failure. In most TLP systems, this functional test is usually implemented by a simple leakage current measurement at a given voltage level with a defined parametric failure criterion, for example, a leakage deviation of one order of magnitude from the original level. For example, in the case of data for the 20 V snapback device presented in Fig. 2.26, the left plot for pulsed I–V shows no peculiarity associated with the irreversible failure that is already observed in the leakage current obtained with the functional test at an I_{T2} current level of ~ 2.5A.

TLP characteristics are very convenient for comparative analysis. They are widely used across this book both to represent the device parameters, pulsed SOA, and for the debugging of the analog circuit product pin characteristics. Pulsed SOA in the ESD time domain (further as ESD SOA) is a SOA measured for specific pulse conditions. In principle, this SOA depends on the specific pulse waveform.

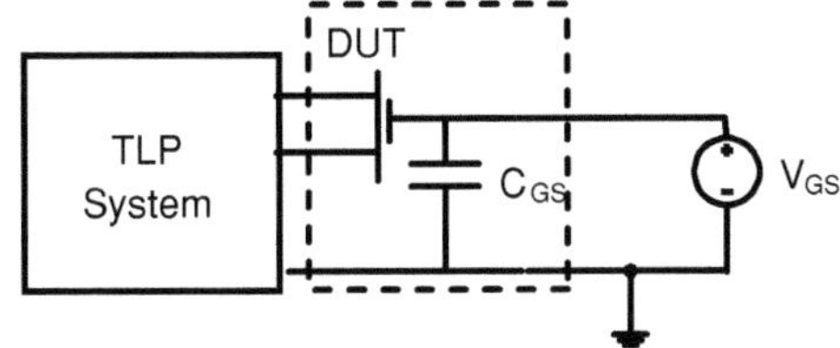

Fig. 2.27 Circuit diagram for setup for pulsed SOA evaluation using TLP measurements

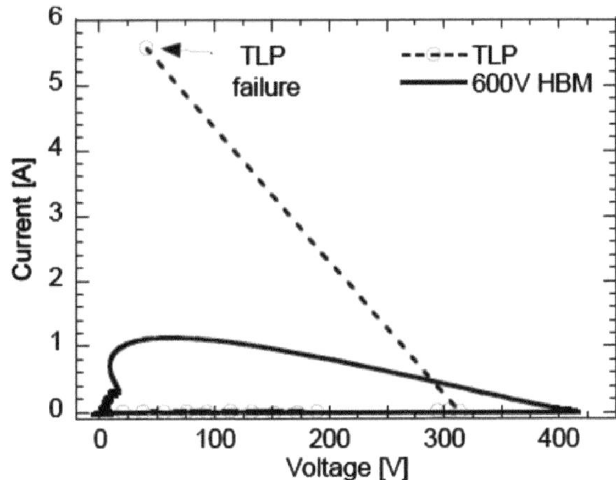

Fig. 2.28 Overlay of TLP I–V and HBM I–V curves taken from the same device for the stress level 0.6 kV

Known miscorrelation expected a different SOA for HBM, MM, and CDM, as well as for IEC and CDE system level test pulses. The most common characterization technique for ESD SOA evaluation combines TLP measurements with constant gate bias (Fig. 2.27). This technique applies TLP pulses to a DUT, for example, NMOS devices under constant gate bias on the gate electrode (Fig. 2.27). The setup includes a constant voltage source to provide gate-source bias, base-emitter bias, or current. TLP stress is applied under different bias conditions to obtain the DUTs SOA.

The impact of the TLP tester impedance on the failure level cannot always be neglected. One example is the so-called very high voltage (VHV) switching devices which can operate at voltages of several hundreds of volts up to more than one kilovolt. *E.g.* triggering a 600 V SCR device into the snapback mode in a 50 Ω TLP setup can result in a current level above 10 A. This may be much higher than the current capability of the device. An example of a device that passes 2 kV HBM stress (1.33 A), but fails the 50 Ω TLP stress is presented in Fig. 2.28. Another important issue for the high-voltage devices is the precise measurement of the holding voltage that is usually hidden by the load line of the 50 Ω TLP tester.

An experimental methodology to overcome this issue is either to increase the source impedance of the TLP tester [38] or to use a multi-level TLP tester [39]. The 100 ns TLP I–V curves of a HV nLDMOS-SCR were obtained with a 50 and a 500 Ω TLP tester (Fig. 2.29). The device snaps back to a higher current value

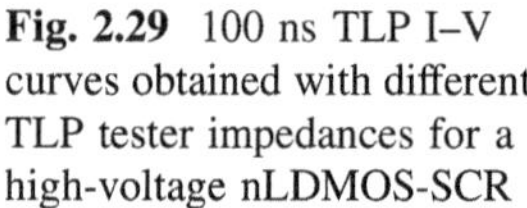

Fig. 2.29 100 ns TLP I–V curves obtained with different TLP tester impedances for a high-voltage nLDMOS-SCR

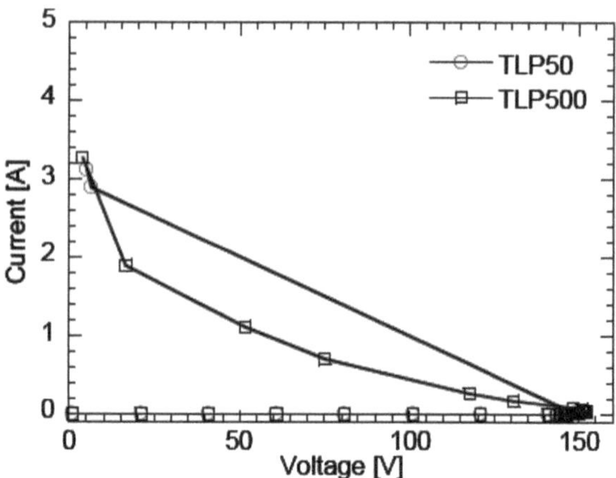

when using the 50 Ω tester. Due to the different load line the current after snapback is lower when using the 500 Ω tester. When using the 500 Ω system even points between the device turn-on around 150 V and the completed snapback around 3 V can be captured.

2.3.2 Very Fast TLP Test Method

The very fast TLP (vfTLP) test methodology has been proposed for the first time in [40]. The motivation for the vfTLP testing is to enable the device pulsed characterization and voltage/current measurement in the CDM time domain. In the standard practice document [41] vfTLP is defined as a TLP stress with less than 10 ns pulse width and 100–500 ps rise time. The fast rise time and short pulse duration allows the measurement of transient device behavior in the nanosecond time domain.

The TLP and vfTLP I–V characteristics of the same device are different due to the pulse length. Conventional 100 ns TLP measurements represent a quasi-static state of a device. The vfTLP I–V curves represent a more dynamic device behavior which is related *e.g.* to the DUT triggering delay (Fig. 2.30).

The quasi-static characteristics of the device are accessed by means of TLP I–V characteristics analysis. TLP waveform analysis enables the study of the transient device behavior during ESD stress. Because of the transient nature of the TLP pulses calibrating TLP and vfTLP measurements is an important step to obtain accurate testing results. For example in on-wafer TDR TLP setups the parasitic elements of the connection cannot be neglected due to distortion of the measurement results. The probe needles contain the main contributing parasitic elements. They are represented by their resistance $R_S < 1\ \Omega$ and their inductance $L_S \sim 10\text{–}20$ nH (Fig. 2.31) [42, 43].

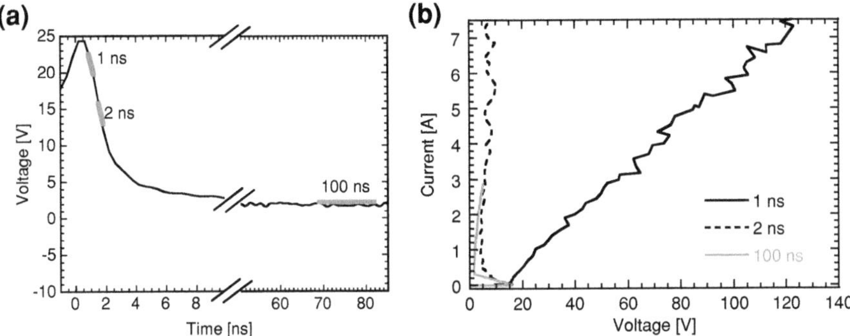

Fig. 2.30 Illustration of different averaging windows during extraction of vfTLP and TLP I–V curves: **a** location of averaging windows and **b** I–V curves based on the different averaging windows

In the high current region of a TLP I–V characteristics the parasitic resistance of the probe needles causes a significant parasitic voltage drop due to the needle resistance R_S. The additional voltage drop brings inaccuracy to the TLP waveform analysis. The impact of the parasitic elements must be calibrated out from the measurement data. To take into account the additional voltage introduced by the needle resistance the TLP tester can be connected to a short circuit element. The I–V characteristic is measured as an initial step of the calibration procedure. The slope of the obtained I–V curve equals to the serial resistance of the probe needles. By introducing the corresponding correction to the measurement data the real I–V characteristics at the DUT can be obtained.

A more complex calibration routine for vfTLP testing setups is related to the significantly faster rise time and shorter pulse duration. The corresponding measurement setup with high bandwidth current and voltage probes results in a need for high cost RF probes. Their main disadvantage in case of on-wafer vfTLP measurements is the non-flexible probe pitch. The layout of the on-wafer test structures has to be designed for available RF probe pitches.

A practical alternative with standard tungsten probe needles requires (Fig. 2.32) a de-embedding of the parasitic from each measured voltage and current waveform using a dedicated calibration/de-embedding methodology [44]. The method is independent of the pulse shape generated by the tester. The vfTLP pulse can feature different durations and rise and fall times.

The de-embedding/calibration methodology uses three loads to characterize the full vfTLP setup: an open circuit, a short and a 50 Ω resistor. Voltage and current are captured from those loads. With the data, a model for the needle parasitic and the loss in transmission line are extracted.

The voltage and current for a short load measured by the oscilloscope were calibrated. The needle inductance causes a voltage overshoot of 23 V (Fig. 2.33a). After calibration this voltage overshoot is completely removed. As a result the

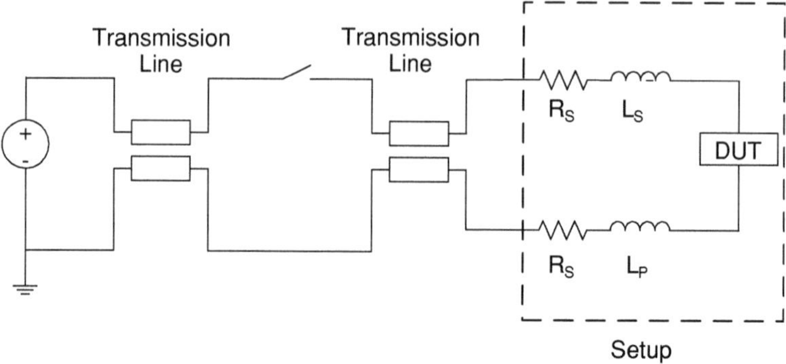

Fig. 2.31 Circuit diagram for on-wafer TLP measurement setup with device under test connection with parasitic resistance R_S and parasitic inductance L_S of the probe needle

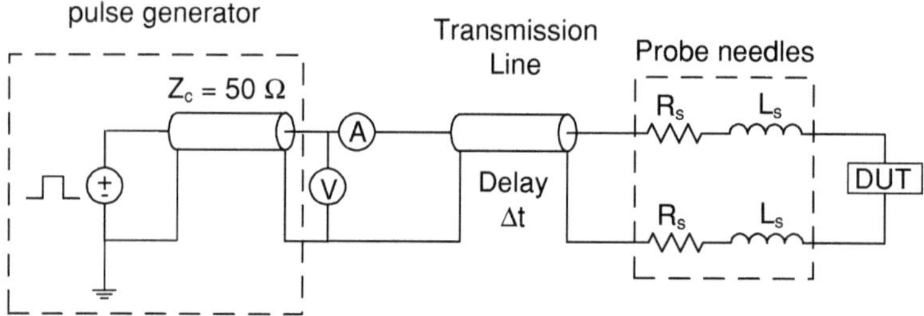

Fig. 2.32 Circuit diagram for an on-wafer TDR vfTLP measurement setup including loss in transmission line and probe needle parasitic

voltage from a short load has eliminated the voltage overshoot due to the needle parasitic (Fig. 2.33b).

The calibration methodology is applied to vfTLP measurement data, obtained from a diode-triggered SCR (DTSCR). The impact of the test setup parasitic is demonstrated for a diode-triggered SCR (DTSCR) by comparing the vfTLP I–V characteristics before and after the calibration (Fig. 2.33c). Without calibration, both the on-resistance and the holding voltage of the DTSCR are higher. The additional voltage drop due to the parasitic is included in the extracted I–V curve. After calibration, the on-resistance is much lower and a more accurate holding voltage is extracted from the I–V curve. The accuracy of the calibration methodology is validated by comparison of the results with RF probe needles (Fig. 2.33d).

For both TLP and vfTLP testing setups a Kelvin type of setup (TDT) can be used with an additional DUT connection to eliminate the need for a calibration procedure related to the probes parasitic [38]. However due to possible different cable lengths of voltage and current measurement an alignment of the obtained voltage and current waveforms might be required even if a Kelvin setup is used.

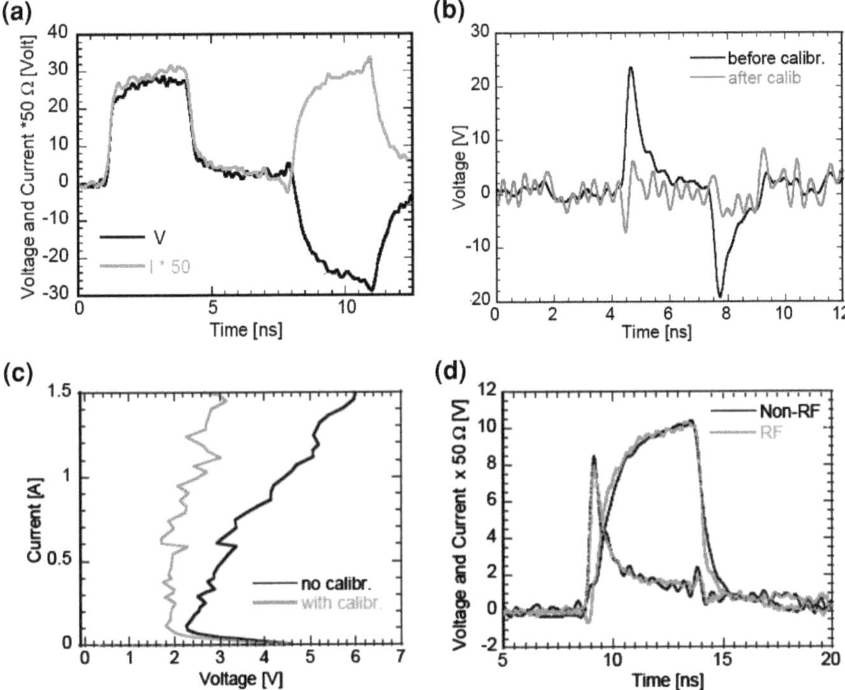

Fig. 2.33 Application of vfTLP calibration routine to measurement data: the voltage and current for a short circuit load (**a**), comparison of the voltage before and after calibration (**b**), 3 ns vfTLP I–V curve of a diode-triggered SCR (**c**) and comparison of measured current and voltage waveforms during 5 ns vfTLP stress after calibration and when measured with RF probes (**d**)

2.4 Transient Waveforms Characterization for ESD Stress

While TLP characterization and the passing level for ESD pulses are the most used tools for practical design, the analysis of the waveforms for the different transient pulses often can reveal additional useful information and help to debug and optimize different ESD solutions. For example using current and voltage waveform capture techniques for HBM pulses the phase-diagram like HBM I–V characteristic can be reconstructed by plotting instantaneous current $I(t)$ over $V(t)$ over the entire HBM time domain (Fig. 2.34).

The analysis of such phase I–V characteristics reveals the region representing the device turn-on, oscillations around the peak current and a stable monotonous part during the remaining 150 ns HBM pulse discharge. The first two regions represent the transient device behavior under ESD stress, whereas the latter represents the quasi-static device pulsed operation.

Similar analysis can be completed to represent the transient characteristics for the system level pulse. To achieve this goal several important aspects must be

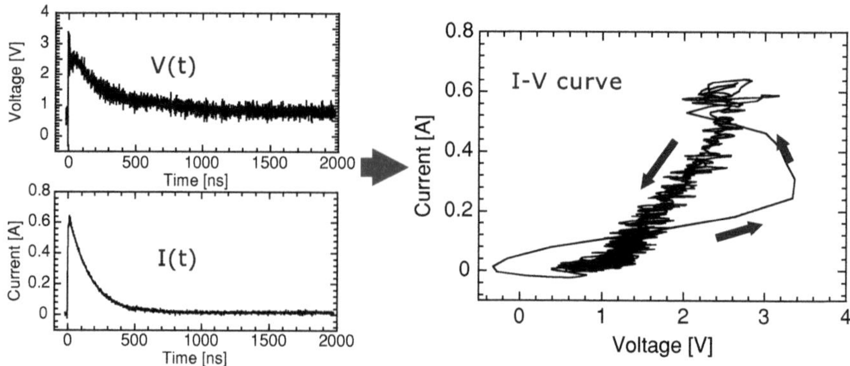

Fig. 2.34 Voltage *V(t)* and current *I(t)* waveforms captured from a ESD diode at a HBM stress level of 1 kV and plotted phase HBM I(t)−V(t) characteristics

taken into account. They are related to the required calibration procedures for the waveform capturing, clamp operation specific and the setup itself. These aspects are discussed in this section.

2.4.1 Calibration of ESD Waveforms

The example of a calibration procedure is described in this section for the case of HBM tester. The methodology can be also applied to other two-pin component-level tests like MM and HMM.

At the standard HBM pulse rise time of ∼2–10 ns the voltage and current waveform measurement is very sensitive to parasitic elements in the setup (Fig. 2.35). The parasitic elements contributed by the probe needles and test setup cause an additional voltage drop around the current peak and to the linear region of the HBM I–V characteristic. They must be eliminated by an appropriate calibration procedure. The current transformers usually used as current probes have a limited bandwidth, which is between 25 kHz and 2 GHz. This results in distortion of the falling part of the measured current waveform.

To remove this low frequency distortion of the current transformer, the transfer function *TF* of the current transformer needs to be determined for the calculation of the real current I_{corr} out of the measured current I_{CT} (2.1).

$$I_{corr} = TF \cdot I_{CT} \tag{2.1}$$

This corresponds to a de-convolution problem, where the determination of an unknown input signal is calculated from the measured output signal if the transfer function of the system is known. The methodology allows to extract the transfer function *TF*, the needle parasitic resistance R_p and inductance L_p. The extracted

Fig. 2.35 Circuit diagram for an HBM on-wafer test setup with indicated parasitic tester capacitance C_{HBM}, parasitic tester inductance L_{HBM}, board capacitance C_B, probe needle resistance R_S and probe needle inductance L_S

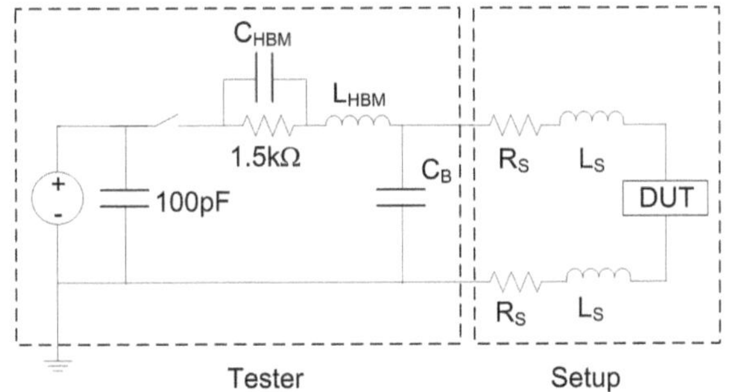

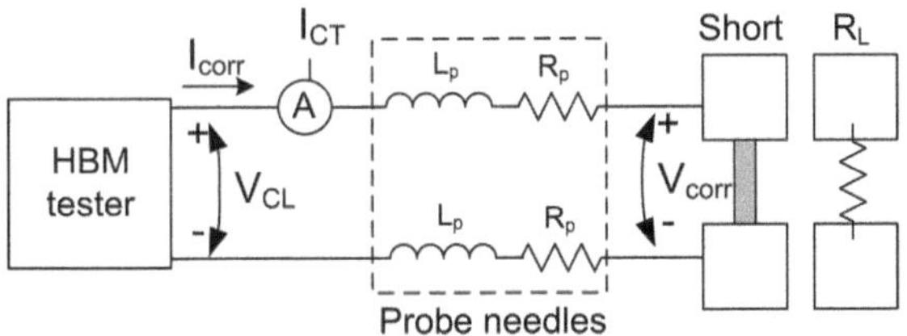

Fig. 2.36 Setup for HBM calibration with indicated corrected 'real' current I_{corr}, distorted current measured by current transformer I_{CT}, measured voltage V_{CL}, corrected 'real' voltage V_{corr}, parasitic needle inductance L_P, parasitic needle resistance R_P and load for calibration R_L

values are calculated from HBM voltage V_{cl} and current I_{ct} waveforms captured for given resistive load R_L and short circuit conditions (Fig. 2.36).

Measured voltage V_{cl} and current I_{CT} are aligned in time and transformed to the frequency domain. Two expressions of the transfer function of the current transformer are obtained - one for the load R_L (2.2) and one for the short (2.3) measurement.

$$TF_{cl}^{load} = \frac{I_{corr}^{load}(\omega)}{I_{CT}^{load}(\omega)} = \frac{V_{cl}^{load}(\omega)}{(R_L + Z_P) \cdot I_{CT}^{load}(\omega)}, \tag{2.2}$$

$$TF_{cl}^{short} = \frac{I_{corr}^{short}(\omega)}{I_{CT}^{short}(\omega)} = \frac{V_{cl}^{short}(\omega)}{Z_P \cdot I_{CT}^{short}(\omega)}, \tag{2.3}$$

$$Z_P = 2 \cdot (R_P + j\omega L_P), \tag{2.4}$$

where Z_P is the impedance of the needles. Both transfer functions are identical as they are obtained with the same current transformer and on the same setup:

$$\frac{V_{cl}^{short}(\omega)}{Z_P \cdot I_{CT}^{short}(\omega)} = \frac{V_{cl}^{load}(\omega)}{(R_L + Z_P) \cdot I_{CT}^{load}(\omega)} = TF. \tag{2.5}$$

From (2.5) Z_p is obtained as

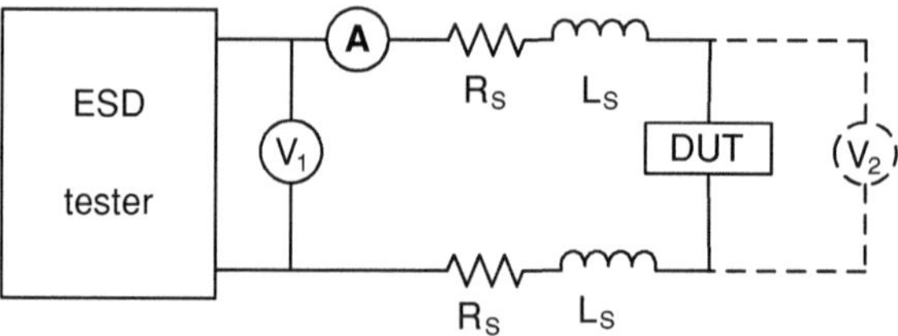

Fig. 2.37 Circuit diagram of an ESD on-wafer test setup including parasitic with option Kelvin setup using direct voltage V_2 measurement, where R_S and L_S are probe needle resistance and inductance, V_1 tested represents standard voltage measurement

$$Z_P = \frac{V_{cl}^{short}(\omega) \cdot R_L \cdot I_{CT}^{load}(\omega)}{V_{cl}^{load}(\omega) \cdot I_{CT}^{short}(\omega) - V_{CL}^{short}(\omega) \cdot I_{CT}^{load}(\omega)}. \tag{2.6}$$

For a typical on-wafer measurement setup the series resistance R_p is ~ 0.5–0.8 Ω and the inductance L_p is ~ 10–15 nH, extracted per a single needle. Finally, *TF* is obtained by substituting (2.6) in (2.2) or (2.3). To obtain the real current through a device under test (DUT), the measured current waveform $I_{DUTmeas}$ is transformed to the frequency domain and multiplied with the transfer function *TF* (2.7). A corrected voltage waveform across the DUT is calculated referring to Eq. (2.8).

$$I_{corr}^{DUT}(\omega) = TF \cdot I_{meas}^{DUT}(\omega), \tag{2.7}$$

$$V_{corr}^{DUT}(\omega) = V_{meas}^{DUT}(\omega) - Z_P \cdot I_{corr}^{DUT}(\omega). \tag{2.8}$$

Due to limited power of the signal spectrum at high frequencies, the numerator and denominator in Eqs. (2.2) and (2.3) becomes very small. The result is unrealistic values at high frequencies that have to be removed before the IFFT operation. To reduce the noise level additional filtering of the obtained data is required. The corrected current $I_{corr}^{DUT}(\omega)$ and voltage $V_{corr}^{DUT}(\omega)$ waveforms are transformed to the time domain. The calibration data is independent of the pre-charge voltage. A calibration needs to be performed only once before a full set of HBM waveform measurements.

The de-embedding of the HBM tester parasitic requires the application of the calibration data to every captured voltage and current waveform. Also the data needs to be filtered to remove the increased noise due to FFT/IFFT operations on the data. As an alternative to the calibration procedure the voltages can be captured in a Kelvin setup [38]. The advantage of the setup (Fig. 2.37) is that the voltage is measured directly at the DUT using a second pair of probes. Since the stress current has a separate path the parasitic voltage drop is not interfering with the voltage measurement thus eliminating the need in calibration.

The comparison of the maximum voltage dependence upon the stress level for the peak voltage with and without Kelvin technique is apparent (Fig. 2.38a, b) and less noisy. It can be further used to validate the calibration procedure too (Fig. 2.38c).

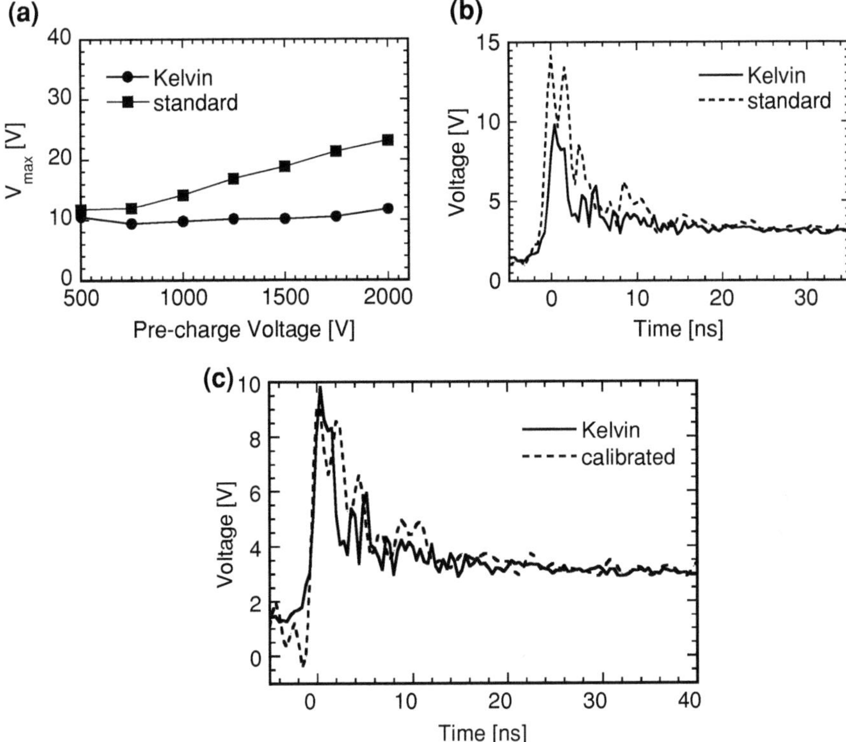

Fig. 2.38 Application of Kelvin methodology for low-voltage-triggered SCR device measurements: maximum voltage versus HBM stress level (**a**) and voltage waveforms for a 1 kV HBM stress (**b**) and comparison with the calibration/de-embedding results

Additionally the frequency response of the current transformer can be modified to improve the bandwidth of the current transformer. Changing the transfer function of the current probe can significantly simplify the calibration procedures. The lower frequency limitation of inductive current probes is changed to a level which is acceptable for the measurements. The equivalent circuit of an inductive current transformer typically used in ESD measurement setups includes the self-inductance and the termination resistance (usually 50 Ω) (Fig. 2.39). The formed L–R filter has a lower frequency limit that can be reduced further by adding a low value parallel resistor between the current transformer and the input of the oscilloscope [45].

For example, by adding a resistor of 5 Ω in parallel to a Tektronix current transformer CT-6, the lower bandwidth limitation is reduced from 250 kHz (datasheet) down to $\sim$40 kHz (Fig. 2.40) impacting significantly the measured current waveforms. As a result the negative current part of the current waveform is strongly reduced (Fig. 2.41)

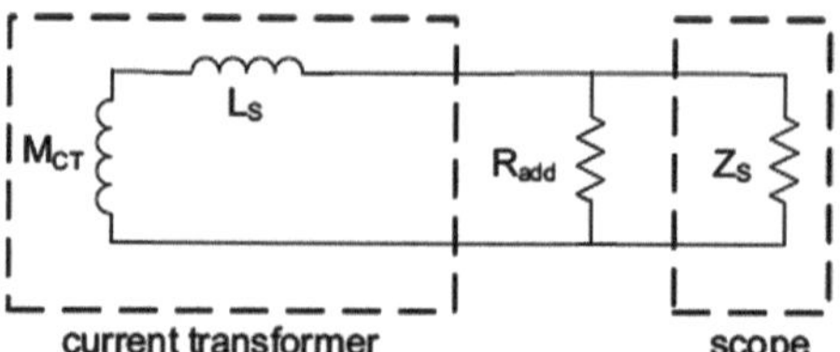

Fig. 2.39 Equivalent circuit diagram of CT probe with added resistor representing the mutual inductance M_{CT}, the self-inductance L_S and added resistor R_{add} with the input impedance oscilloscope Z_S

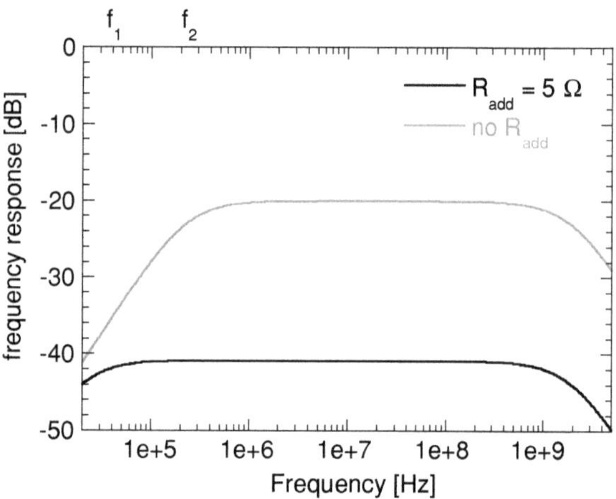

Fig. 2.40 Simulated frequency response of a Tektronix CT-6 current transformer with and without added parallel resistor; f_1 lower frequency limit with added resistor of 5 Ω, f_2 lower frequency limit without added resistor

The main advantage of modifying the current probe is the reduced effort during the HBM tester calibration. If a modified current probe is used together with a Kelvin setup no HBM tester calibration is required when measuring voltage and current waveforms during HBM stress. Consequently no FFT is applied to the measurement data and the increase of the noise in the measurement data is prevented.

Once the waveform capture setup is done, the measurements of the transient characteristics can be applied for comparative analysis of the ESD solutions. Such application can be demonstrated on the example of a low-voltage SCR local clamp operation as a function of the driver circuit design. The clamp design includes a diode triggered SCR device. To evaluate the clamp performance an nMOS transistor has been added as a gate monitor device in parallel to the clamp (Fig. 2.42).

Three different variations A, B and C of the clamp design were used [44] for comparative analysis. The type A was representing the baseline clamp composed from the SCR device and the small width reference diodes. The diode controlled only small local Anode-G2 junction (Fig. 2.42) of the SCR N-base. Thus the diode

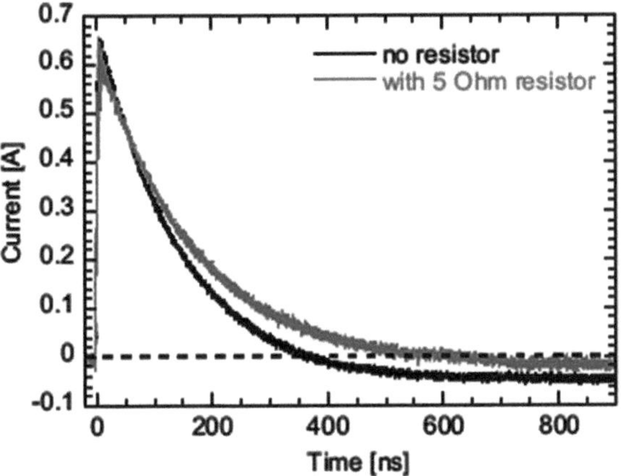

Fig. 2.41 Current waveform into a short load, measured with a current transformer Tektronix CT-6 with and without added 5 Ω resistor; HBM stress level: 1 kV

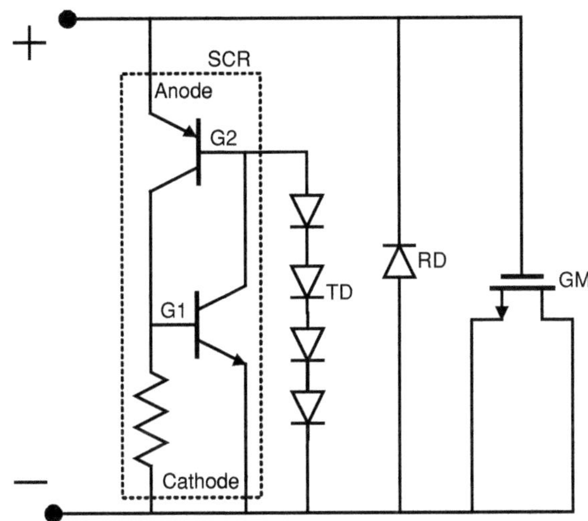

Fig. 2.42 Schematic of diode triggered silicon-controlled rectifier with gate-monitor in parallel

pull-down circuit was triggering the SCR locally and the delay of the turn-on was expected to provide relatively large clamp voltage overshoot due to the turn-on delay. Type B was representing an improved turn-on speed. This was achieved by wider trigger diodes to control the entire width of the SCR N-base junction connected to the Anode-G2 terminals [47].

Finally the further turn-on speed improvement was in implemented in the Type C version, when the conventional shallow-trench-isolation (STI) diodes in the low side reference circuit were replaced by poly-bounded diodes. Due to better on-state performance of the poly-bounded diodes over the STI diodes the SCR triggering speed was expected to be the best among the three types.

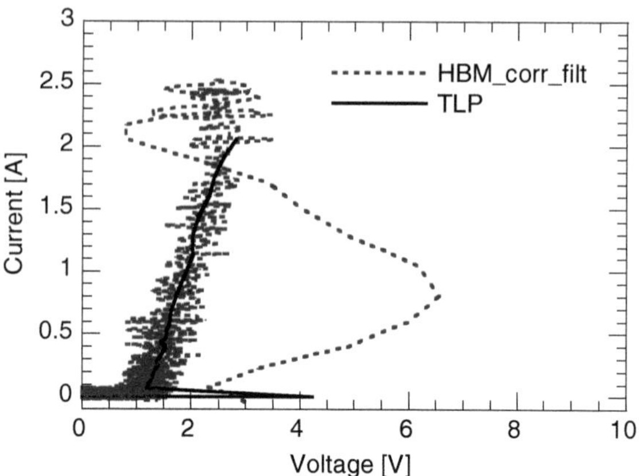

Fig. 2.43 Overlay of 4 kV HBM I–V and 100 ns TLP I–V curves, obtained from device type A

To emphasize the informative value of the waveform capture analysis value the TLP measurements were also performed for all three SCR clamp types with no gate monitor present. As expected, no difference for the three clamp types can be observed in the TLP I–V characteristics. There is a clear matching between the TLP I–V curve and the linear part of the HBM I–V curve. The TLP characteristics of each clamp were similar to the TLP I–V curve of type A (Fig. 2.43). A different result for maximum TLP current was recorded when the gate monitor was connected demonstrating the best performance for clamp type C.

This fact points on a critical role of the voltage overshoot that was damaging the nMOS device gate oxide. The effect can be made visible by an overlay of a calibrated HBM phase I–V characteristics (Fig. 2.43). Unlike in TLP I–V curves the voltage overshoot is visible in a HBM I–V plot.

The comparison of the HBM I–V characteristics for the three clamp design types demonstrated the expected level of voltage overshoot depending on the clamp design (Fig. 2.44). Device type A is the slowest device and provides the highest overshoot voltage. The combination of the DTSCR clamp type *A* and the gate monitor fails at ∼1.9 kV HBM.

The faster turn-on speed of the type *B* design results in an increased passing level of 2.6 kV when the gate monitor is connected. Finally the biggest improvement is achieved for clamp type *C* with poly bounded reference diodes. It generates the lowest overshoot due to fastest turn-on. This leads to an increased HBM robustness in the gate monitor experiment of about 4.6 kV.

2.4.2 Transient Characterization of HV Circuits

The high-voltage (HV) ESD devices include longer drift and blocking junction regions in comparison to LV CMOS devices. Therefore a longer carrier transient time from the anode to cathode is involved in the initiation of the conductivity

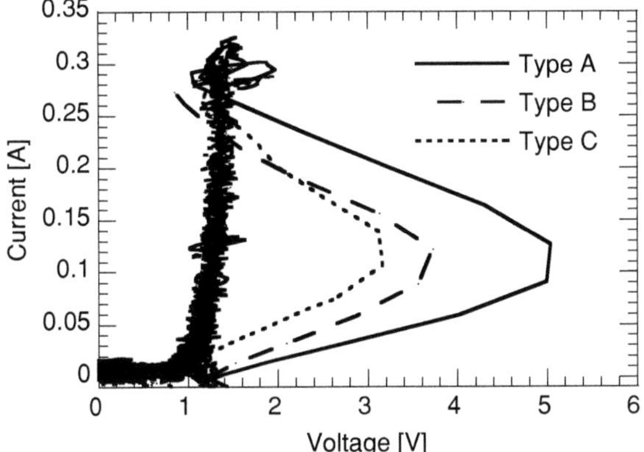

Fig. 2.44 HBM I–V curves obtained from the three SCR types for the same HBM stress level of 500 V

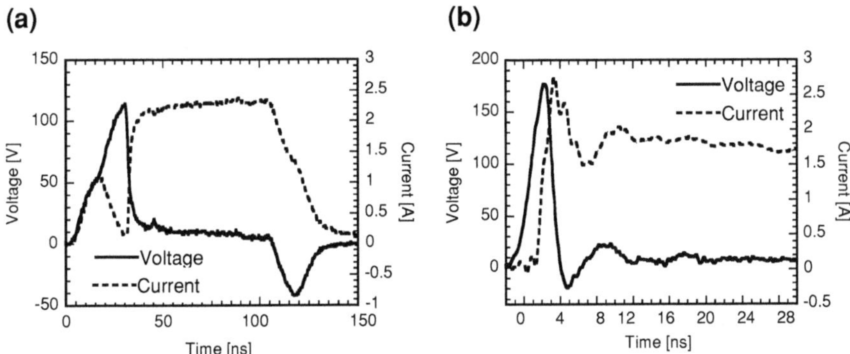

Fig. 2.45 Calibrated TLP (**a**) and HBM (**b**) voltage and current waveform at a stress level of ∼2 A for 100 V VDMOS-SCR device

modulation process. This results in slower device reaction during ESD (over-) stress, non-uniform triggering, and filamentation [48, 49]. An example of the HV devices realized in a 100 V BCD process technology is discussed below to demonstrate the peculiarities. The voltage and current waveforms were captured using the described calibrated methodologies for TLP and HBM ESD pulses. The DUT was a vertical double-diffused MOS SCR (VDMOS-SCR) device stressed with about 2 A equivalent current (Fig. 2.45).

A comparison of the TLP and HBM waveforms shows that during HBM testing current and voltage overshoots are occurring but not during TLP testing. The TLP trigger voltage is about 120 V (Fig. 2.45a), whereas in HBM an overshoot voltage of about 180 V (Fig. 2.45b) is measured. Additionally, in HBM a delay between voltage and current peak is observed. Through the measured device flows a HBM current with a peak ∼2.7 A. This is much higher than the nominal expected current

Fig. 2.46 Equivalent schematic of HBM (**a**) and TLP (**b**) setups during device stress: C_{HBM} parasitic tester capacitance, L_{HBM} parasitic tester inductance, C_B test board capacitance

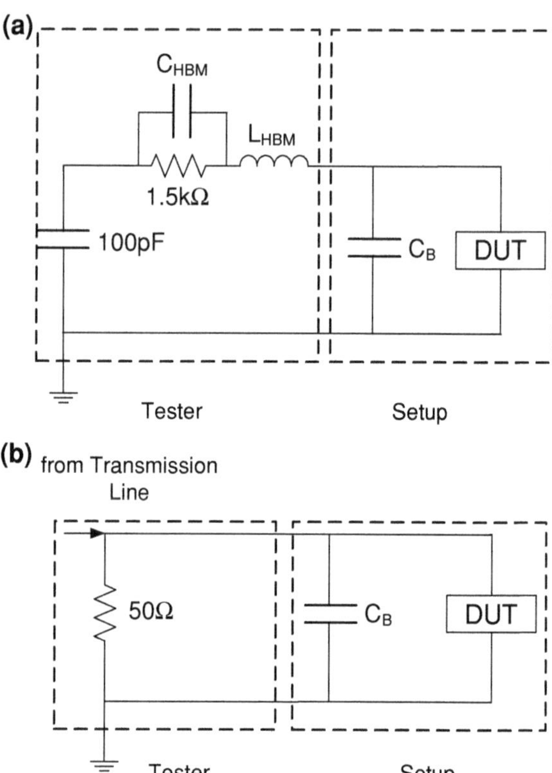

of 2 A for 3 kV pre-charge voltage. What are the underlying reasons for these results? The different testing environment in TLP and HBM influences strongly the transient behavior of the ESD clamp. In a HBM test setup the 100 pF discharge capacitor is in parallel to the board capacitance C_B. This results in a much larger equivalent capacitance in parallel to the DUT in the HBM testing setup. The much larger equivalent capacitance in the HBM setup interacts with the DUT and causes the described behavior. The equivalent schematics for TLP and HBM measurement setups show different ways how the C_B capacitor is charged and discharged (Fig. 2.46). In the HBM setup the capacitor is charged to the level of the triggering voltage. During the fast transient triggering of the HV SCR device in high conductivity state, the capacitor C_B releases the major charge into the SCR device. This generates CDM-like high current peak. The faster rise time of this initial peak is the likely reason for the higher voltage overshoot. In case of TLP stress the C_B capacitor is driven by the 50 Ω load that limits the current through the capacitor.

It is important to note that the capacitive loading of the voltage probe impacts the measured HBM waveforms. Typically commercially available voltage probes have an input capacitance in the 8–10 pF range. These values in general cannot be neglected. For example in case of the above devices the voltage probe capacitance influences the device triggering and results in a growth of the current amplitude

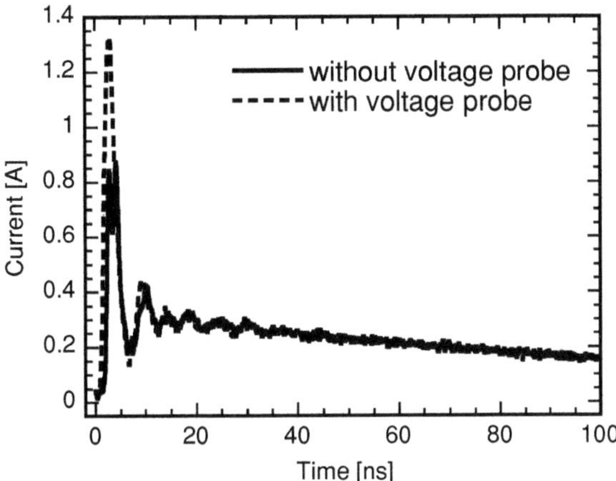

Fig. 2.47 HBM current waveforms obtained from the same device with and without connected voltage probe, stress level: 500 V (equivalent to 0.33 A)

from ~0.8 A without voltage probe to 1.4 A with connected voltage probe (Fig. 2.47).

To evaluate the influence of the capacitive load of the voltage probe on the HBM waveform measurements, additional capacitors with different values are placed in parallel to the DUT (Fig. 2.48a). In the described setup for HBM waveform capturing the VDMOS-SCR device peak voltage and current depend on the total capacitance across the device (Fig. 2.48b). The maximum voltage obtained at a HBM pre-charge level of 4 kV remains practically unchanged with increased capacitance, while the increasing capacitive load increases significantly the current overshoot.

2.4.3 Transient Characterization with On-Wafer HMM Setups

On-wafer HMM testing setups can be used not only to measure the ESD passing level of the components for this pulse. Similar to the HBM methodology, described above, this setup can be also modified to enable the voltage waveform capturing. This enables both the preliminary validation of IC pins during ESD stress and the research and design of ESD devices for system-level ESD protection designs.

Due to the fast rise time of the HMM pulse first peak a more sophisticated setup design is required in comparison to HBM waveform capturing. If the voltage measurement is made across the probe needles a calibration and de-embedding of the test setup parasitic required extracting the real voltage waveforms at the tested device. The much larger voltage amplitudes during HMM stress combined with the faster rise time in general results in much higher voltage overshoots. During voltage capturing the scaling of the oscilloscope is usually set to capture the full

Fig. 2.48 Simplified circuit diagram for the HBM measurement setup with additional capacitor C_{add} (**a**) and peak voltage and current during an HBM stress on a VDMOSSCR device at pre-charge voltage: 4 kV upon the capacitor C_{add} variation (**b**)

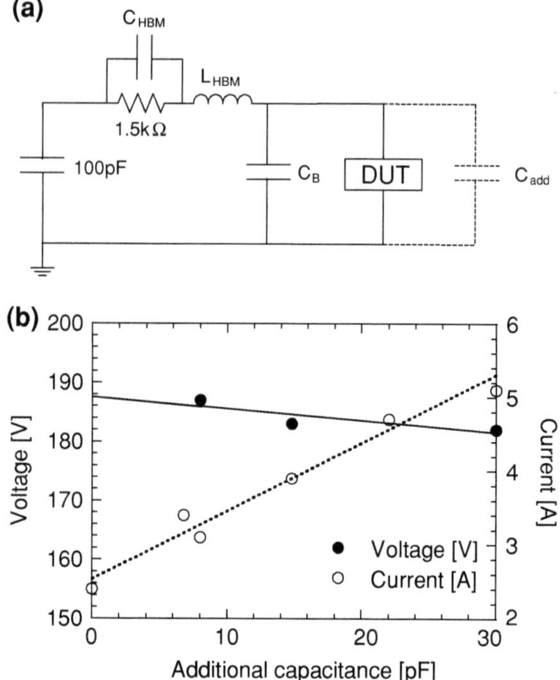

Fig. 2.49 Simplified circuit diagram for on-wafer HMM setup with voltage measurement in Kelvin configuration

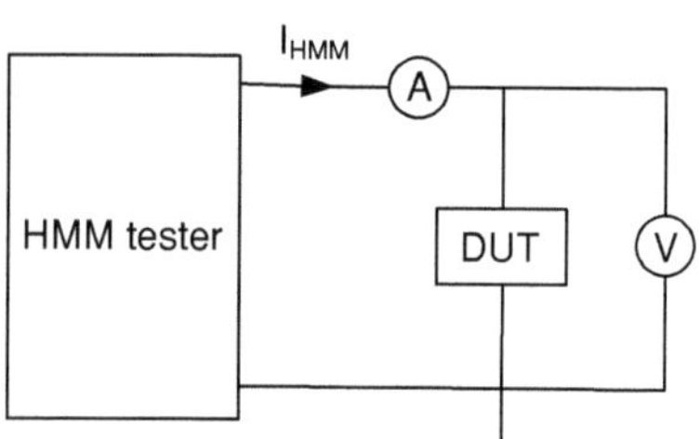

amplitude of the voltage overshoot. This reduces the measurement resolution for the part of the voltage waveform where the device is in the on- or holding state after the first 5–10 ns time domain of the pulse. Hence, the preferable setup for HMM on-wafer measurements is the Kelvin setup (Fig. 2.49). The voltage probe is connected through a second pair of probes to the probe pads of the DUT which decouples it from the HMM current path. This limits or even eliminates the corresponding waveform distortion.

For example, the voltage and current waveforms for on-wafer HMM stress of a nLDMOS-SCR measured in a Kelvin setup provide the realistic values for the triggering voltage of the nLDMOS-SCR, holding voltage and the device turn-off (Fig. 2.50).

Fig. 2.50 Voltage and current waveforms during on-wafer HMM stress; device: nLDMOS-SCR; 1.5 kV HMM stress level with Kelvin setup voltage measurement

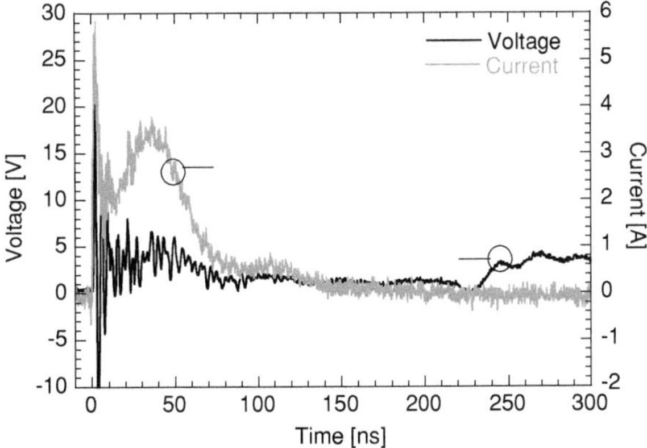

2.5 HMM Tester Correlation

The HMM standard practice document [30] outlines methods how system-level ESD stress should be applied to components. "Components" are defined as both single devices and integrated circuits. Two main tester concepts are proposed: HMM testing setup using the IEC 61000-4-2 standard discharge circuit and 50 Ω HMM pulse generators. The 50 Ω HMM pulse generators create the stress waveform by shaping a rectangular pulse with transmission lines and filters. In spite of a good correlation reported for 50 Ω HMM (HMM-50) and ESD gun based testers [50] for standard and advanced LV CMOS technologies, this section demonstrates a miscorrelation case study. The impedance of HMM-50 tester can lead to false results in comparison to testers which use a discharge circuit of the IEC 61000-4-2 type.

2.5.1 Test Setup and Device Characterization

The HMM testers used for the comparison are connected to on-wafer test setups. The gun based HMM tester is a HANWA HED-5000 M (HMM-IEC, Fig. 2.51a) tester which uses an IEC 61000-4-2 RC type of discharge circuit to create the stress current [51]. By design, the discharge module of this HMM-IEC tester does not generate the typical electromagnetic fields produced by ESD guns during the discharge. The tester form factor is compatible with on-wafer setup integration which allows a short connection between the discharge module the DUT on the wafer. The second tester is a 50 Ω HMM tester HPPI 3010C/3011C (HMM-50, Fig. 2.51b) which is based on a modified Transmission Line Pulse tester [52]. The transmission lines and rise time filters are modified in a way that allows the sourcing of a stress current which has a shape within the specifications of the IEC 61000-4-2 standard.

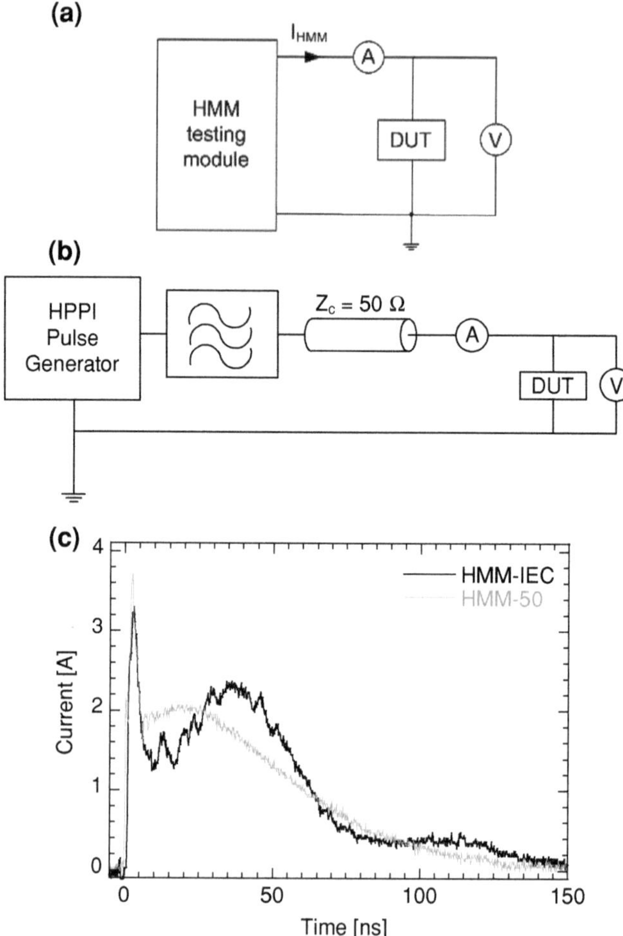

Fig. 2.51 Block diagrams for HMM-IEC (**a**) and HMM-50 (**b**) on-wafer test setups with connected voltage and current probes and comparison of their short circuit current waveforms for 1 kV HMM stress equivalent (**c**)

In spite of the difference in the output impedance, both testers provide current waveforms into a short circuit load that are compliant to the IEC 61000-4-2 standard. In general the real tester pre-charge voltages cannot be used for the comparison of failure level due to the different methods of creating the ESD stress current. The use of the first peak of the HMM current waveform is neither an adequate figure of merit for the comparison. The amplitude of this first peak strongly depends on the parasitic load of the measurement setup, the DUT impedance and the accuracy of the pulse source. In addition there can be a variation from pulse to pulse when repeating zaps at the same stress level and in the same test setup. For tester comparison, the current after 30 ns is a more reliable

Table 2.4 Devices under test and process technologies used

Device name	Device type	Technology
LV N-SCR	nLDMOS-SCR	90 nm CMOS
Diode	ESD diode, forward	100 V BCD process
HV N-SCR	nLDMOS-SCR	100 V BCD process
PNP	Lateral PNP	100 V BCD process
NMOS	Grounded-gate NMOS	5 V analog CMOS

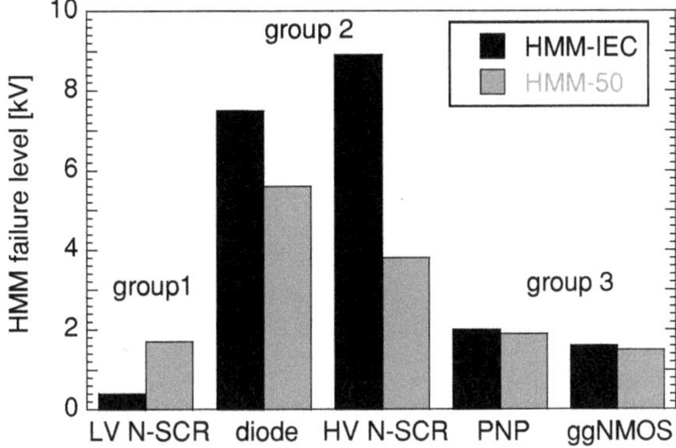

Fig. 2.52 Summary of obtained measurement results; three device groups: *group 1* lower failure level during HMM-IEC testing, *group 2* higher failure level during HMM-IEC testing, *group 3* similar failure level when using HMM-IEC and HMM-50 tester

figure of merit. To comply with the standard the current level after 30 ns is expected to be ~ 2 A/kV, although with ± 30 % variation.

To minimize waveform distortions the voltage waveforms were captured with a Kelvin setup. The selected DUTs are typical ESD protection devices for analog, high-voltage-tolerant and high-voltage/smart power applications (Table 2.4).

The experimental results obtained from all devices can be subdivided in 3 groups based on their correlation between the HMM-IEC and the HMM-50 testing results (Fig. 2.53). The lateral PNP and the NMOS show similar failure level for both testers (group 3). The diode and HV-SCR devices fail at two times lower passing level when using the HMM-50 tester (Group 2, Fig. 2.52). In contrary the LV N-SCR device fails at substantially lower level when using the HMM-IEC tester.

The difference in failure level for the LV N-SCR (0.5 kV vs. 1.8 kV) can be explained by the gate oxide breakdown effects due to a faster rise time of the HMM-IEC tester. To support these conclusions mixed-mode simulations with DECIMM^{TM} [19] are carried out. The models of the two HMM tester are implemented as well as the device TCAD model. The simulated voltage waveforms (Fig. 2.53) for the same equivalent stress level show a smaller voltage amplitude and a slower rise time with the HMM-50 tester suggesting less gate oxide stress.

To compare the impact of the HMM tester on the GOX stress the vertical electrical field across the gate oxide (GOX) was also extracted from the device-

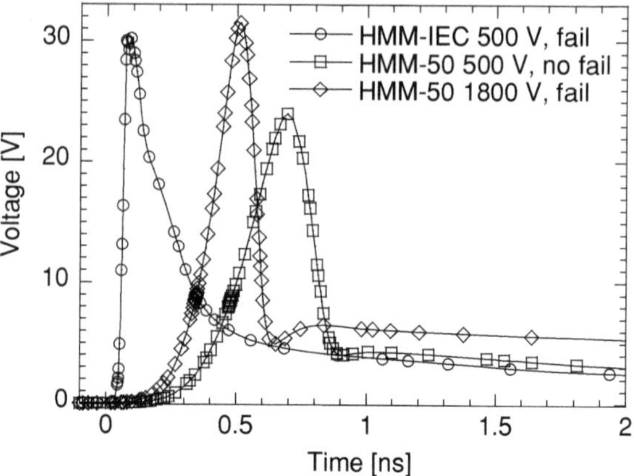

Fig. 2.53 Simulated voltage waveforms across the low-voltage LV N-SCR during triggering when stressed with different HMM tester; stress level: equivalent stress level

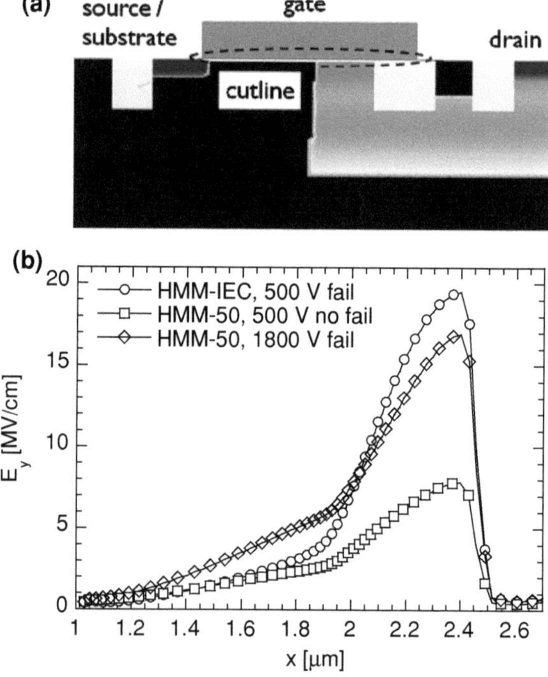

Fig. 2.54 Simulated cross section of the LV N-SCR (**a**) and simulated vertical electrical field through GOX for different HMM testers and equivalent stress level (**b**)

circuit mixed-mode simulations. The simulated cross-section of the LV N-SCR (Fig. 2.54a) was used to extract the maximum vertical electrical field across the GOX at different stress levels. For HMM-IEC tester pulse the electrical field is significantly exceeding the value of the HMM-50 tester for the same stress level. The peak value correlation of the 500 V HMM-IEC is matching the corresponding

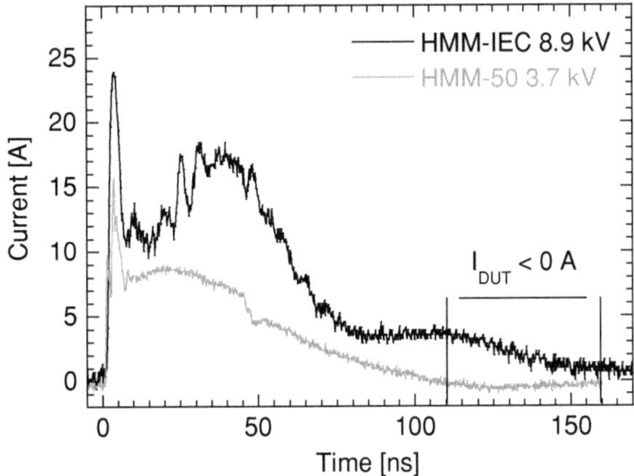

Fig. 2.55 Measured HMM current waveforms at (equivalent) failure level for the high-voltage ESD diode when stressing with both HMM testers

value for 1.8 kV HMM stress which supports the conclusion of the failure cause. GOX breakdown occurs because of the faster rise time of the HMM-IEC tester.

A different reason for the miscorrelation was derived for the HV ESD protection diode for 100 V applications. The device passed 8 kV HMM-IEC tester stress, but failed below ~ 3.7 kV during the HMM-50 tester stress. This unexpected result was explained by examining the current waveform. A negative current through the diode at the end of the pulse was observed as a result of reflections at the 50 Ω HMM transmission line. The low on-state resistance of the diode causes an impedance mismatch with the 50 Ω source impedance of the tester. Part of the current through the DUT is reflected back into the HMM-50 tester [52]. The observed negative current forces the diode to switch fast from the forward into the reverse biased state. When switching from forward to reverse bias, the excess minority carriers, collected and stored in the diffusion capacitances during forward conduction, need to be discharged. This takes a finite time for the reverse recovery. It is visible in the negative part of the measured current waveform as a constant negative current flow (Fig. 2.55). During this reverse recovery time the diode conducts although the voltage across the device is much lower than the reverse breakdown voltage of 150 V. The amplitude of the reverse recovery current is high enough to cause a failure of the diode since reverse biased diodes can only withstands very small current densities.

Mixed-mode TCAD simulations are done for the same geometrical parameters and doping profiles to understand and visualize the reverse recovery and resulting stress in the device. During reverse recovery, the diode junction is in the transition to the reverse biased state. The external negative voltage enforces the built up of a high electrical field across the junction (Fig. 2.56a), the electrical field stimulates impact ionization Fig. 2.56b) and at certain HMM stress levels the impact ionization leads to avalanche breakdown. The resulting current flow results in a current density which exceeds the limits of the reverse-biased diode and causes device failure.

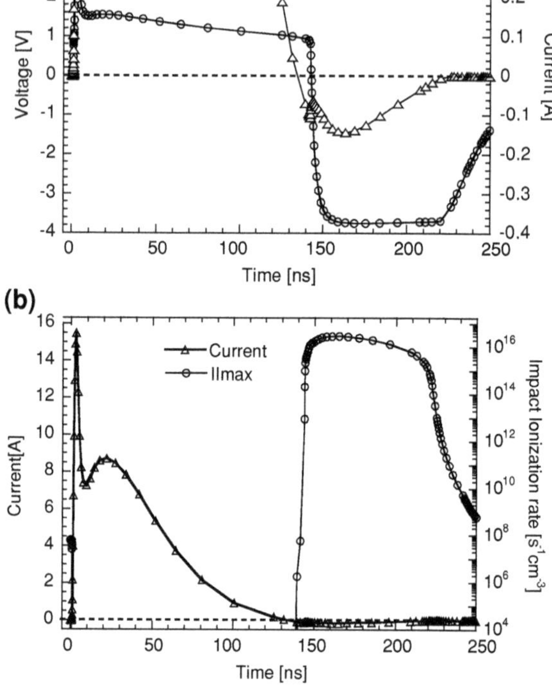

Fig. 2.56 Mixed-mode simulation of ESD diode during HMM stress applied with the HMM-50 tester for the current and voltage (**a**) and the current and impact ionization rate (**b**) at HMM stress level 3.7 kV (equivalent)

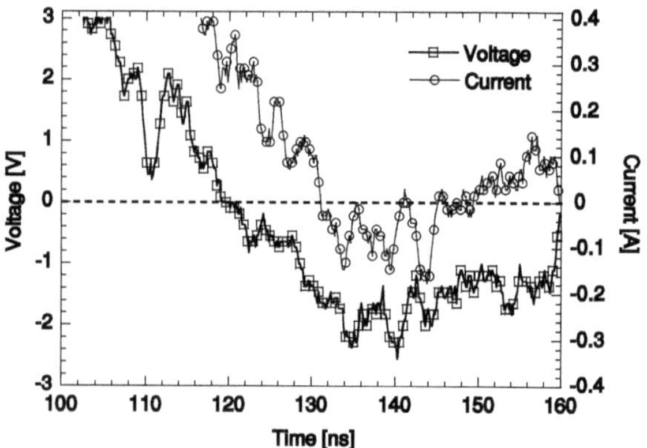

Fig. 2.57 Measured voltages and current through nLDMOS-SCR when stressed with HMM-50 tester, zoom-in on moment of reverse recovery; stress level: 5.6 kV (equivalent)

The 100 V tolerant HV nLDMOS-SCR has failed at 5.6 kV HMM-50 and at 7.5 kV HMM-IEC stresses. Similar to the ESD diode case described above the likely reason of miscorrelation is the negative currents generated by HMM-50 tester (Fig. 2.57) and the corresponding reverse recovery effect.

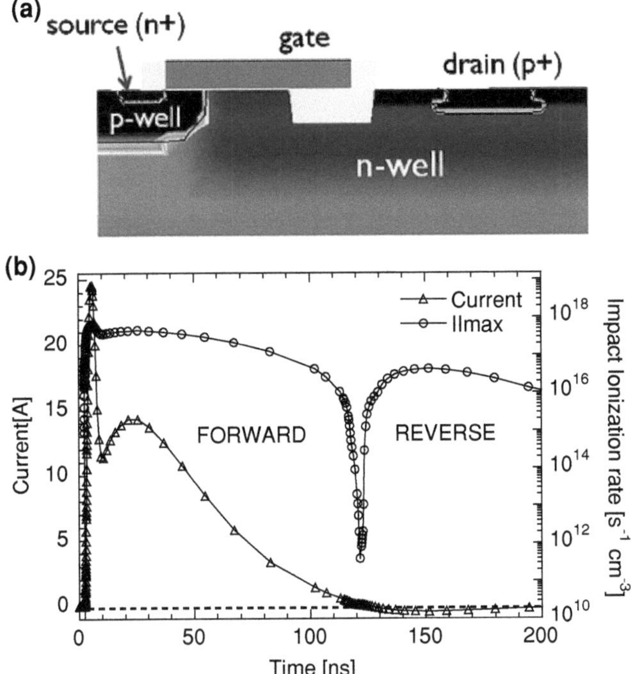

Fig. 2.58 Simulated cross-section of a nLDMOS-SCR (**a**) and simulated HMM current and maximum impact ionization rate in the device during 5.6 kV (equivalent) HMM-50 stress (**b**)

The mixed-mode simulation with the nLDMOS-SCR device cross-section (Fig. 2.58a) in DECIMM^{TM} [19] was used to analyze the lower failure level during HMM-50 stress. The simulated HMM current and the impact ionization rate of the nLDMOS-SCR at the failing stress level (Fig. 2.58b) demonstrate the significant increased impact ionization rate at t $\sim$ 150 ns in spite of a full decay of the HMM current. The increased impact ionization rate after the decay indicates reverse recovery in the nLDMOS-SCR device.

The location of the highest impact ionization during HMM stress during forward conduction mode is the drain side of the device (Fig. 2.59a). During reverse recovery, the impact ionization is concentrated mainly at the source side (Fig. 2.59b). The negative external voltage forces the n+-to-p-body junction on the source side of the nLDMOS-SCR into reverse bias thereby building up a strong electrical field. This electrical field stimulates impact ionization along the junction. The impact ionization does not spread uniformly across the junction. Instead there are two hot spots where the impact ionization rate is locally higher (Fig. 2.60). This induces locally a higher carrier and current density.

The impact ionization is also concentrated around the gate oxide of the nLD-MOS-SCR. During reverse recovery, some of the free hot carriers move towards the gate. They get trapped when they enter into the gate oxide. With increasing

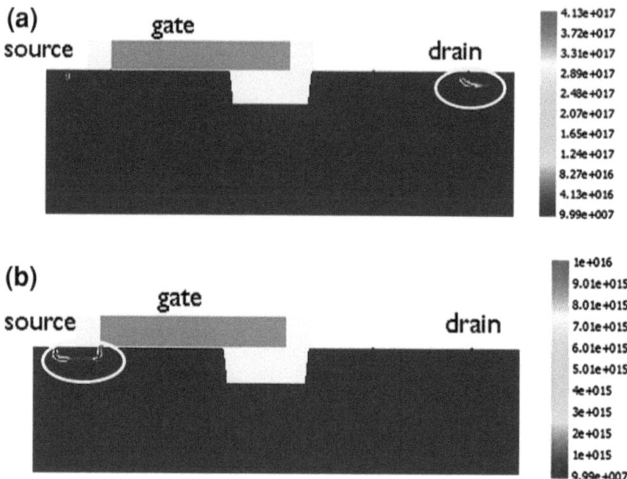

Fig. 2.59 Location (*circle*) of impact ionization peaks during HMM-50 stress: **a** after 20 ns (*forward conduction*) and **b** after 150 ns (*reverse recovery*); *stress level* 5.6 kV (equivalent)

HMM stress levels, more carriers are trapped in the gate oxide which results in a degradation of the gate oxide and an increase of the device leakage similar to the hot carrier degradation effect at much longer stress in normal operation conditions.

The evolution of the device leakage current during HMM stress on the nLD-MOS-SCR for both the HMM-50 and the HMM-IEC testers is different (Fig. 2.61). On the contrary, an abrupt failure is observed during the HMM-IEC tester stress. This supports the conclusion that different failure modes occur when stressing the HV nLDMOS-SCR with the two different HMM testers. The device fails thermally when using the HMM-IEC tester. Hot carrier stress induced by reverse recovery causes a gate oxide failure during stress with the HMM-50 tester.

2.5.2 Impedance Matching and Impact on Failure Level

The observed device failures are directly related to the pulse source impedance of the particular HMM-50 tester. The on-state resistance of the tested devices is much lower than 50 Ω. This causes mismatching with the source impedance of the HMM tester. If a series 50 Ω resistor is added to the HMM setup (Fig. 2.62) the reflections are minimized and failure due to reverse recovery is suppressed.

The repeated HMM-tests with the series 50 Ω resistor show different results (Table 2.5). The equivalent failure level when stressing the two devices from group 2 with the HMM-50 tester becomes comparable.

A small negative current is measured when stressing the ESD diode with the 50 Ω resistor. This forces the diode into reverse bias and the reverse recovery

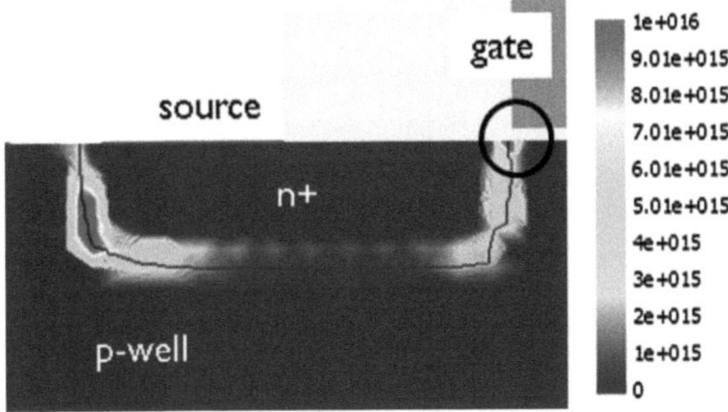

Fig. 2.60 Zoom-in on locations of impact ionization (after 150 ns) in device cross-section, circle: impact ionization hotspot close to gate oxide; *stress level* 5.6 kV (equivalent)

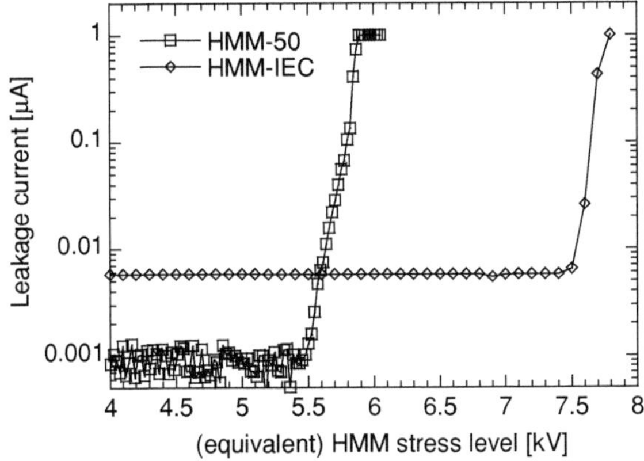

Fig. 2.61 Device leakage current evolution (at 100 V) of 100 V nLDMOS SCR during HMM testing with HMM-50 tester and HMM-IEC tester

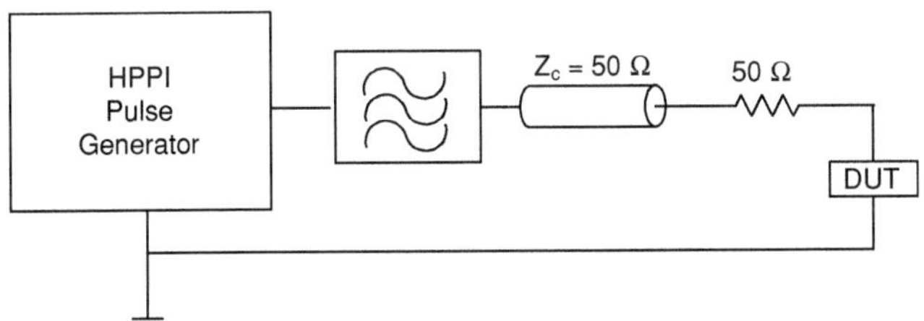

Fig. 2.62 Modified HMM-50 setup with added 50 Ω resistor

Table 2.5 Equivalent failure level of ESD diode and nLDMOS-SCR when stressed with HMM-50 tester: with and without added 50 Ω series resistor

Device	No resistor	With 50 Ω resistor
ESD diode	3.7 kV	3.8 kV
nLDMOS-SCR	5.6 kV	5.4 kV

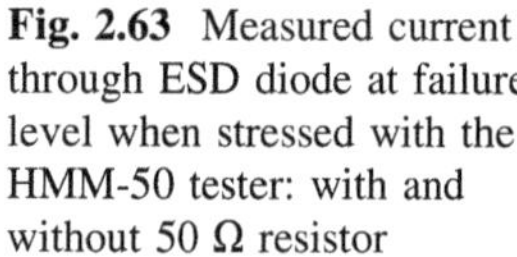

Fig. 2.63 Measured current through ESD diode at failure level when stressed with the HMM-50 tester: with and without 50 Ω resistor

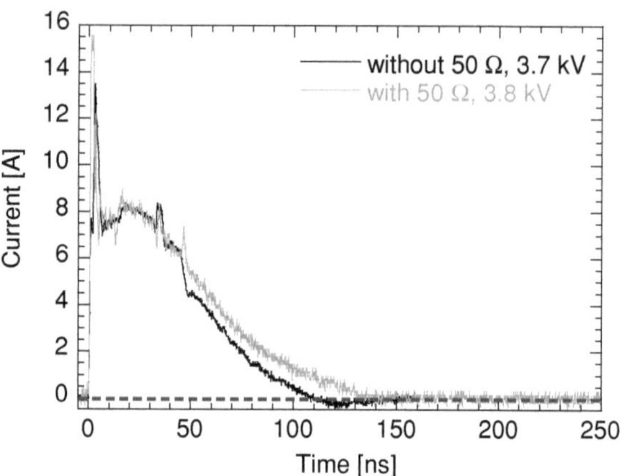

occurs. Although the amplitude of the negative current is much smaller in comparison to the stress without added resistor the resulting failure level is similar to the case when the resistor is added. This is attributed to a different stress current shape (Fig. 2.63) for the measured HMM current through the ESD diode with and without added series resistor.

This larger forward current charges more the diffusion capacitances of the diode. When the diode goes into reverse recovery, more carriers need to be swept away to get a stable reverse-biased state. The higher amount of free carriers, together with the impact ionization due to reverse recovery creates a similar current density in the diode than without series resistor.

The additional series resistor prevents negative voltages and currents when the nLDMOS-SCR is stressed. Consequently it should prevent reverse recovery in the n+-to-p-body junction on the source side of the nLDMOS-SCR. However, small negative voltages and currents are still measured. This indicates that reverse recovery still occurs in the device when stressed with the HMM-50 tester. The measured HMM current through the nLDMOS-SCR with and without added series resistor (Fig. 2.64).

A larger current flows through the device when the resistor is added to the setup. More excess minority carriers are stored in the diffusion capacitances of the device's junctions. Due to the larger amount of minority carriers during forward conduction, a similar number of carriers are trapped in the gate oxide. This results in a similar degradation of the gate oxide in comparison to the case when no resistor is added to the setup (Fig. 2.65).

Similar failure level are obtained when stressing the lateral PNP and the ggNMOS with both the HMM-50 and HMM-IEC tester. PNP devices conduct ESD stress with a higher on-resistance. This results in fewer reflections when the stress is applied with the HMM-50 tester. Consequently only a small negative current flows after the decay of the HMM pulse. Furthermore, a PNP device in reverse behaves like in forward operation but with a different I–V curve. It is

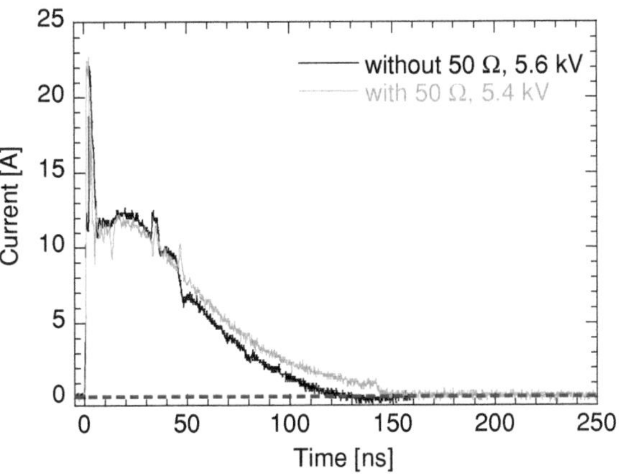

Fig. 2.64 Measured current through nLDMOS-SCR at failure level when stressed with the HMM-50 tester: with and without 50 Ω resistor; equivalent HMM stress level are given

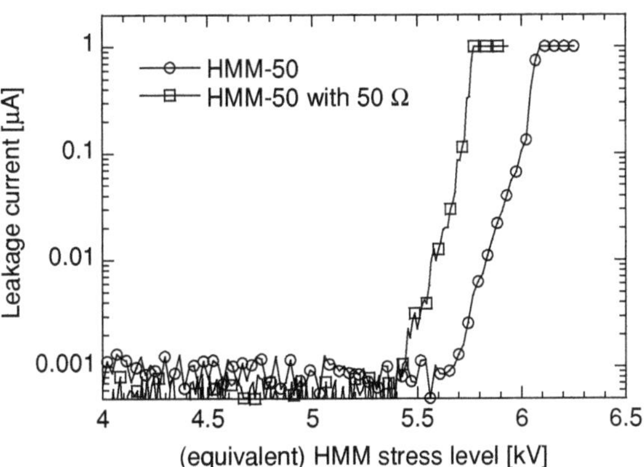

Fig. 2.65 Leakage current evolution of 100 V nLDMOS-SCR (at 100 V) during HMM-50 test with and without added 50 Ω resistor

expected, that the PNP can conduct at least the same ESD stress as during forward conduction.

Thus correlation issues exist for IEC 61000-4-2 type of HMM tester and a 50 Ω HMM testers. They might need to be address to gain an appropriate confidence in component level test results. In general 50 Ω HMM test systems should be used taking into account the reverse recovery and negative current generation effects.

2.6 Summary

The demand on highly integrated SoC and SoP ICs created a design paradigm shift towards the on-chip system level ESD protection. The significant gap between component and system level test methods and standards has been bridged by the introduction of the HMM method and component level ESD gun test as well as on-wafer methodologies. These methods provide stress pulses for the component level close to system-level standards pulse waveforms like for example IEC 61000-4-2. As a result, today, not only the final system blocks, but also modules, PCB and IC designs can be evaluated for system level ESD stress.

Both from technical realization perspective and from an overall methodological point of view, the on-chip system level ESD design requires a good understanding of the test standards and procedures, especially for their adaptation to the IC component level and on-wafer verification. The aspects of such understanding include the measurements of the pulse waveforms, transient characteristics, and the correlation factors between different tester types.

The physical aspects of the key test methodologies and their application specific support of on-chip ESD system level design are related to ESD gun testing itself, the test board design, and the tester and ESD pulse correlation factors. The gun system level tests according to the commonly used IEC 61000-4-2 and ISO 10605 standards and their realization in the ESD laboratory environment are described above.

The approach for Human Metal Model (HMM) testing as a major practical component level emulation of the system level ESD stress is based on the understanding of the major standards and test procedures, as well as board and wafer level testing methodologies.

A significant progress has been made in on-chip system level IC development by the implementation of the component and on-wafer testing methods. These methods simulate the system level ESD discharge pulse waveforms at least with some acceptable accuracy for the initial design. In particular, understanding the IEC 61000-4-2 standard requirements led to the development of a component-level test method with the same stress waveform. While the system level IEC and ISO ESD standards for the contact discharge define the tests for the system ports, the HMM methodology, on the contrary, is primarily targeting the evaluation of the IC pins robustness.

A similar approach has been successfully applied for evaluation and comparative analysis of the standalone ESD solutions on test chips. This is done based on the assumption that, if the HMM pulse waveform repeats the system level waveform, then some correlation of the passing level can be expected. The follow-up studies have demonstrated that in many cases the correlation between standard system level gun tests in the corresponding laboratory setup has a direct correlation with the passing level during HMM stress. In spite of a number of reported miscorrelations, today, HMM likely represents the most practical and widely used approach for on-chip ESD design. This will be broadly demonstrated in the examples of the following Chapters.

The primary goal of any ESD design is to meet the required passing level during component and system-level ESD qualification. However, relying only on the passing level would be rather difficult during the development of the ESD solutions and the conduction of ESD case studies. For that the capturing of voltage and current during ESD stress is required. The exact implementation of suitable measurement setups is not an easy task. Therefore calibration and de-embedding methods for the removal of distortions by tester parasitic have been presented and applied to examples. As an alternative, Kelvin setups are used for ESD on-wafer measurement setups. Both, calibration and the Kelvin setups are required to capture the real transient device and circuit response during ESD stress.

Chapter 3
On-Chip System Level ESD Devices and Clamps

Design of mixed-signal analog integrated circuits (ICs) often involves a co-design of the internal functional analog circuit blocks, on-chip ESD solutions and even the process integration in case of power optimized technology. Protection of the pins with system-level specification requires an in-depth understanding of a number of rather cross-disciplinary subjects. They include ESD device operation principles in the breakdown state, injection and conductivity modulation modes, clamp layout design, integrated process technology options, latch-up, safe-operation area and self-protection capability of standard power devices, ESD network and analog internal circuit blocks. The goal of this chapter is to present these cross-disciplinary subjects in a logical and condensed format. This is achieved by compilation of a high-level introductory material, important for further understanding, followed by a presentation of the ESD devices and clamps design principles with emphasis for system-level on-chip protection.

3.1 Important Introductory Material for On-Chip ESD Design

3.1.1 Local Clamp and Rail-Based Protection Network

The ESD protection requirements are imposed to guarantee that integrated components or systems are able to withstand a specified ESD pulse level. For the mixed-signal analog IC chip, the ability to withstand a certain level of ESD current is achieved by implementing an on-chip ESD protection network. This network is a combination of ESD clamps, high-current metal routing with fully or partly self-protected internal circuit blocks. Historically, conventional IC protection specifications only included the so-called component-level pulses (HBM, MM, and CDM). The emergence of portable systems generated a demand in additional protection on the so-called system level.

V. Vashchenko and M. Scholz, *System Level ESD Protection*,
DOI: 10.1007/978-3-319-03221-4_3, © Springer International Publishing Switzerland 2014

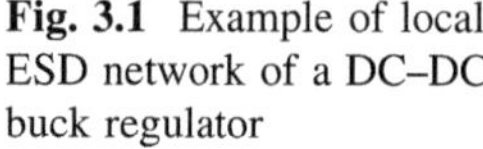

Fig. 3.1 Example of local ESD network of a DC–DC buck regulator

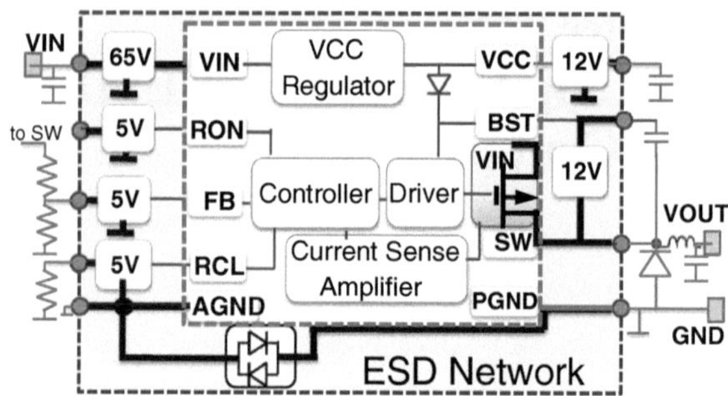

In systems, the ESD protection network can be designed with transient voltage suppressors (TVS) and additional passive components on the printed circuit board (PCB). In this case, TVS can be treated as an equivalent to discrete ESD protection clamps used to form a multi-stage protection network. In most cases, an analog IC or a system represents a combination of internal functional circuit blocks and an additional embedded ESD pulsed-power circuit. In general this virtual pulsed-power circuit cannot always be de-coupled from the analog circuit blocks [5].

The principal functionality of the on-chip ESD components is to limit the voltage at the IC pin below the absolute maximum rating of the internal circuit blocks in the pulsed regime. Thus, on-chip ESD protection is applied at the IC pins. As described in Chap. 1, it can be either rail-based or implemented using local clamp approach. An example of a local clamp protection network is an asynchronous buck DC-DC voltage regulator (shown in Fig. 3.1). The high voltage (HV) DC power input VIN pin is protected by a HV local NLDMOS-SCR 65 V clamp, and each low voltage (LV) analog control pin is protected locally by a 5 V tolerant snapback NMOS clamp. 12 V tolerant isolated field-oxide snapback clamps are protecting the BOOST pin and VSS regulator pins (Fig. 3.1), while the SWITCH pin relies on the self-protection of the large power array.

An alternative rail-based network can be applied to the control pins by combining the 5 V active core clamps and rail ESD diodes.

In spite of many similarities, the ESD functionality of the discrete TVS components is different from the on-chip ESD protection clamps. TVS are typically applied at the system periphery to the port location. When an ESD stress is applied at the system port, the TVS limits the electrical overstress voltage, preventing the excessive current from propagating across the board. In general PCB design, it is difficult to foresee all possible coupling scenarios (Fig. 3.2). Therefore sometimes additional TVS components are even used as a last moment patch solution to help the system to pass to the ESD stress level. An additional major difference is that TVS cannot always limit voltage precisely, so the system design in general relies on a multistage network.

Thus, on-chip system-level protection is generally not a replacement for system-level on-board design with TVS. However, continuing trends in portable systems minimization and cost reduction as well as the demand for high-reliability

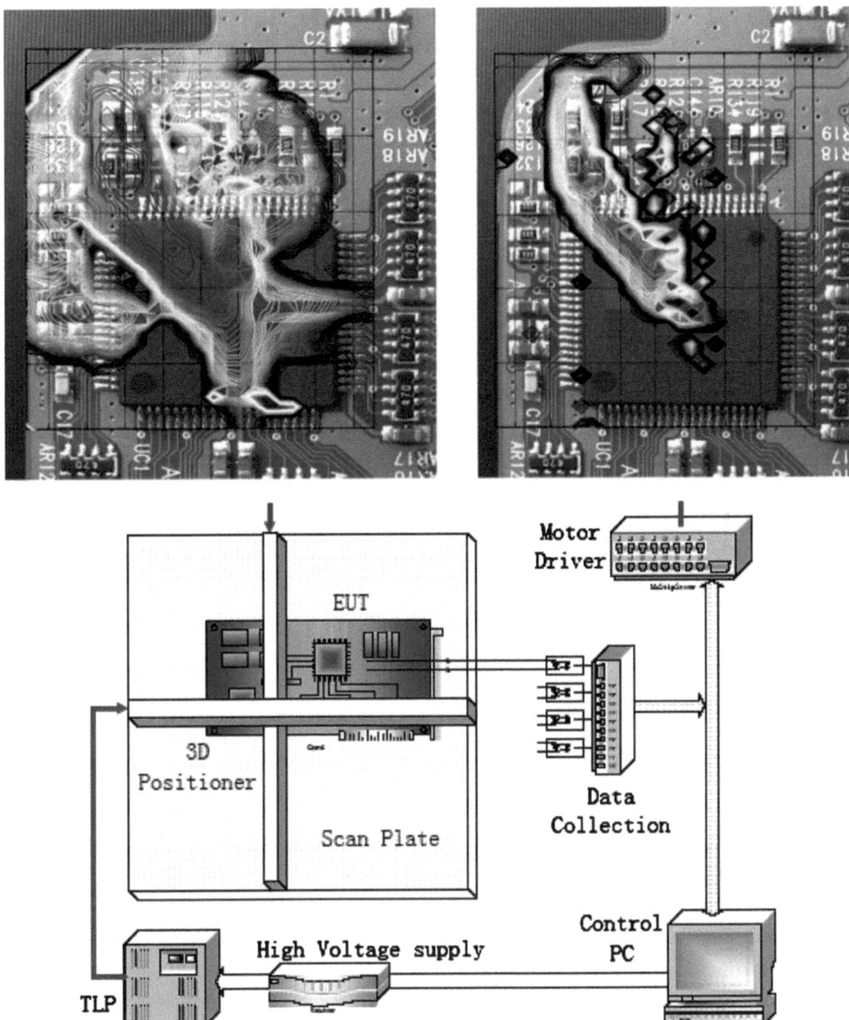

Fig. 3.2 Measured response on system-level stress across PCB area for two different designs (**a**) obtained using EMC/ESD Scanning System (**b**) by Amber Precision Instruments (2010)

automotive and aerospace electronics allow on-chip system-level solutions to evolve with significant pace.

The ESD protection network both on-chip and on-system-PCB provides a dedicated high pulsed current path for every pin-to-pin combination, while simultaneously limiting the voltage below the critical regimes of the devices in the internal circuit blocks that are interfacing with the protected pin.

System-level stress specification is defined in several standards. The commonly used standard is IEC 61000-4-2 (Chap. 2). This Standard combines both the contact and the air gap discharge. When system level stress is applied, a more complicated ESD current propagation scenario is realized in comparison with the component level ESD stress conditions. The system typically should tolerate the discharge applied to an arbitrary user accessible location in addition to the ports. The external ports are typically subject to contact and air gap IEC specification stress. Other isolated system locations may need to withstand the air gap discharge, for example an LCD screen.

In the simplest cases, the IC with system-level-capable pins is directly connected to the system ports—for example, in USB, CAN and LIN receivers. In such cases, the system can avoid relying on the on-chip protection and completely bypass the use of a number of PCB components, including the TVS.

On-chip protection for analog circuit blocks can tolerate the rail-based ESD protection network approach only in some rare cases. Local ESD protection clamps with system-level pulse current clamping capabilities represent the mainstream solution. The complexity of local ESD protection is related to the generally unknown conditions of the internal control electrodes in the devices interfacing with the pad during the ESD pulse. Typically, the ESD clamp is understood as a triggering circuit that, under certain conditions, can conduct a high current in the ESD pulse time domain and provides a certain clamping voltage. The high current conduction is usually activated either above some critical voltage level or at transient voltage detection at the protected node.

3.1.2 Conductivity Modulation in Semiconductor Structures

Although in some cases local clamps can be composed using the same principles as active clamps, the most conventional approach is to use ESD devices. The ESD devices are essentially pulsed power devices with an added scalable high current conduction capability. Small footprint ESD devices operate in a certain conductivity modulation mode realized in isothermal or, more precisely, adiabatic conditions. In these conditions, physical mechanisms due to electrical (rather than thermo-electrical) conductivity modulation determine the high current operation and the protection capability of the device. Therefore, self-heating of the device can be neglected until the high-current mode while the heat generation is localized to a few microns of the device cross-section. In spite of a broad variety of ESD protection devices and clamps, there are only four known basic electrical conductivity modulation mechanisms of practical value to Si microelectronics: *avalanche breakdown, avalanche injection, double avalanche injection* and *double injection* [4] (Table 3.1).

These conductivity modulation mechanisms are used to produce high-current density in ESD devices at relatively low clamping voltage. In avalanche breakdown conditions, both the internal and external dependence of current upon voltage occur at positive differential resistance (PDR). The remaining three

Table 3.1 Conductivity modulation mechanisms realized in lateral ESD devices for high-current density

Conductivity modulation mechanism	Typical ESD devices based upon the mechanism	Typical lateral current density (mA/μm)
Avalanche breakdown	Avalanche diodes; blocking junctions, PMOS, PNP	0.01–0.1
Avalanche injection	Snapback NMOS; NPN, field oxide devices	0.1–3
Double avalanche injection	P-i-n, M-i-n diodes	0.1
Double injection	LVTSCR, SCR, bipolar SCR, LDMOS-SCR	10–100

conductivity modulation mechanisms in case of Si material produce at least the internal negative differential resistance (NDR).

One the major non-linear physical phenomena associated with NDR is spatial instability of the current distribution along the structure width. It occurs when the characteristic dimension of activation (positive feedback phenomenon) in the device is much smaller than the characteristic dimension of the inhibition (negative feedback phenomenon). The non-uniform current distribution may result in current crowding and formation of solitary or multiple current filaments [4]. Unless the increase of current density in conductivity modulation mode is limited across the distributed ESD structure width, the high-current capability is not suitable for on-chip ESD protection.

To limit the positive feedback produced by the conductivity modulation, structure-level negative feedback must be implemented as part of the ESD device design. This measure typically involves embedding current saturation regions, which provide a compensating voltage drop at high current density. The current density can become sufficient for conductivity modulation of the saturation region, formation of an electric field maximum with a significantly lower melting temperature than Si at the contacts. At these conditions, conductivity modulation mechanisms causing thermal carrier generation and current instability are still responsible for structure burnout.

The conductivity modulation phenomena can be understood as a non-linear change of the structure conductivity as a result of applied voltage or current increase. The definition is based upon external structure parameters. Using internal structure parameters, the effect can be defined as a change in the conduction properties of particular structure regions as a result of change in carrier space charge balance due to avalanche and injection processes. Thus, conductivity modulation is the result of an internal avalanche-injection process initiated by change in the applied current or voltage, rather than the change of structure conductivity due to injection/extraction carriers from the base contact region or creation of accumulation/depletion regions by the field control electrode.

From a practical ESD device design point of view, the most interesting cases are concerned with a non-linear change of conductivity that results in the NDR effect. The NDR can be defined as a decrease in the voltage drop on the structure

Fig. 3.3 Elementary semiconductor n-p-n, p-n-p, p-i-n and p-n-p-n structures for numerical experiments with conductivity modulation

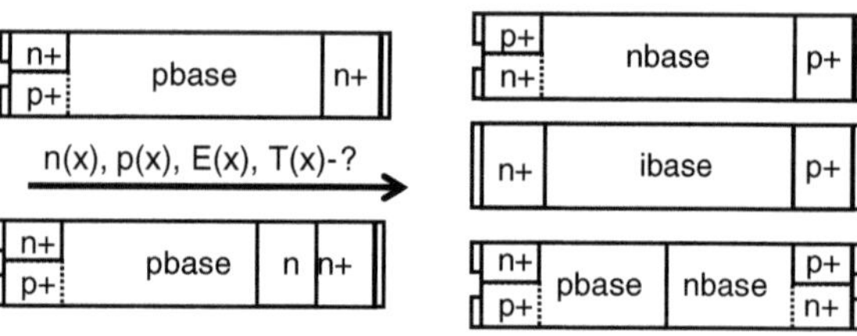

terminals under a current increase through them. This effect is usually observed as formation of S-shape I–V characteristics, although in some cases the additional voltage drop on the device current saturation regions can compensate and "hide" the internal NDR [5].

During operation with external loading, when the device achieves the NDR regime at some critical voltage, a self-turn-on or self-triggering is observed into the corresponding high current state according to the load characteristic. Among ESD application engineers and circuit designers, this phenomenon is usually referred to as *snapback*. Local ESD clamps based upon snapback ESD devices are called snapback clamps.

The spatial current instability phenomenon can be understood as an uncontrollable increase of local current in time after certain critical conditions are achieved in originally uniform current distribution. Both in experimental physics and numerical simulation the change of the uniform state is enabled by fluctuations. A small deviation from the uniform state can be introduced by the local inhomogeneity as well. In the numerical simulation analysis the finite accuracy of numerical solution of the finite-difference scheme in each mesh point provides the numerical noise source equivalent to physical fluctuations.

The spatial current instability process occurs due to growth of current fluctuations. It arises when the total current applied to the structure corresponds to the NDR regimes. In these conditions, local current increase results in local voltage drop and vice versa. Essentially, ESD devices are designed to support current density levels that correspond to the maximum current filament amplitude. At the optimal ESD device design at high current the filament simply occupies the width of the entire structure.

Understanding of the basic conductivity modulation mechanisms is critical for a confident ESD device engineering. It can be effectively achieved using numerical simulation analysis. Using the following 2D basic semiconductor structures (Fig. 3.3), comparative analyses of *E(x)*, *n(x)*, and *p(x)* distributions for Si material can be done for different current level regimes:

(i) *p-n diode structure* at reverse bias in avalanche breakdown mode.
(ii) *n-p-n* and *p-n-p triode structures* in avalanche injection conditions
(iii) *p-n-p-n thyristor structure* in double injection conditions
(iv) *p-i-n diode structure* in double avalanche injection conditions.

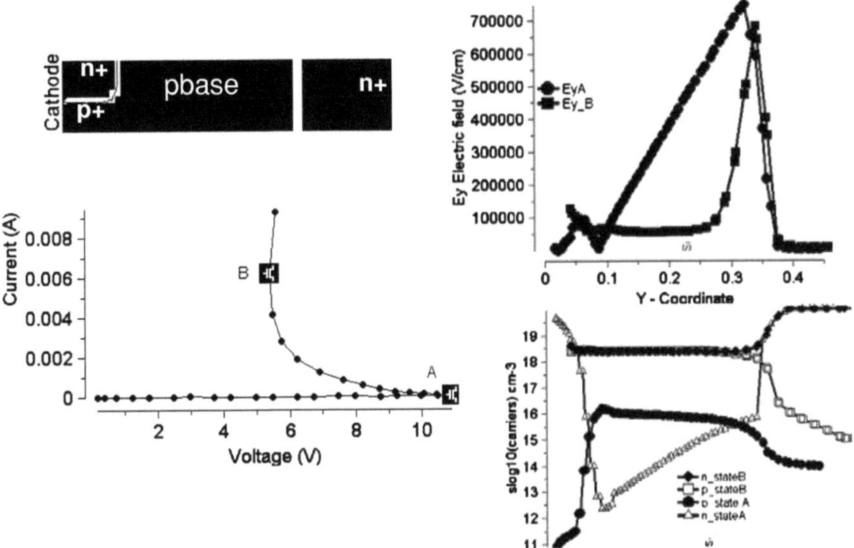

Fig. 3.4 Simulation analysis of the avalanche-injection conductivity modulation in the n-p-n structure: I–V characteristics (**a**) and comparison of the electric field, electron and hole distributions (**b, c**) for the avalanche breakdown state "A" and the avalanche injection state "B"

An example of such analysis is presented in Fig. 3.4 for the avalanche injection in the n-p-n structure. In the low current avalanche breakdown state "A", the electric field is distributed along the structure according to the significant imbalance between the electrons and holes along the entire p-base. The corresponding high total voltage drop is an integral value of the electric field distribution. In the high current state "B", the quasi-neutral area is formed with $n \sim p$ in a major part of the p-base. This results in a reduction of the electric field in the base, including at the collector junction where intense impact ionization occurs. Similar analysis can be done for other elementary ESD devices.

The next level of physical understanding of structures with negative differential resistance and the resulting phenomena can be gained using quasi-3D numerical simulation analysis. The current filamentation phenomenon in distributed structures is the result of spatial current instability of the uniform solution, due to NDR conditions in the transient operation mode. As mentioned above, a numerical solution for the current filament becomes possible due to growth of the numerical fluctuations of the finite-difference scheme.

The spatial instability is realized in the transient electrical regime with the current density that corresponds to the NDR state. At some critical electrical regime a uniform current distribution becomes unstable due to the growth of numerical fluctuations in non-equilibrium conditions. The high fluctuation growth increment is realized due to local positive feedback between the local current density and the local voltage drop. It results in the formation of one or more

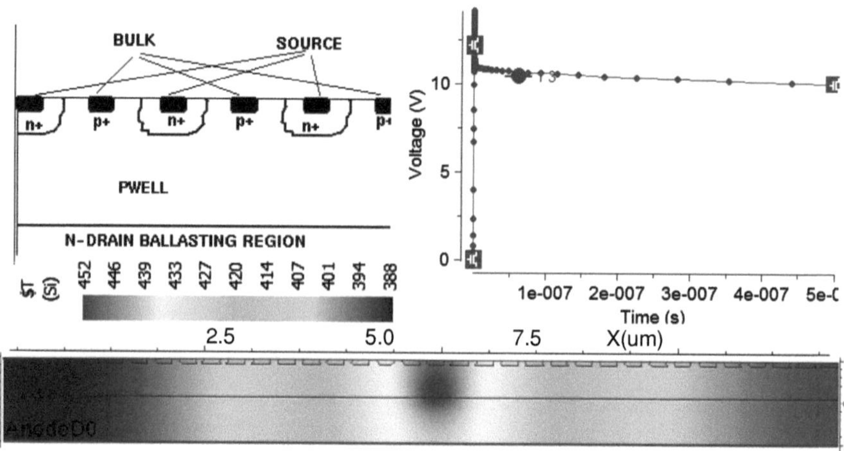

Fig. 3.5 Simulation analysis of the avalanche-injection conductivity modulation in a distributed n-p-n structure: I–V characteristics (**a**) and comparison of the electric field, electron and hole distributions (**b, c**) for the avalanche breakdown state "A" and the avalanche injection state "B"

current filament states. In the example of the p-n-p distributed structure (Fig. 3.5), the thermal coupled solution with the heat equation was enabled in addition to the drift-diffusion model. In this case, local self-heating and electrical current filamentation can be observed too.

For the 15 kV air gap IEC system level ESD pulse, a local ESD clamp is needed to sustain transient currents up to ~ 30 A. In the most optimized case, the structure's capability is below ~ 10–30 mA/μm. Therefore, the clamp layout is typically accomplished in the form of either multifinger arrays or racetrack devices around the bond pad, rather than just a very long single finger.

Similar to elementary diode structures, a distributed ESD device array with NDR can also experience instabilities in the uniform current density distribution. To compensate for this non-linear effect, negative feedback can be implemented at both the array and clamp levels to contain the current density within safe operating conditions. For example, in a snapback NMOS ESD clamp, positive feedback can be realized both on the structure level with a saturation resistor formed by the drain ballasting region, and on the cell level with additional back-end ballasting to eliminate the undesired effect of non-even finger turn-on. At high system-level clamp currents, the current ballasting measures become even more efficient.

When the peak current density in a snapback ESD device is limited down to appropriate levels, the pulsed operation of the ESD device is expected to be fully reversible. An opposite scenario occurs in standard devices with NDR. With some exceptions, standard devices do not provide a reversible operation in NDR conditions and snapback mode. Instead, NDR limits the pulsed safe operation area (SOA) due to a local burnout of the device structure at the location of the high amplitude current filament.

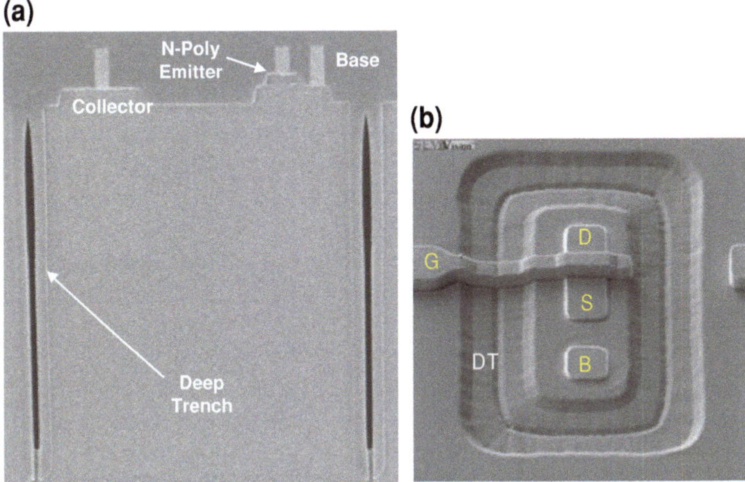

Fig. 3.6 Cross-section and *top view* of the 20 V NPN structure with deep trench isolation (**a**) and *top view* of deep-trench (DT) isolated NMOS device (**b**)

The physical limit of the operation regime is not always related to the semi-conductor part of the device. The backend metallization and contacts should also be taken into account, especially in high voltage devices. In high power regimes with high self-heating, rapid melting and electromigration of the metal-semiconductor alloy can also be a limiting factor. These effects will be discussed below.

3.1.3 ESD Related Specifics in Integrated Process Technology

Mixed-signal power analog circuits are usually designed with extended voltage CMOS (Complementary MOS), CBiCMOS (Complementary Bipolar CMOS) or BCD (BJT-CMOS-DMOS) integrated process technology platforms. A large selection of process integration options and modules is used to optimize the production cost by reducing the mask count and process steps, depending on the specification targets of a particular product.

For optimal power performance, integrated HV devices are usually implemented in the form of lateral drain-extended MOS devices (DeMOS) or self-aligned lateral double diffused MOS devices (LDMOS). Both N-channel and P-channel MOS devices are usually supported to provide proper power train design options. Extended voltage CMOS processes can support voltages up to 40 V, while HV BCD processes can support voltages over the entire 100 V range.

The LV CMOS modules in HV processes typically support dual gate oxide options, with the lowest voltage logic blocks and higher voltage analog components used for power applications. The BJT modules can combine regular diffusion emitter vertical and lateral BJT or rather high-performance poly-emitter devices based on different self-aligned emitter-base architectures.

The main differences between CMOS and BCD processes are the starting substrate material, epi growth and the substrate isolation scheme of HV devices. In analog processes, different devices typically share a substantial amount of mask layers to reduce process costs. An important part of on-chip ESD device design is the different mask options and their alignment.

Analog CMOS processes are usually based either on the low cost lightly doped substrate material or the more expensive initial substrate with a few-micron thick preliminary grown pEpi region. The lightly doped substrate is critical in reducing the induction current in the substrate if high-Q on-chip inductors are required for high speed product design. The drawback of the low doped substrate is a bigger sensitivity to the latch-up effects (Chap. 4). This is especially critical at system-level ESD events in power-up conditions where significant current is injected in the substrate. The substrate-generated carriers and spread of the injected current into the substrate can significantly change the characteristics of the parasitic devices at ESD conditions or pulsed SOA limits.

To relax latch-up spacing rules for guard rings in custom analog processes and to reduce the space needed for lateral latch-up isolation on the chip, a highly doped substrate material with a low doped pEpi growth ~ 2–4 µm thick is usually used. An alternative approach in advanced BCD processes involves a deep trench isolation formation (DTI) (Fig. 3.6). A more costly BCD process version is based on full substrate isolation using silicon-on-oxide (SOI) wafers. It can support both CMOS and BCD architectures, completely eliminating latch-up between the isolated epi pockets.

In analog CMOS processes, substrate isolation of the high side devices is achieved by triple well architecture with a high-energy deep N-well implant (DNWELL). The DNWELL dose and energy due to material defects and implanter tool capabilities limit the voltage tolerance of the HV devices and imply some latch-up limitations. These limitations are relaxed in analog BCD processes using N-buried and P-buried isolation layers implanted prior to nEpi growth. Interaction of the vertical substrate isolation regions with ESD device regions is an important part of ESD solutions design.

In the case of the advanced polyemitter bipolar module with a low impedance subcollector, highly doped subcollector N-sinker and P-sinker regions are usually available. In combination with the buried layers, these regions can be used for stronger latch-up isolation and formation of active ESD device regions.

A comparison of generic active devices in CMOS and BCD processes is presented in Table 3.2. Practically digital and analog CMOS components as well as high or extended voltage devices are fabricated using similar process steps.

A recent trend is to use low-cost foundry technology for LV analog 20–180 nm CMOS processes provide an advantage in a more accurate mask alignment needed to integrate digital CMOS components with small feature dimensions. This feature is widely used for in extended voltage device design where non-self-aligned body

Table 3.2 Comparison of the standard devices in generic 0.18 μm CMOS and BCD processes

Device types	0.18 μm 30 V CMOS	0.18 μm 30 V BCD
Digital (LV) conventional CMOS	2 V NMOS & PMOS	2 V NMOS & PMOS
Analog (LV) conventional CMOS	5 V NMOS & PMOS	5 V NMOS & PMOS
Power and high voltage CMOS	12 V NDeMOS & PDeMOS 20 V NDeMOS & PDeMOS 30 V NDeMOS	5 V NLDMOS & PLDMOS 12 V NLDMOS & PLDMOS 20 V NLDMOS & PLDMOS 30 V NLDMOS
Bipolar devices	20 V Lateral NPNP & PNP	20 V Vertical Polyemitter NNP & PNP;10 V JFET

region architecture can be represented by low-voltage pWell replacing the original double diffused body implant. The power-optimized extended voltage devices are represented by drain-extended MOS (DeMOS). The architecture of the device contains a drain drift region that provides depletion in all operation regimes with a high voltage, overcoming the limiting factor of the gate oxide voltage tolerance. In DeMOS devices, neither the drift region nor the body mask is self-aligned.

A similar HV device architecture is used in BCD processes to create a so-called lateral double-diffused MOS HV device (LDMOS). The name of this device comes from the self-aligned vertical DMOS architecture that is broadly used in discrete DMOS components. The major physical difference between Lateral DMOS and DeMOS is in a self-aligned pBody implant. This implant has sufficiently low energy to be shaded by Poly region. The pBody implant is further diffused so it overlaps the drain contact diffusion implant and thus forms the channel of the CMOS part of the device. However, in BCD processes, the pBody mask is often not self-aligned, relying on accurate alignment of masks in modern BCD processes such as Lateral DeMOS. The naming convention is also not so strictly followed across the industrial companies. Therefore, in further material, we will not differentiate lateral DeMOS and DMOS devices, referring to both as LDMOS unless otherwise noted. In addition, in spite of the usual CMOS design practice to refer to digital CMOS as low voltage devices and to analog CMOS as high voltage devices, we will refer to these as digital and analog, respectively, to differentiate them from high voltage LDMOS power devices.

A major paradigm of on-chip design is the implementation of ESD devices with target parameters free from additional process steps. This "free" ESD device design approach is realized under the expectation that standard process devices provide a sufficient range between maximum operating voltages specified by process design rules and the real absolute maximum voltage rating in the ESD pulse time domain. In other words, the ESD protection voltage window can tolerate non-self-aligned free ESD devices, unless large standard device arrays can provide self-protection.

Fig. 3.7 Comparison of the generic BCD (*left*) and CMOS process fabrication flows

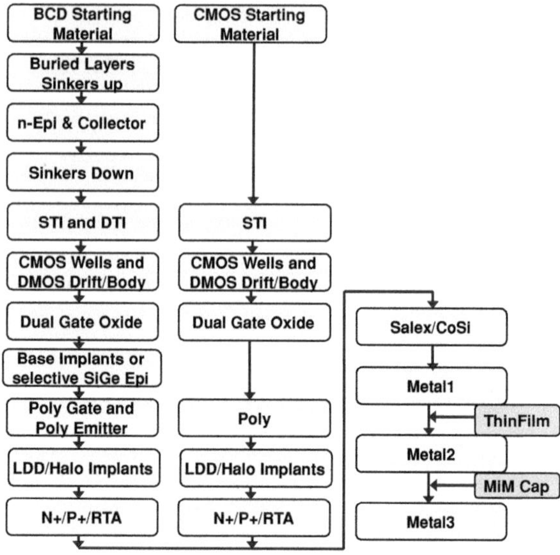

In the free ESD device design approach, a major development strategy is to re-use standard device architecture and regions. The blocking junctions, RESURF schemes and overall design are replicated as close as possible to minimize technology variation and mask alignment impact. Certainly, free ESD devices must be implemented without violating the physical minimum technology process rules.

In line with the advanced analog circuits and systems design, the process technology development field has experienced a tremendous evolution over the last decade. Major milestones include scaling of technology nodes from sub-microns down to 180–20 nm, integration of the lateral high voltage power devices, complementary Si-Ge BJT devices, shallow and deep trench isolations, and multiple front-end modules for on-chip non-volatile memory and passive components. A number of books in the field describe the peculiarities of the CMOS and BCD processes. This section compiles only a minimal introductory material, which is directly related to the subject of this book and is necessary to understand the terminology used in following text.

The differences in HV BCD and extended CMOS processes (Fig. 3.7) are the front-end steps before and after nEpi growth and optional advanced polyemitter BJT formation. Essentially, the CMOS process can simply be treated as a module of BCD process. The difference between these technology platforms is in vertical and lateral isolation schemes. An additional in-depth pRESURF region is often formed in BCD processes, as in the case of power-optimized in-depth NLDMOS design. The lateral surface isolation with STI is standard for both platforms. Optional deep trench isolation step can be introduced in advanced HV BCD processes with high performance BJT devices. A selective base-emitter epi growth in BCD and BiCMOS processes is used for complementary Si-Ge BJT devices, with a poly-emitter for high-gain or a grown mono-emitter for low noise.

The N-epitaxial region and heavily doped buried layers enable ESD device implementation specific to several system-level solutions. In analog CMOS processes, the HV DeMOS devices can often be implemented as partially free devices, reusing the conventional digital and analog CMOS well implants and minimizing new drift implants for proper depletion. In power-optimized BCD processes, considerably more implants are usually used. For example, Lateral DMOS devices are power optimized for breakdown versus on-state resistance physical limit of the silicon material.

The device level optimization of the DMOS involves multiple RESURF features delivered by poly field plates, in-depth P-RESURF implant connected or disconnected from the body as well as more sophisticated super-junction device architecture. In-depth understanding of the process technology options is important for optimal on-chip ESD device design and minimization of non-self-aligned device regions.

Typically, high side devices in CMOS process are isolated via the deep nWell (DNWELL) implant. Due to high energy, the photoresist mask layer is rather thick with corresponding large feature dimensions. In BCD processes, highly doped n- and p- buried layers (NBL and PBL, respectively) are implanted before the nEpi growth and laterally and vertically diffused during the nEpi growth (depending on its thickness). To avoid high spreading of NBL into the nEpi region, a slower-diffusing antimony (Sb) is often used instead of phosphorus or arsenic species.

After the vertical regions are formed, pWell and nWell regions are developed in both processes (Fig. 3.8). For these regions, multiple implants with different energies and doses are used to form a box-like profile. The surface implants are tuned for a proper threshold voltage level for CMOS devices and arranged in an anti-punch-through profile for better CMOS latch-up immunity. LDMOS drift and body regions, unless partly reused from well profiles, are implanted at the same stage.

The nDrift regions in nLDMOS devices typically need to be optimized for proper depletion in the DMOS on-state and off-stage high voltage operation regimes. In pLDMOS, the pDrift region acting implant must be depleted in the same regimes. These complementary HV BCD processes are usually provided by a pair of HV blocking junctions formed by the drift and body acting regions, which can be incorporated into free ESD device solutions together with poly field plate and in-depth pRESURF regions.

Formation of the well profiles is followed by the dual gate oxide (DGO) CMOS and self-aligned BJT process steps. Typically, two complementary pairs or complementary wells are used to support the most optimal diffusion profiles in the thin-gate-oxide digital CMOS devices and thick-gate-oxide analog CMOS devices used in LV analog blocks. Additional combinations with shallow implants can be used to integrate a variety of CMOS devices, for example high- and low-threshold voltage or leakage devices, as well as power optimized LV analog DMOS devices. These optional implants can help to achieve optimal parameters for free ESD devices. Significant advances in such design are achieved with physical process and device simulations using Technology CAD (TCAD) tools. Advanced TCAD methodology is described in Sect. Sect. 1.5.

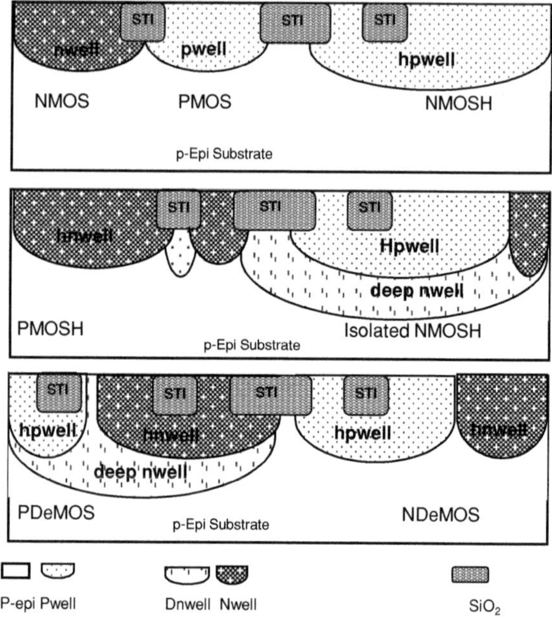

Fig. 3.8 Process flow cross-sections after the pWELL and nWELL implant steps for the CMOS devices

Formation of the dual gate oxide is accomplished by thermal oxidation to produce a thick gate oxide for analog LV CMOS devices. This step is followed by a gate oxide masking layer, oxide by etch and a high-quality thin gate oxide growth for digital CMOS devices. Then, a polysilicon (poly) layer deposition and etch (Fig. 3.9) forms the poly gates, poly-RESURF regions, poly-capacitors and poly-resistors. Finally, an optional high poly resistor mask is applied.

Shallow n- and p-lightly-doped drain (LDD) regions with corresponding opposite doping type phalo and nhalo implants for LV CMOS devices are implanted after the poly layer etch. These implants are self-aligned to the poly and STI edges. The nitride deposition and etch that forms the spacer region is followed by heavily doped shallow nplus and pplus contact diffusion regions. The regions are self-aligned to the corresponding edges of spacer and STI region etches.

Reference components for ESD clamps can be formed as avalanche diodes based on surface nldd-pplus or pldd-nplus junctions. It is important to take into account that contact diffusion regions are designed to maintain a normal operation current density. These regions have minimum process rules dimensions and might not be optimal for the extremely high current densities, especially those realized under system-level ESD current conditions. Experimentation loops for each particular case are needed to determine the most optimal diffusion region length, number of contacts, or the need for an additional silicide blocking region.

After the rapid thermal anneal (RTA) process activates the shallow implants, silicide deposition, the silicide exclusion mask and silicide etch are developed to remove silicidation from the desired diffusion regions and polyresistors. The

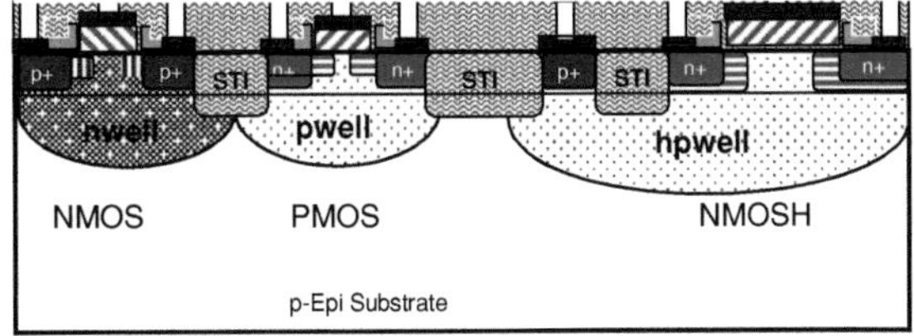

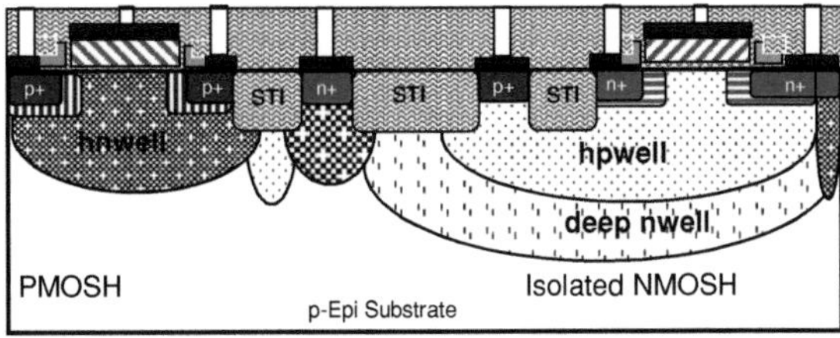

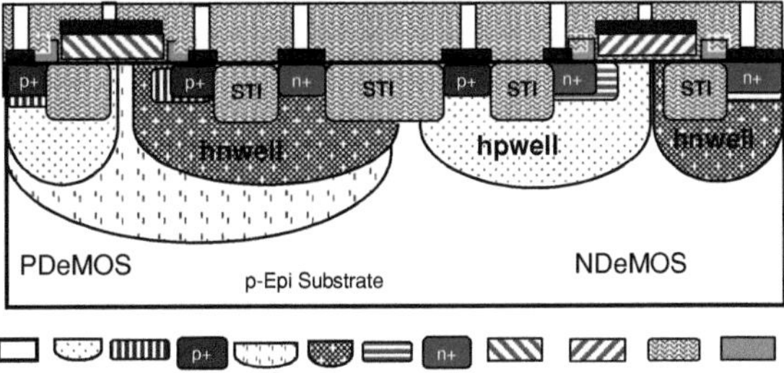

Fig. 3.9 Process flow cross-sections after silicide formation and oxide deposition and etch for contacts in the CMOS process for digital and analog CMOS and DeMOS devices

silicide block option is critical for implamentation of the device-level negative feedback in ESD structures, which provides local current ballasting and spacing between the semiconductor-metal composition with a low melting temperature and regions with high local heat generation.

The front-end process steps form the contacts, vias and metal layers. The current capability of the metal layers at the system level is another potentially limiting factor in optimization of the silicon structure. For proper cell and clamp design, the metallization scheme of the ESD device array requires a rather careful and detailed strategy.

There are numerous "tricks" for innovative ESD device design that rely on an in-depth understanding of processes technology physics and features. For example, highly diffused NISO, NBL, and PBL implants can be patterned with minimal mask dimensions to obtain partially diluted layers and either control of gain of the parasitic BJT structures engaged in the conductivity modulation process or to tune the breakdown voltage of the free ESD structures [53, 54]. The subcollector and deep regions can source a number of implementation possibilities for different parasitic SCR- and BJT-type structures, whose in-depth current conduction is favorable for higher ESD performance due to optimal in-depth heat dissipation.

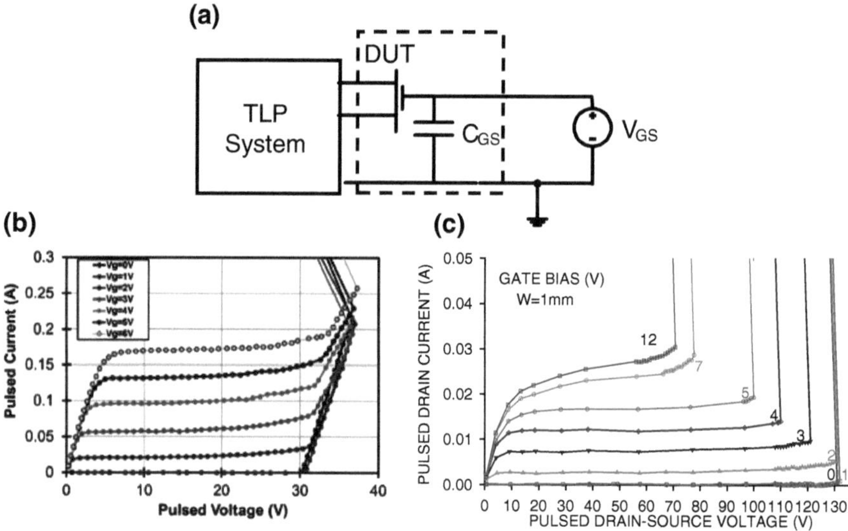

Fig. 3.10 Experimental setup for pulsed SOA measurements (**a**), pulsed drain-source characteristics of 20 V NLDMOS device with high (**b**) and 100 V NLDMOS with low (**c**) self-protection capability

3.1.4 SOA in ESD Pulse Domain and Self-Protection

Standard device operation above some critical electrical regimes is irreversible and results in instantaneous burnout. Physical limitation of electrical regimes in the ESD pulse domain is usually represented by pulsed save operating area (SOA). The pulsed SOA for a given device can be measured using a TLP system combined with an additional DC voltage source in the setup (Fig. 3.10a). For example in case of NMOS devices the pulsed drain-source I–V characteristics can be measured at constant gate bias until either reversible or irreversible triggering. In the last case a fresh sample of similar device is taken for each next gate bias measurements.

In the example of NLDMOS device (Fig. 3.10) the drain-source electrical regimes are simply limited by the boundary-defining maximum drain current as a function of the drain-source voltage. Depending on the device design, different shapes of the pulsed SOA boundary can be realized and plotted by connecting the critical points of the drain-source I–V characteristics family (Fig. 3.10b, c).

Another important figure of merit associated with SOA is the self-protection capability. It can be defined as a minimum critical current that the device can sustain in pulsed regime. In NLDMOS devices, this current is typically realized in off-state conditions in the avalanche breakdown mode. This current is the critical current before the formation of a negative differential resistance region, followed by irreversible device operation. An example of a 20 V NLDMOS with high self-protection capability is shown in Fig. 3.10b. Conversely, a device with low self-protection capability (Fig. 3.10c) demonstrates practically no self-protection current levels before snapback on the example of 100 V NLDMOS.

In general there are two scenarios where the device can provide self-protection: (i) substantial avalanche current before snapback and (ii) reversible snapback into the high current mode. The majority of standard devices cannot sustain reversible operation in snapback mode. In some rare cases, LV NMOS and PMOS devices may exhibit reversible snapback operation; however, this is often accompanied by high sensitivity to process variation and TLP-ESD pulse miscorrelation.

Thus, the current instability boundary for standard devices essentially represents the irreversible pulsed SOA [4, 5]. In this case, self-protection becomes important as it allows width scaling of the critical avalanche current, as in the case of large power arrays (Fig. 3.10b). While often suitable for component-level ESD pulse spec, the high current magnitude at system-level ESD events rarely allows any practical utilization of this approach. The voltage level of standard devices during ESD stress must be limited by the ESD protection network, though self-protection remains important for the second stage current.

Essentially, the pulsed SOA represents a physical limitation of the semiconductor devices' electrical regimes due to a two-stage positive feedback phenomenon. In the first stage, the device achieves a uniform current distribution along the contact width with a total NDR regime. The NDR is formed due to lateral conductivity modulation specific to the structure type. For standard active n-channel MOS devices, conductivity modulation of the pBase of the parasitic n-p-n BTJ structure is observed. Respectively, in the p-type active device, nBase conductivity modulation of the p-n-p BJT can take an effect. However, due to the difference in electron and hole mobility in Silicon, positive feedback in the n-type device with parasitic n-p-n is much stronger and provides NDR, while p-type devices typically show no, hidden, or rather small NDR.

Similar to phenomena in elementary semiconductor structures, when a device with some load resistance in the circuit reaches snapback, the uniform current distribution becomes spatially unstable and the device transitions into a state with non-uniform current and solitary or multiple current filaments. Further device operation depends on the current level, which is determined by the load resistance, local current ballasting and pulse duration. If there is structure-level negative feedback that limits local current density in the conductivity modulation mode, then the filaments will be able to expand until a new fully-uniform high-current state is reached. In these conditions, the snapback is reversible. When there is no device-level inhibiting factor to counterbalance the spatial current instability, the device experiences irreversible local burnout due to unlimited local current density increase.

The second scenario of self-protection, mentioned above, is a reversible operation in snapback mode. This type of operation is an essential design goal of snapback ESD devices. Similar to standard devices, the area of operational regimes is limited by structure-level positive feedback due to electrical conductivity modulation mechanisms, which forms the current instability boundary.

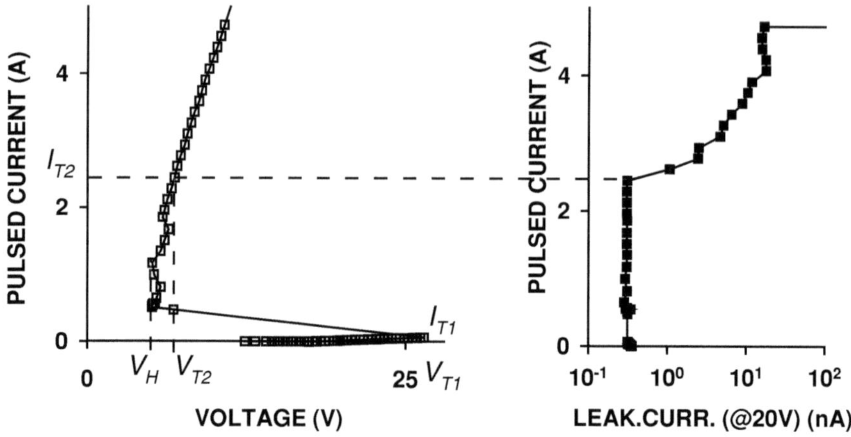

Fig. 3.11 Example of TLP characteristics for a 20 V tolerant nLDMOSSCR ESD device

However, this limitation is reversible. Due to specific ESD device architecture, structure-level negative feedback phenomena limit the increase of the local current density in the device until it exceeds the level of next-stage irreversible conductivity modulation of the contact regions or the self-heating and related electromigration limits. To verify reversible operation, S-shape reversible TLP characteristics can be measured in combination with the functional leakage test at every step (Fig. 3.11).

Since the duration of system-level ESD phenomena is short, structure heating is localized in the vicinity of a few microns. Thus, first stage positive feedback phenomena are mainly of an electrical nature, rather than an electro-thermal one. When a local ESD protection clamp is added to the pin, the absolute maximum voltage is limited by the ESD clamp triggering voltage, rather than the SOA of standard devices used in the internal circuit blocks connected to the pin.

3.2 Low Voltage ESD Devices for System-Level Protection

In system-level protection, an important engineering task is implementing a solution that combines energy balance inside the ESD device structure, uniform linear current density across the distributed cell array, and achieves specified electrical characteristics in the smallest possible footprint. The last two requirements include characteristics in ESD pulsed and normal operation regimes. These characteristics are the triggering and holding voltages, appropriate DC voltage tolerance, leakage current, high-side isolation and parasitic capacitance, and resultant transient ESD pulse voltage waveforms.

To minimize sensitivity of the ESD device to process technology variation and avoid yield impact, a key method of ESD device design is transforming the original standard device into a free ESD device. This method will be demonstrated in the following sections of this chapter for LV and HV on-chip ESD devices. Under this approach, the current instability that limits pulsed SOA of the standard device becomes the first stage of the triggering operation for the free ESD device. The ESD device is obtained by adding new regions and transforming the semiconductor structure to provide reversible operation in conductivity modulation mode. The conductivity modulation produces a positive feedback used to obtain the high current. However at some high current a device-level negative feedback must be implemented in the structure to balance the current instability and limit the current density below the safety limits. Thus, unlike in the case of standard devices, the ESD device operation in snapback mode is reversible.

3.2.1 Non-snapback Solutions

ESD devices and clamps for the component level specification are broadly presented and classified in [5] as well as in many other books in the ESD field [55–59]. To some extent a number of known component level ESD solutions can be width-scaled up to the appropriate current level and used for system pin protection. For example, system-level protection for input and output pins in multiple pin count, low-voltage products can reuse the rail network by combining a core active clamp with properly oversized ESD diodes arranged from the system level to the rails.

An example of serial data line pins that can tolerate a 2 kV cable discharge event (CDE) for Serializer and Deserializer IC over a single differential pair is presented in Fig. 3.12. This device is specified to work with cable up to 10 m for automotive video displays, with an LVDS standard compliant to ISO 10605 "Test methods for road vehicles—electrical disturbances from electrostatic discharge". The product also can tolerate a rather high cable discharge level, as verified by ISO gun ± 10 kV contact and ± 30 kV air discharge when an on-board 0.5 μF bypass capacitor is added.

From practical point of view (with some exceptions), compact device-level solutions are favored because they target an optimal use of chip space and cost levels. The level 4 IEC 61000-4-2 standard with an 8 kV contact and 15 kV air gap discharge requires a pulsed current capability of ~ 30 A for ESD clamps, at the desired footprint of 4×10^4 μm^2. This pulsed current capability can be achieved at linear current density above 10 mA/um, and thus requires the use of ESD devices operating in conductivity modulation mode.

To achieve this goal, LV device solutions with a voltage tolerance below 5–7 V typically require an SCR-type device based on double-injection conductivity modulation in the p-n-p-n structure. Both avalanche diodes and devices based on avalanche-injection conductivity modulation of parasitic n-p-n and p-n-p structures are not an adequate design approach.

Active clamps also typically cannot be used because they must pass system ESD stress. In power-on conditions during system level stress, the active clamp is

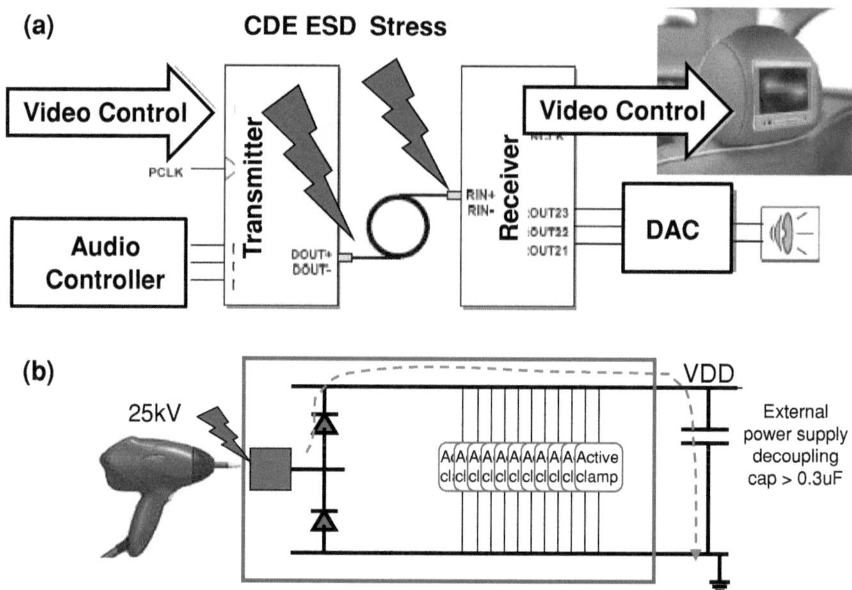

Fig. 3.12 Example of the interface application network (**a**) and ESD protection scheme with application of a board bypass capacitor (**b**)

practically disabled due to fully charged RC-timer capacitor. Therefore the clamp can provide a discharge current path only in the transient voltage range substantially above the power supply level.

In certain conditions, triggering of the snapback clamp in the on-state during system level ESD event under dc voltage at the pin may result in transient latch-up (Chap. 4). This design challenge may be resolved in some cases either with circuit design shutdown measures or may require the ESD device holding voltage to be above the power supply level. ESD devices with high holding voltages are also discussed in this chapter.

Avalanche diodes for of 5.5–7 V voltage range with reasonable leakage levels after the system-level width scaling can be designed as surface devices. The breakdown voltage reduction below 5 V level is limited by the high leakage current generated due to a band-to-band tunneling effect. To target the low voltage spec below 5 V the SCR devices can be used. They are described in the following section.

A major engineering challenge is the design of a self-aligned device. To avoid impact of mask alignment on the device characteristics the active junction of the device can be design using implants overlap. For example an appropriate doping level for the self-aligned lateral LV avalanche diode is represented by a combination of a lightly doped drain and shallow contact implants. For digital or analog

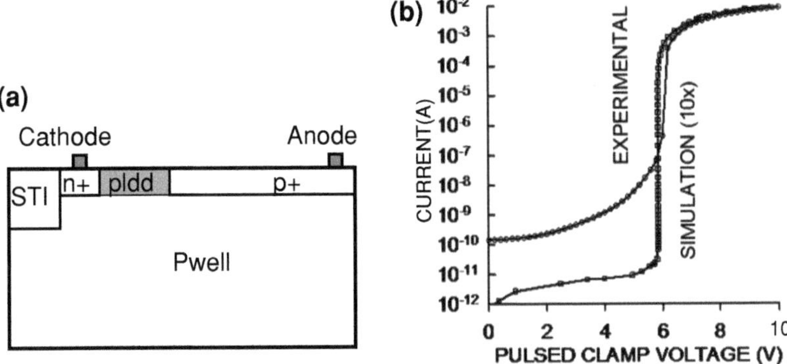

Fig. 3.13 Simplified cross-sections for a n+-pldd surface avalanche diode in CMOS process (**a**) and the corresponding experimental I–V characteristics (**b**)

CMOS devices the device can be implemented as a self-aligned surface diode formed by n+-pldd, p+-nldd implant (Fig. 3.13) with some mask overlap between.

Alternatively, a quasi-self-aligned diode can be made using poly-bounded architecture with a small gate region. The poly-gate region provides for quasi-self-alignment (with nldd-pldd), and the corresponding halo implants form the diode junction under the floating poly region (Fig. 3.14). In a poly-bounded diode, the forward injection current path is directly at the surface between the two diffused anode and cathode regions. This current path is much shorter than the path around STI in a conventional diode. So, the forward bias operation has a lower on-state resistance (Fig. 3.14c). In foundry processes, obtaining such a device poses a technical difficulty due to the automatic generation of LDD mask layers by the CAD package.

TVS components present another alternative to the lateral high holding voltage structures. Discrete TVS components are based upon vertical device architecture. These devices are reviewed in the last section of this chapter. The devices are not likely to be integrated in power-optimized mixed-signal BCD or extended voltage CMOS processes. Co-packaging of the TVS, on the other hand, can be done instead.

3.2.2 SCR and LVTSCR Devices

An on-chip free SCR design is the most effective approach to realizing system-level ESD solutions. It consists of a "parasitic" p-n-p-n structure that supports the double injection conductivity modulation. This type of free ESD device can be obtained by forming parasitic p-n-p in an n-type standard device that already has a parasitic n-p-n structure. Alternatively, a parasitic n-p-n can be formed in a p-type standard device with an existing parasitic p-n-p.

Fig. 3.14 Comparison of the STI (**a**) and poly-bounded (**b**) avalanche diodes process simulation cross-sections and their experimental I–V characteristics (**c**)

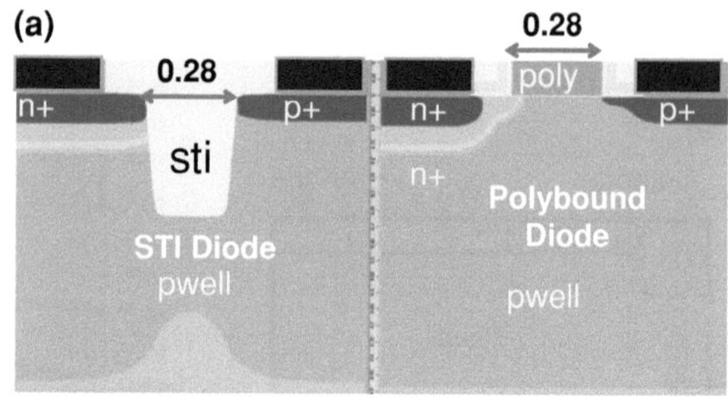

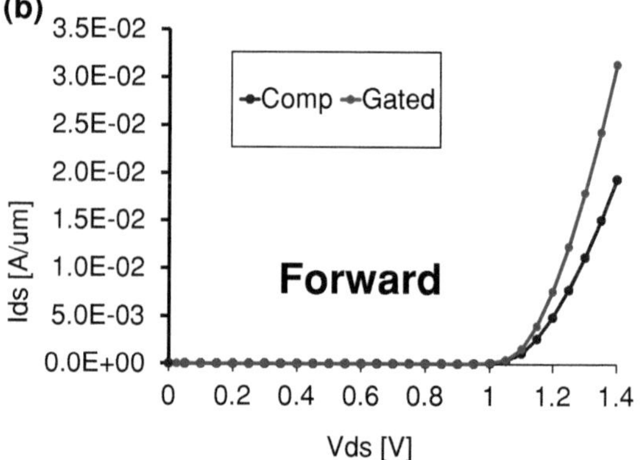

The device-level engineering goals are to achieve the proper gain and multiplication coefficient in this embedded structure. The gain and multiplication factors are adjusted by the dimensions of the device region, with implant profiles chosen to target the desired triggering voltage, high current capability and sensitivity to the control electrode conditions.

In general, operation of an SCR device represents a domino effect scenario. At first, the critical current for instability is reached either by avalanche multiplication of the relatively high channel current at the low multiplication coefficient or by multiplication low off-state leakage current at rather high multiplication coefficients. Then, avalanche-injection of the n-p-n (p-n-p) structure in the n-type- (p-type-) based SCRs results in an additional substantial current increase. In n-type SCR devices, the current increase is accompanied by negative differential resistance, while p-type SCR devices maintain a positive differential resistance at this stage. At some critical current, an internal voltage drop above ~ 0.7–1 V on the n- (p-) base-emitter junction of the p-n-p (n-p-n) causes the junction to open. The positive feedback between n-p-n and p-n-p structures results in double injection conductivity modulation. In these conditions, the structure triggers according to load characteristics into the low holding voltage state.

In n-type-based SCRs, the negative differential resistance stage is often hidden. Alternatively, double S-shape I–V characteristics can be realized for the device with certain parameters. It is hard to differentiate between the first two stages in the p-type-based SCR device, unless the SCR is compared to the lateral avalanche diode with the same blocking junction regions design.

At high injection current, the SCR structure provides an internal negative feedback that balances the current density across the entire structure width. At the transition from avalanche-injection to double-injection mode, the avalanche current component is substituted by the injection current component. As a result, on-state avalanche multiplication in the device is realized at a very low rate, with a corresponding low electric field. The forward injection from opposite n- and p-emitter junctions makes the device practically equivalent to a forward biased p-n diode with positive differential resistance. Thus, isothermal spatial current instability is fully inhibited until high current levels are reached. When heat generation becomes dominant a different electro-thermal spatial instability phenomena results in local structure melting followed by accelerated electro-migration in the contact regions and other irreversible effects.

When additional channel or base current is an available device design option, the triggering characteristics of the SCR device can be passively or actively controlled by the clamp components. One of the first SCR ESD devices to utilize this capability is so-called Low Voltage Triggered Silicon Controlled Rectifier (LVTSCR) [60]. This device is one of the most conventional practical solutions for LV system-level protection. The ancestor device of the LVTSCR is a corresponding standard LV analog or digital NMOS device.

The transformation target for the device is the addition of the parasitic p-n-p structure. The major steps of this process consist in creating three new regions: the pplus emitter diffusion, the nplus floating drain diffusion with the pEmitter vertical and lateral isolation by the nWell region. The nWell essentially represents the nBase of the SCR (Fig. 3.15a).

Similarly, standard LV PMOS structures can be transformed into a p-type SCR by embedding the appropriate n-p-n structure. In this case, the corresponding nEmitter diffusion region is isolated by the pWell (here representing the pBase) (Fig. 3.15b). The pWell can also be isolated from the substrate by a Deep nWell region in CMOS processes or by a nEpi and n-buried layer (NBL) in BCD processes.

The most critical parameters of the structure are those that define the gain of the embedded n-p-n and p-n-p structures. In LVTSCR, the bases are typically shorted to the corresponding emitter and only the gate is used as a control electrode. In this case, the gain is mostly defined by the internal base-emitter well resistances R_{NW} and R_{PW} (Fig. 3.15a). The internal p-n-p and n-p-n base resistors R_{NW} and R_{PW} are controlled by the base-emitter spacing and region length. To some extent, linear parameters of the structure can be used to control the holding voltage. These parameters include the floating drain region length L_F, the pEmitter length L_P, and corresponding spacing L_{FP} and L_{PD} (Fig. 3.15a). Varying these parameters produces greater isolation of the pEmitter.

Excluding the substrate and the nEpi terminals, the final nLVTSCR ESD structure is a five-terminal device. To preserve original NMOS devices' terminal

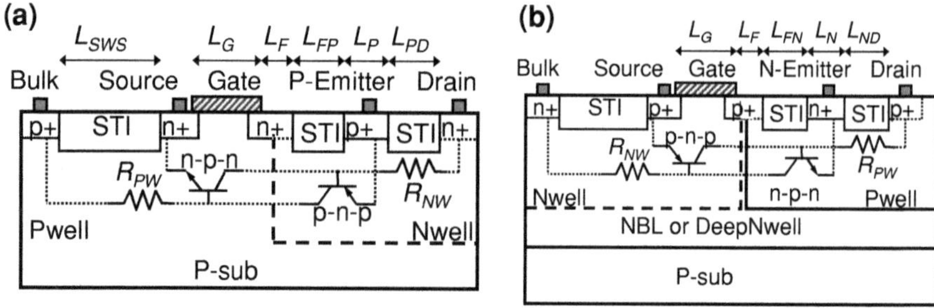

Fig. 3.15 Simplified cross-section for n-type (**a**) and p-type (**b**) LVTSCR devices with critical structure parameters and parasitic BJT and resistor structures

nomenclature, the following terminology is often used: n-Drain (as nBase of SCR); p-Emitter (as pEmitter of SCR); Gate; nSource (as nEmitter of SCR); and pBody (as pBase of SCR).

Thus, the device in general provides one field control electrode for initial channel current multiplication at the avalanche-injection stage and two base control electrodes that can be used to inject minor carriers through the base-emitter junctions. These control channels are widely used in design of SCR clamps with specified triggering and holding characteristics.

The simplest LVTSCR clamp circuit with grounded gate is presented in Fig. 3.16. At fast transient ESD pulse and large gate resistance, a gate coupling effect is used to reduce the triggering voltage substantially. Due the dV/dt effect the gate potential increases above the threshold voltage. If the parasitic n-p-n in the LVTSCR is designed to support the falling dependence of the triggering voltage upon the gate bias the LVTSCR triggering voltage in high conductivity state is realized below the DC breakdown voltage level.

For passive LVTSCR triggering control the large gate resistor can impact the signal at the pin. At fast transient characteristics the dV/dt coupling in normal operation regime might be undesirable due to a parasitic channel current. A more appropriate clamp design involves active reference circuits. A reference circuit can potentially be added to all three voltage and current control electrodes in order to bring the clamp triggering characteristics into the desired triggering voltage range.

Examples of typical voltage- and current-referenced clamps are presented in Fig. 3.17a–d. For example a small contact width avalanche diode can be used as a high-side voltage reference component to actively bias the gate above the diode avalanche breakdown voltage (Fig. 3.17a). Similarly, for p-LVTSCR, a low-side voltage reference can be implemented based upon this principle (Fig. 3.17a). The current reference for the base electrodes can be realized with a forward biased diode-triggered SCR (DTSCR) [61] (Fig. 3.17c, d).

The principle of operation of the clamp utilizes the waning dependence of the triggering voltage on either the voltage of the field control electrode or the current through the base electrode. This condition is satisfied rather often, but not

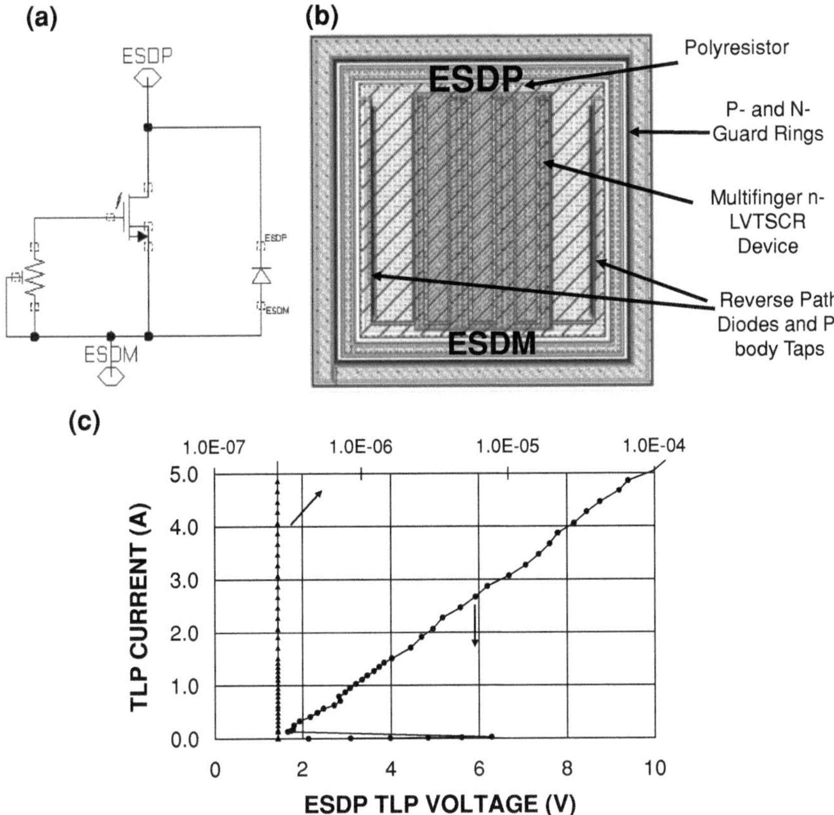

Fig. 3.16 Circuit (**a**), cell layout topology (**b**) and measured TLP characteristics (**c**) of the grounded gate LVTSCR clamp

guaranteed. In order to properly engineer the clamp, a three-terminal characterization of the device (similar to a pulsed SOA) is informative.

A more sophisticated mixed-device-circuit ESD solution, which was first proposed in [62], is based upon the driver-controlled triggering characteristics of the device under ESD and normal operating conditions.

LVTSCR devices possess the same voltage tolerance limitations due to long-term gate oxide reliability as NMOS devices. If the voltage tolerance needed for system-level input and output pins is above the analog CMOS module operating voltage, a field-oxide SCR (FOXSCR) can be implemented following a simple transformation of the LVTSCR device. The transformation is just a replacement of the poly-gate region with a minimally-sized STI region (Fig. 3.18a).

In the resultant FOXSCR device, both the breakdown and triggering voltages are determined by the diffusion-to-well breakdown voltage, which is usually relatively high $\sim$10–13 V. In regular SCR devices, the well-to-well breakdown voltage can be even higher, in the range of 16–24 V (Fig. 3.18b). Often, dV/dt

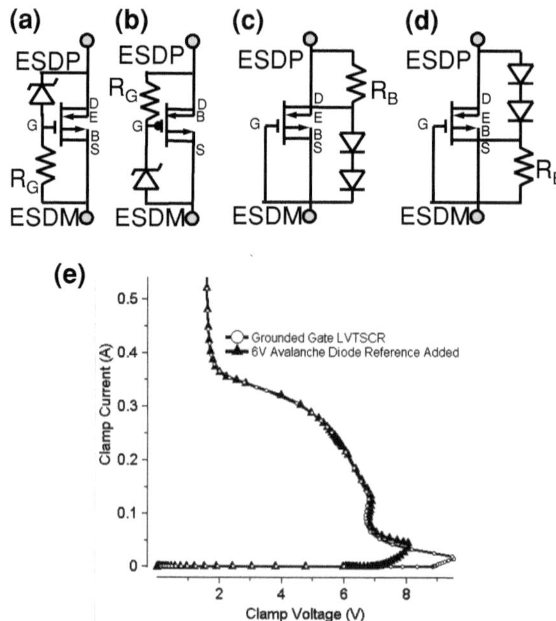

Fig. 3.17 Clamp circuits for nLVTSCR (**a**) and pLVTSCR (**b**) with an avalanche diode reference; nLVTSCR with a forward diode string n-base (**c**) and p-base (**d**) current reference and simulated I–V characteristics for nLVTSCR clamps with a 6 V avalanche diode gate voltage reference (**e**)

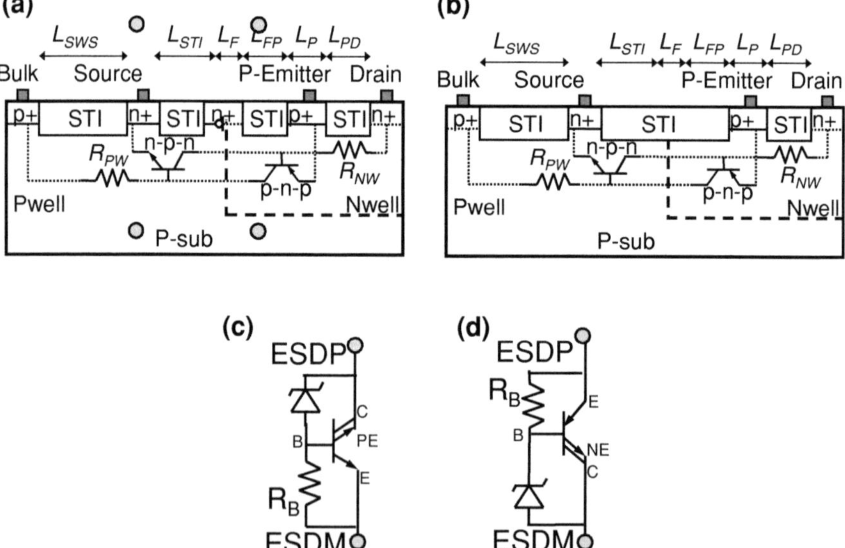

Fig. 3.18 Cross-sections of the FOXSCR (**a**) and conventional SCR (**b**) and the clamp circuits with high-(**c**) and low-(**d**) side avalanche diode voltage reference

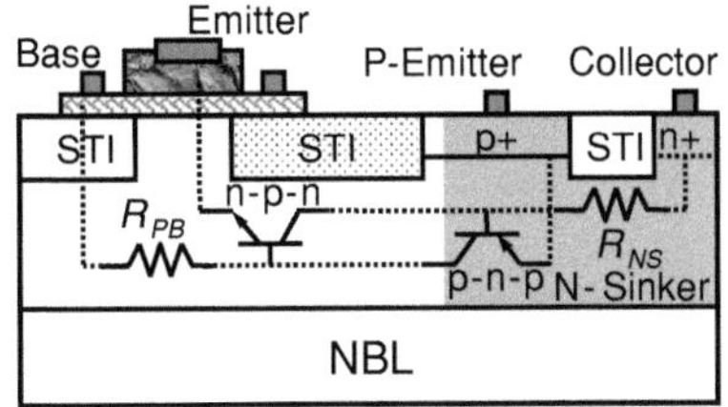

Fig. 3.19 Simplified cross-section of the Bipolar SCR in a Si-Ge n-p-n BiCMOS process

triggering due to gate electrode coupling or the displacement current effect in the parasitic n-p-n and p-n-p structures are undesirable, due to load dependence, interference with the pin signal or transient latch-up jeopardy. To trigger the FOXSCR at a lower voltage, avalanche-diode-referenced clamps can be implemented, thus providing base current injection after the avalanche diode breakdown (Fig. 3.18c, d).

Another type of efficient SCR devices called *Bipolar SCR* (BSCR) devices can be designed for advanced, high-speed BCD or BiCMOS processes with gain poly-emitter BJT devices [63, 64]. The principle of operation of BSCR devices and the major device regions is similar to SCRs implemented in CMOS processes. The major difference in the architecture is the original transformation of the high-gain poly-emitter n-p-n BJT rather than NMOS device. This device transformation is achieved by embedding an additional p-emitter diffusion region between the pBase and the nCollector (Fig. 3.19).

The pEmitter isolation by nSinker implant is rather high. Therefore in small signal operation, the BSCR gain is similar to the ancestor n-p-n device. This allows reuse of standard compact models. The current instability criteria for the SCR is $\alpha_{NPN} M_N + \alpha_{PNP} M_P > 1$ [5]. When the n-p-n gain is much higher than the p-n-p gain, $\alpha_{NPN} \gg \alpha_{PNP}$, this can be simplified to $\alpha_{NPN} M_N > 1$. Thus, in a properly designed BSCR, the instability boundary can be achieved similarly to the ancestor n-p-n BJT device.

3.2.3 High Holding Voltage SCRs

When system-level ESD stress is required to pass in power-on conditions and the voltage source at the stressed pin has low impedance, transient latch-up may occur. The most reliable way to avoid latch-up is by implementing an SCR clamp with an on-state holding voltage above the power supply level. Alternatively, a special power supply with a fast shutdown and reset circuit features can be used.

When the emitters of the SCR device are not excessively isolated by base diffusion, the holding voltage in the on-state is typically in the 1.5–2 V range. While this might be adequate for LV digital or RF inputs and outputs with an operating voltage of 1–1.5 V, the analog domains with 2, 3.3 or 5 V voltage require a different approach.

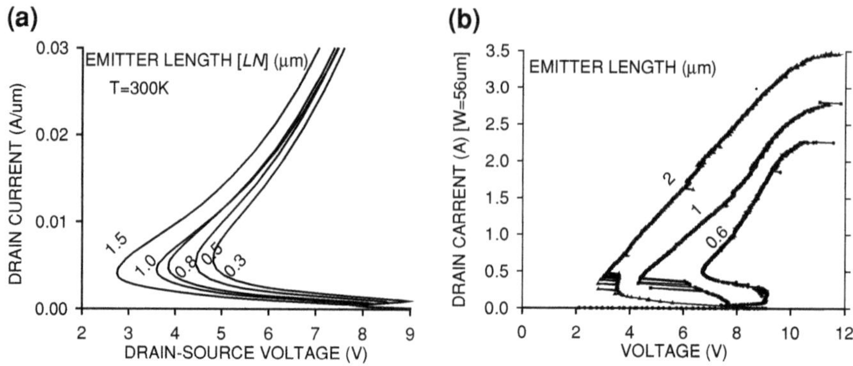

Fig. 3.20 Simulated (**a**) and measured TLP (**b**) LVTSCR I–V characteristics for different p-emitter lengths [5]

The desired 3–5 V holding voltage can be achieved either on the device level or on the clamp level. On the device level, the holding voltage of an SCR can be controlled by the structure parameters that impact the gain of the internal n-p-n and p-n-p structures (Fig. 3.20). These control parameters are related to the level of isolation of the emitters. Unfortunately, these parameters control the triggering characteristics and the carrier balance at high current operation too. Thus, this straightforward approach in increasing the holding voltage has limitations and may result in both undesirable significant increase of the triggering voltage and loss of the structure's high current capability. Overall, increase in the holding voltage on the structure level reduces the ESD protection capability of the SCR devices, due to the increase of the avalanche current component required to maintain the carrier imbalance for the relatively high electric field in the structure and higher heat generation.

Ultimately, it is clear that if an nLVTSCR pEmitter is completely isolated, for example, due to reverse positioning of the emitter and drain regions the device is essentially physically equivalent to a snapback NMOS. In this case respectively higher holding voltage is realized due to different conductivity modulation mode as well as higher triggering voltage is the result of current saturation in the n-Well.

Nevertheless, some progress to voltages above the 2–3.3 V range can be made by experimental variation of the semiconductor process specifics. A properly designed LVTSCR ESD cell with the same width and the same holding voltage as snapback NMOS cell still provides a significantly higher current capability per contact width (Fig. 3.21).

An alternative method [65] for clamp-level design uses the self-adjusted p-emitter de-biasing circuit. Here, p-emitter injection is controlled with high-current serial de-biasing diode structures (Fig. 3.22a). On the clamp level, the p-emitter de-biasing diodes can be either internal LVTSCR device regions (Fig. 3.22b) or external isolated clamp components. The holding voltage of the resultant clamp is controlled by the number of diodes (Fig. 3.22c). The pEmitter de-biasing circuit

Fig. 3.21 Comparison of
NMOS and LVTSCR with
the same holding voltage; the
last point of each *curve*
corresponds to the beginning
of soft-leakage degradation

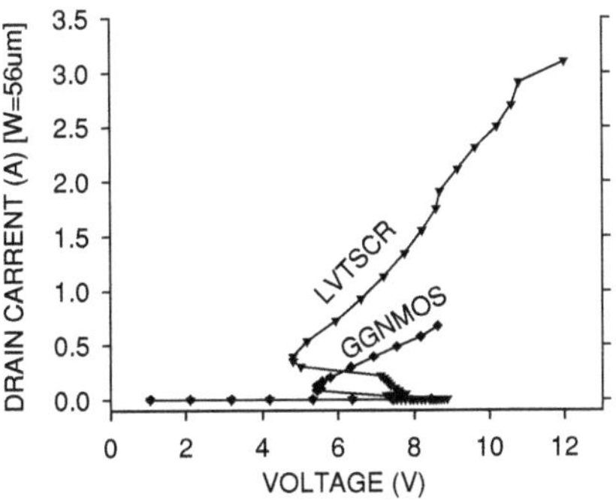

creates an additional voltage drop in the pEmitter circuit, which prevents pEmitter
injection below the corresponding voltage level (Fig. 3.22c). The current capability of this solution significantly exceeds that of the topological and device-level
solutions that target internal gain reduction.

3.2.4 Low Voltage Dual-Direction Devices

A small footprint and dual-direction functionality are often specified as targets for
many analog applications. These applications include display column drivers in
portable electronics that are subject to system-level stress, RF inputs, common-
mode voltage regulators, and interface applications.

One of the main reasons for the dual direction tolerance is namely system
specific. When an electrical connection is established between two remote systems, the ground potentials for each system may be different. In this case, the
presence of a diode at the input-output to the ground of such systems cannot be a
tolerable design. In this case dual-direction ESD protection becomes the design
solution of choice. The dual-direction requirement usually refers to the case where
the IC pin must tolerate low leakage more than just the forward-biased diode
voltage drop (>1 V) bias conditions both above and below the ground potential.

The design of dual-direction devices is one of the most sophisticated challenges
when system-level current capabilities, a small footprint and high holding voltage
are among the design targets.

Apparently, the dual-direction solution can only be implemented when pWell
isolation from the pSubstrate region is available as a process option. A simplified
dual-direction clamp can be composed by back-to-back stacking of SCR or n-p-n
BJT unidirectional clamps (Fig. 3.23a). The clamps can be joined by connecting
the respective high-side terminals to the floating middle node. Apparently, the
middle node must be floating to avoid forward-biased diode formation on the
substrate at either positive or negative pin bias. At positive (relative to the ground)
ESD stress, the clamp connected to the ground node provides high clamping
voltage, while the remaining stacked components supply the forward body diode

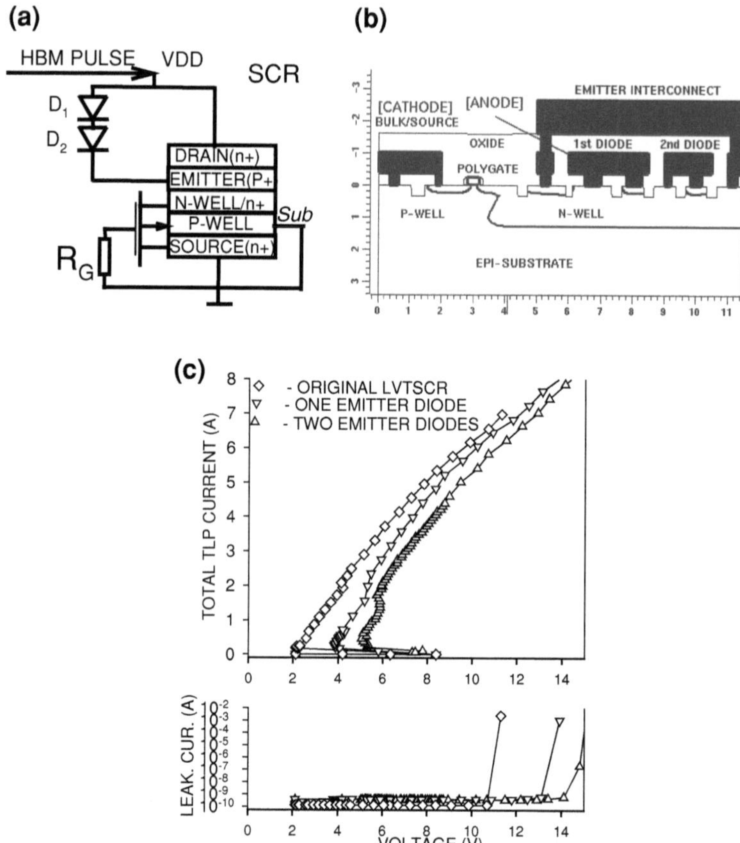

Fig. 3.22 Simplified equivalent circuit for a high-holding-voltage SCR with emitter high current de-biasing (**a**); cross-section of the device with embedded diodes in the N-base region (**b**) and experimental TLP I–V characteristics for the clamp with different numbers of diodes (**c**)

current path. At negative pad bias, the upper clamp provides the high clamping voltage, while the clamp directly connected to the ground provides the body diode. If the clamp has no embedded body diode, the dedicated diode must be added. This case is presented in Fig. 3.23a.

A version of the back-to-back stacked dual-direction n-p-n BJT clamp (Fig. 3.30a) can be implemented using layout-isolated components. A more advanced approach can be realized on the common cell layout. A cell with merged sub-collector back-to-back BJT devices combines four alternating groups of fingers, represented by four circuit components (Fig. 3.23a). The base-collector junction of an n-p-n BJT combined with shorted base-emitter terminals (BJT1 and BJT4) forms the reverse path diodes. At a positive ESD pulse stress, BJT2 remains passive and the current path is provided by BJT3 in the snapback mode and BJT1 acting as a forward-biased collector-base diode. During negative ESD pulse, the ESD current path is formed through BJT2 in snapback mode and BJT4 in the diode mode.

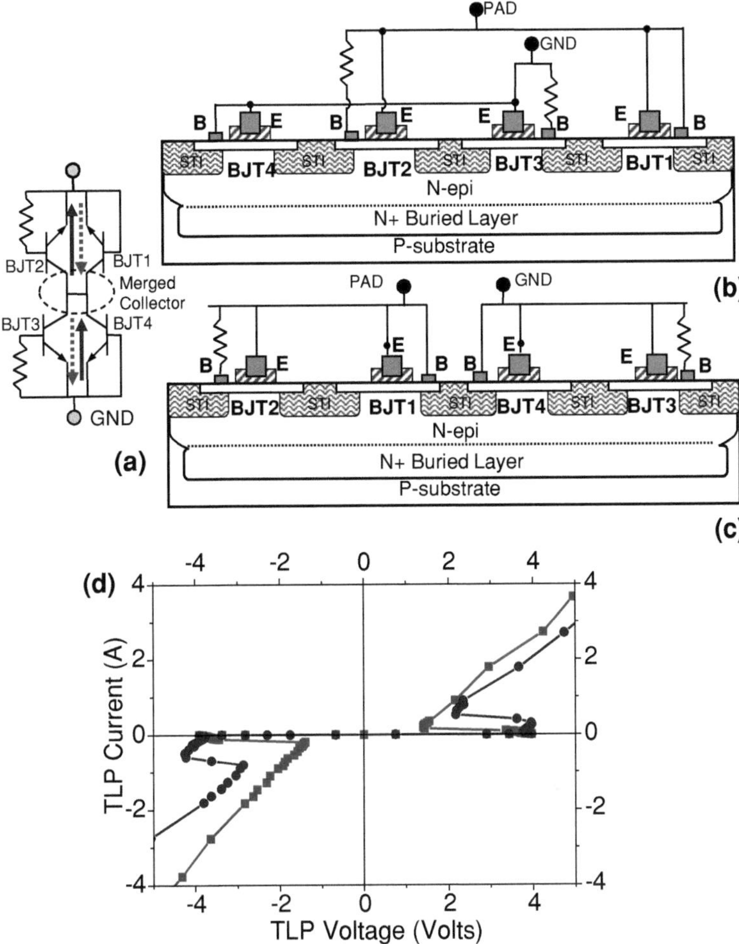

Fig. 3.23 Equivalent circuit (**a**) and cross-sections of merged collector dual-direction BJT structures with distanced (**b**) and close (**c**) arrangements of the BJT1/BJT4 pair and measured pulsed I–V (TLP) *curves* for these two architectures in *diamond* and *square* symbols, respectively

In the arrangement sequence (Fig. 3.23b) or when separation of the base-emitter regions of the diode path BJT devices is large (Fig. 3.23b), the clamp characteristics simply repeat in each direction for the summation of the layout-isolated NPN BJT clamps. However, when the fingers are re-arranged according to (Fig. 3.23c) and the space the between BJT1 and BJT4 device regions is reduced to a minimum, a dual-direction SCR effect becomes visible. In this case, the p-Bases of BJT1 and BJT4 act as SCR emitters and form an SCR current path with the corresponding n-Emitters. The SCR effect is observed on the I–V character-istics as a lower holding voltage for this configuration (Fig. 3.23d).

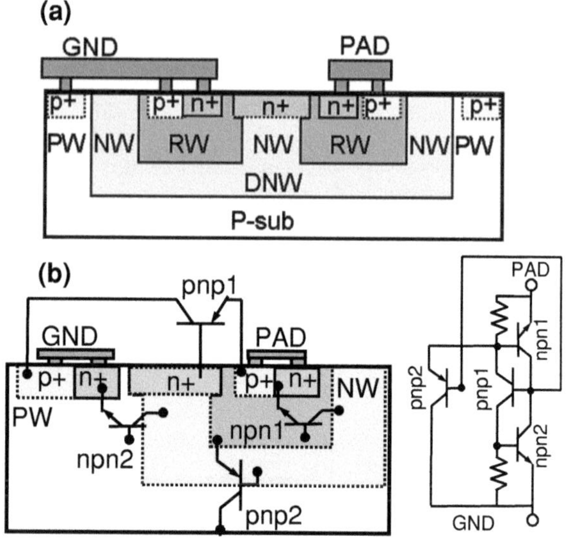

Fig. 3.24 Typical dual-direction SCR structure; PN-n-NP structure with the n+-pWell (**a**), the compact asymmetrical version of the PN-n-PN device (**b**) and its lumped component circuit [66]

This device example presents a good methodological transition toward understanding a more compact dual-direction SCR design. The middle portion of the device (Fig. 3.23c) with BJT1 and BJT4 base-emitter regions and a floating common sub-collector essentially represents a particular case of dual-direction device DIAC architecture. A CMOS-based DIAC for a non-silicided 0.5 μm 5 V CMOS process with DeepNwell isolation is presented in [66]. Below, two versions of this device represent a full substrate isolation design (Fig. 3.24a) and a more compact design (Fig. 3.24b) with only the pad-side device region isolated.

The DIAC device combines two pairs of n- and p- contact diffusions with double blocking Pwell-Nwell junctions. The contact diffusions act as electron and hole injectors and collectors. These interchangeable roles depend on the ESD pulse polarity that forms the elementary p-n-p-n structure that supports double-injection conductivity modulation mode. For positive overstress, the pad-side pplus diffusion provides the pEmitter and the nplus-diffusion plays the role of the nBase, while the ground connected nplus- and pplus- diffusions connected to the ground act as the nEmitter and pBase, respectively (Fig. 3.24c).

The sequence of emitters and bases at each polarity determine the level of injected carriers, since base-emitter resistances are different in each case. The resultant space charge neutralization level impacts both the resultant holding voltage and the high current withstanding capability. Several cell layout options with varying injector region lengths and arrangements can be explored experimentally to identify the necessary holding voltage range at the appropriate high-current performance for a given technology process.

The compact version may seem like a logical choice because it does not require ground-side pWell isolation (Fig. 3.24b). However, in CMOS processes, the characteristics of this version of the device are significantly impacted by the pSubstrate

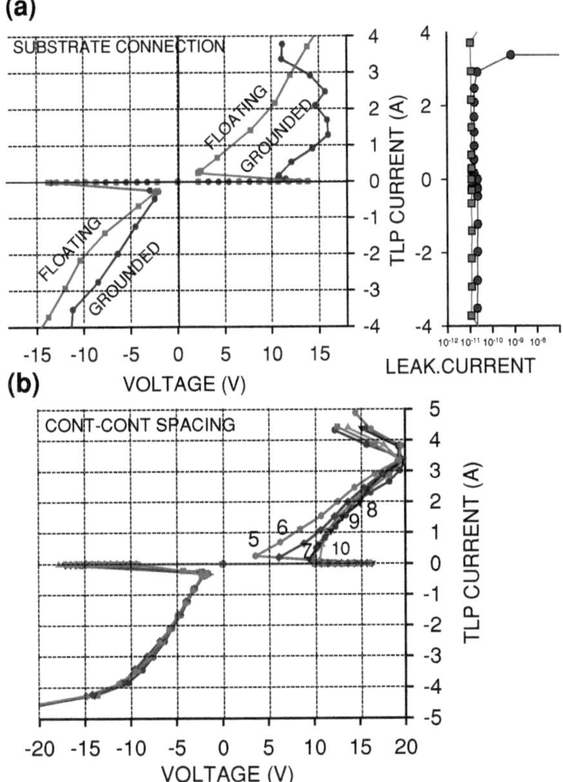

Fig. 3.25 TLP I–V characteristics for symmetric PN-n-NP structures with different substrate connection conditions (**a**) and with scaling of the device characteristics and reducing the contact-to-contact space (**b**)

holes collection and injection effect. At negative pad stress, the grounded pSubstrate acts as an additional vertical pEmitter, providing holes injection and additional positive feedback for a low holding voltage. On the contrary, at positive pad stress, the pSubstrate acts as a collector, providing the path for partial escape of the injected holes from the main SCR current path. This negative feedback results in holding voltage increase and a general reduction of the high-current capabilities [66]. At the same time, understanding these types of positive and negative feedback mechanisms realized on the structure level (Fig. 3.25a) is beneficial in controlling the level of the space charge neutralization between the injected carriers, and thus the corresponding holding voltage (Fig. 3.25b).

3.3 High-Voltage ESD Devices for System-Level Protection

ESD protection of high-voltage (HV) pins with system-level requirements has an elevated complexity not only due to the high current requirements. The most challenging performance specification is avoidance of the transient latch-up and chip-level high voltage latch-up. The second effect will be discussed in Chap. 4. The most straightforward way to safeguard against transient latch-up is to implement a device with a minimum on-state holding voltage above the power supply voltage level at the protected pin. This requirement also maintains system stability against short-term electrical overstress (EOS) events above absolute maximum voltage limits.

Depending on the product design, generally both high holding voltage and low holding voltage solutions can be used. For HV control pins or slow power pins, snapback devices with a holding voltage below the power supply level can be used, as the triggering and holding current cannot be supported by the power at the pin, resulting in a circuit-level reset and shutdown. However, for HV fast transient pins, hot plug-in requirements and system level specification the on-chip HV ESD device solution might be required to provide the high current characteristics under the holding voltage above the power supply level. This section is focused on major HV integrated on-chip devices and clamps with voltage tolerances of ~ 10–200 V, with the aim to provide understanding of system-level solutions.

3.3.1 High Voltage Active Clamps

HV active clamps are based on a driver circuit and a NLDMOS array. In general, HV active clamps are not a space-efficient protection solution due to the very large area required for HV LDMOS arrays, with lengthy drift region under rather low saturation current of ~ 0.4–0.6 mA/μm. Another impact to the area is from the high-voltage capacitance (Fig. 3.26). On the contrary, HV snapback ESD devices, with active regions that are only a few microns longer, can deliver current capability of above 10 mA/μm. Therefore mostly only highly integrated and high-pin-count analog circuits can tolerate integration with the active clamps with the required high holding voltage.

The active clamps are implemented either using RC-based drivers or as a high-side voltage reference circuits to turn on a NLDMOS array. An example of the first design approach with the RC triggered driver is presented in Fig. 3.26a using DECIMM [19] mixed-mode analysis. The operation principle of the clamp is based on fast transient turn-on of the high-side NLDMOS driver component ($M1$) while the capacitor in the $R1C1$ circuit charges (with a rather short time constant of <100 ns).

After $M1$ turns on, the fast charge on the array $M2$ gate through the $M1$ current path is realized before $C1$ is fully charged. With the voltage increase, $M1$ turns off, while $M2$ provides the discharge path until the gate is fully discharged through $R2$.

This clamp can be controlled by enabling the circuit driving the $M3$ and $M4$ LV devices (Fig. 3.26). Even if the design is capable of accommodating a very large $M2$ array, this active clamp design is not likely to have any practical use when system-level stress applied in power-on conditions. In this case, the capacitor $C1$ is

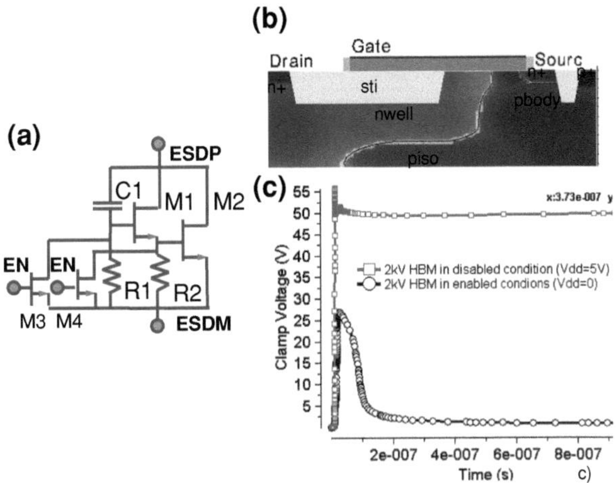

Fig. 3.26 RC-triggered HV active clamp with a shutdown option for fast switching nodes (**a**), NLDMOS FEM device cross-section (**b**) and 2 kV HBM waveforms in enabled and disabled states (**c**)

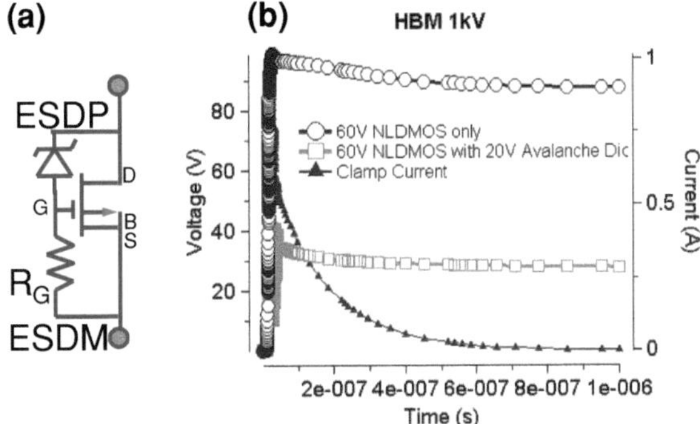

Fig. 3.27 HV Active clamp circuit with an avalanche diode reference (**a**) and 2 kV HBM waveforms with and without a reference avalanche diode (**b**)

fully charged to the pin voltage and the clamping capability is disabled below this pin's bias level.

A more practical design for system-level protection with a large footprint is an active clamp circuit in which the gate pull-up array achieved by a high-side referenced circuit with an avalanche diode (Fig. 3.27). A low-current avalanche diode or stacked components with an appropriate total breakdown voltage can be used for the pull-up driver.

Here, the clamp's principle of operation is rather simple. When the voltage at the protected node becomes higher than the breakdown voltage of the high-side reference avalanche diode circuit, it results in gate pull-up of power array, turning the array on. Depending on the particular product, the NLDMOS array can provide 2–3 times higher current in gate overdrive mode than in saturation mode. A drawback of the clamp is the relatively high clamping voltage resulting from the summation of the breakdown voltage and the overdrive gate bias.

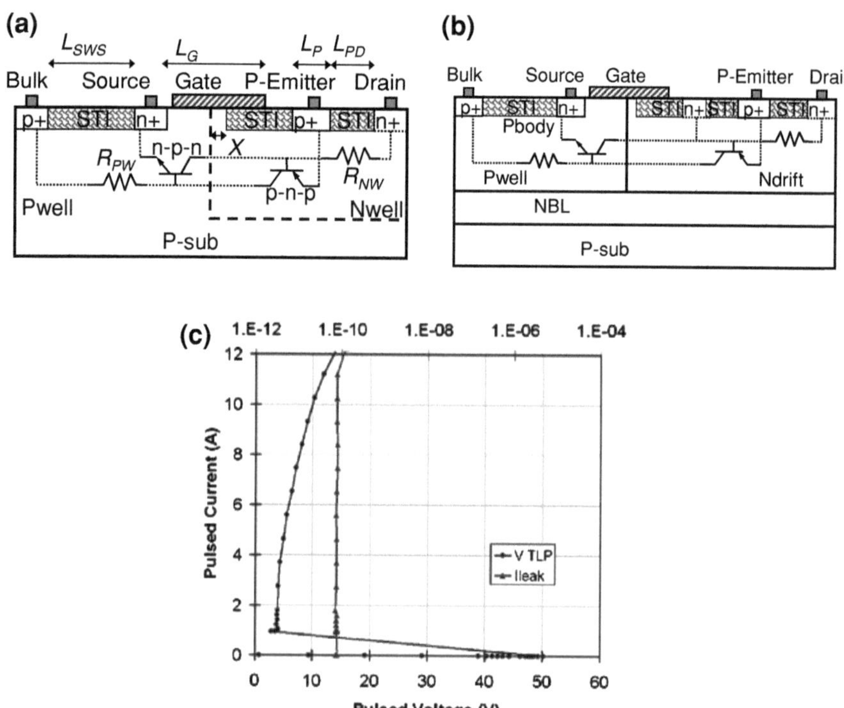

Fig. 3.28 Schematic cross-sections of the nDeMOSSCR (**a**) and nLDMOSSCR (**b**) for extended voltage CMOS and BCD processes, respectively and the experimental TLP I–V characteristics for nDeMOS-SCR for 40 V ESD protection [67]

3.3.2 LDMOS-SCR Devices

The principle of operation, device architecture, and transformation steps for LDMOS-SCR are similar to LVTSCR devices. The starting point standard LDMOS devices are used instead of standard LV MOS components. The physical difference is mostly in HV capability of the n-p-n or p-n-p embedded structures.

Similar to the LVTSCR described in Sect. 3.1, the LDMOSSCR baseline architecture includes a floating drain and pEmitter and contact drain regions reused from the original corresponding either nDeMOS or nLDMOS devices. Simplified cross-sections for nDeMOSSCR (Fig. 3.28a) or nLDMOSSCR devices (Fig. 3.28b) demonstrate this basic device architecture. The experimental TLP characteristics of the nDeMOSSCR for a 0.5 μm extended voltage process are presented in Fig. 3.28c.

The floating drain region may be eliminated in some particular cases, when the proper level of pEmitter isolation is provided by available in the process nWell-like regions. Similarly, depending on the well profiles, reverse positioning of the connected drain and pEmitter may still provide a high current capability of the device. These particular cases are not always optimal for arbitrary nWell or nDrift doping profiles and pEmitter lengths, and may result in miscorrelation effects for system-level pulsed conditions.

In general, nLDMOSSCR operation is based on the same three-stage domino effect that is described in Sect. 3.1 for LVTSCR ESD device. It involves (i) channel current avalanche multiplication up to some critical current for avalanche-injection conductivity modulation and NDR formation in the internal n-p-n structure, (ii) current instability in n-p-n structure that results in snapback according to external load characteristics and (iii) turn-on of the p-n-p emitter-base junction when sufficient current density is generated in avalanche injection mode to open the internal pEmitter-nBase junction. The positive feedback between parasitic n-p-n and p-n-p structures complete the double injection conductivity modulation in the formed p-n-p-n SCR current path.

The avalanche-injection instability in a high voltage n-p-n structure plays a key role in reversible device operation. If the n-p-n device cannot provide an appropriate current level for p-n-p base-emitter junction turn-on, nLDMOSSCR operation in snapback becomes irreversible and the device irreversibly fails immediately, after triggering according to TLP results.

One of the causes of this phenomenon is the low source-body resistance implemented in the original nLDMOS to achieve the highest self-protection capability and avalanche energy. Butted source-body layout placement is often used for this purpose, with a common silicided active region and an additional heavily doped pDeep implant under the n-Source. An HV parasitic n-p-n with these design features cannot usually support any high current because it operates in highly non-linear conditions with high multiplication coefficients accompanied by spatial instability with high-amplitude local current filaments, which are potentially damaging to the device. The first design step in making nLDMOSSCR reversible is to separate the source and body diffusions, thus obtaining the internal source-body resistance that will allow internal biasing of the source-body junction for appropriate levels of electron injection.

The floating drain in high-voltage NLDMOS devices terminates the drift region electric field in low current conditions in order to avoid premature punch-through turn-on of the p-n-p structure. In reverse emitter-drain positioning, this task is accomplished by the drain itself, but as it was already mentioned, this architecture has drawbacks.

The NLDMOSSCR also possesses the major regions standard to a process technology NLDMOS. For obtaining a NLDMOSSCR, the transformation process includes changing the drain diffusion region to accommodate the floating drain and pEmitter, as well as separating the pBody and n-Source diffusion regions, or even partially replacing the pBody by a pWell.

If the NLDMOSSCR is designed to operate in a grounded gate clamp circuit, triggering of the structure can be adjusted below the voltage range of the protected NLDMOS pulsed SOA. This can be done by reducing the drift region or changing the other blocking junction parameters.

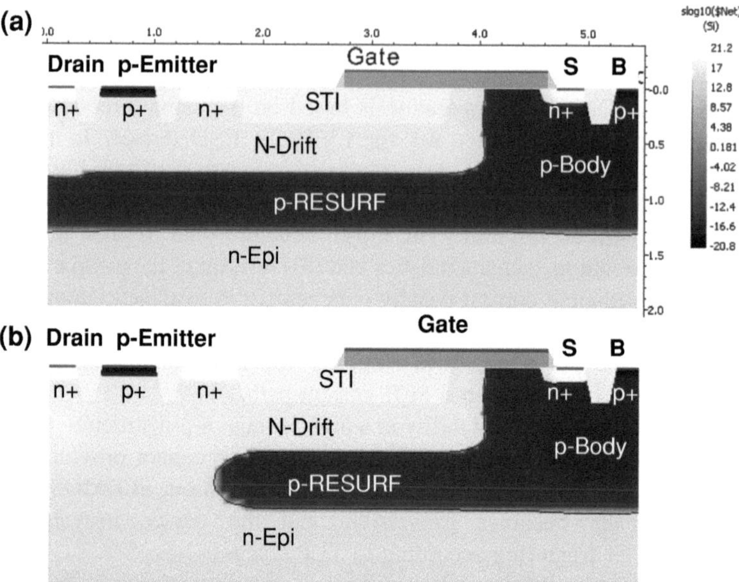

Fig. 3.29 DECIMM cross-section for a NLDMOSSCR device with full (**a**) and partially blocked (**b**) pRESURF region

Often, power-optimized HV NLDMOSSCR device architecture is based on an in-depth pRESURF region. This region forms as a vertical junction that provides depletion for the relatively highly doped nDrift region, in addition to a poly-field-plate RESURF effect. The pRESURF region is electrically connected to the pBody. When a pEmitter is embedded in the device, a vertical p-n-p structure is formed by the pEmitter, nDrift and pRESURF, in addition to the desired lateral p-n-p structure. The maximum breakdown collector-emitter voltage of this vertical p-n-p structure is usually much lower than the targeted voltage tolerance for the HV nLDMOSSCR. Therefore, other design measures with partial blocking of the pRESURF under the pEmitter region are required.

The HV SCR (Fig. 3.29) or PLDMOSSCR (Fig. 3.30b) can be designed similarly to the LV SCRs. The advantage of a PLDMOSSCR device is the potentially high triggering current, which can be useful in avoiding a transient latch-up when the bias at the protected pin is supplied by a high impedance voltage source.

NLDMOSSCR devices are usually preferred to HV SCRs due to the presence of a MOS control electrode. In HV process technologies, designing a NLDMOS that simultaneously has a low dV/dt effect and a high current capability can often become challenging. This is mainly because the same positive feedback avalanche injection processes determine both the triggering and the high current operation conditions.

To resolve this design dilemma, a clamp with a substantial level of channel current before triggering is composed from a high-side avalanche diode in the

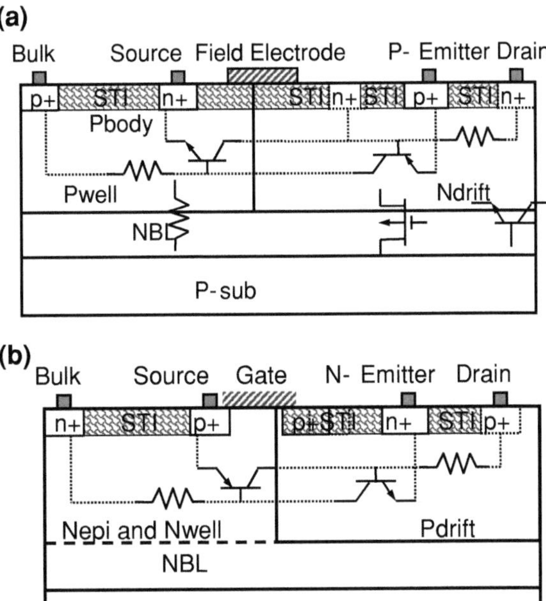

Fig. 3.30 Simplified cross-sections for high-voltage SCR (**a**) and pLDMOSSCR (**b**) devices

NLDMOSSCR gate circuit. In this clamp, the critical regime for snapback is achieved at a relatively low multiplication coefficient. Additionally, the clamp possesses a high channel current after gate biasing occurs due to a voltage drop on the resistor above the avalanche diode breakdown voltage (Fig. 3.31).

Similarly to the LV BSCR, the HV BSCR is obtained by transforming a HV BJT (Fig. 3.32a). The triggering voltage is adjusted to be inside the ESD protection window by changing the internal blocking junction through different base-collector lateral dimensions. As well, adjustments are made at the clamp level by adding a high-side avalanche diode reference component to generate a base current after diode breakdown, according to TLP I–V characteristics (Fig. 3.32b).

3.3.3 High Holding Voltage HV Devices: Avalanche Diodes

With powered-on system-level stress, an SCR structure inevitably turns-on into a high current state with a low holding voltage. A transient latch-up may occur if the protected pin's biasing circuit can supply a current above the minimum holding current of the SCR.

If low holding voltage SCR solutions are not applicable, design choices are limited to lateral avalanche diodes, lateral PNP devices, stacked clamps, or attempts to implement an HV SCR with a higher holding voltage achieved by properly isolating the emitters. Experimental results show that these HV SCR

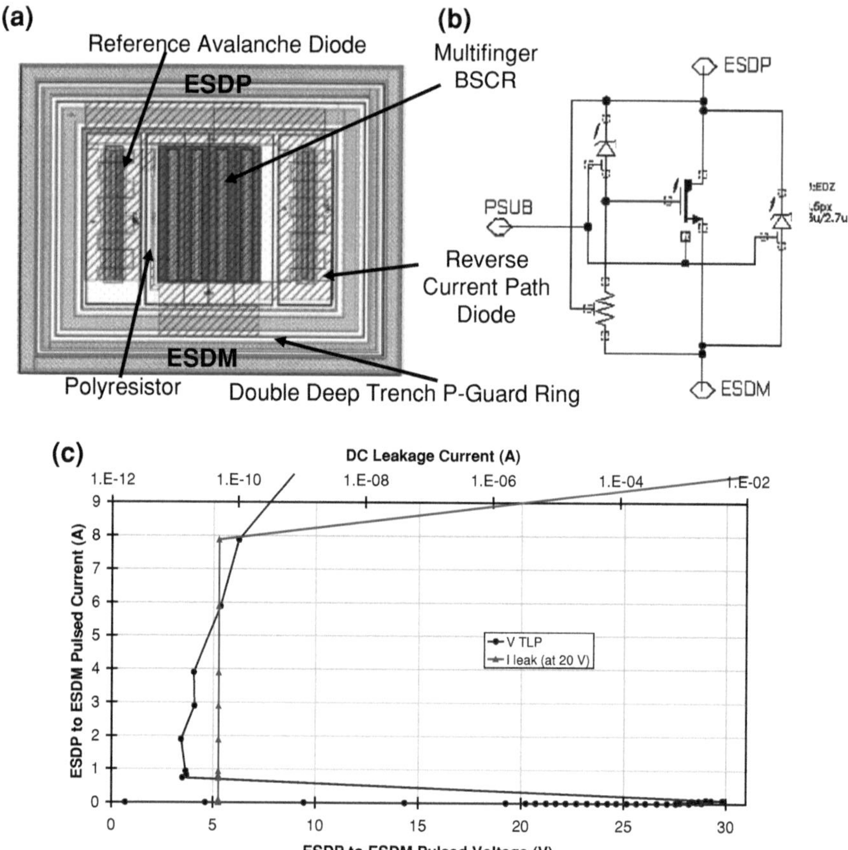

Fig. 3.31 Example of a 20 V NLDMOS-SCR clamp layout (**a**) and schematic (**b**) views and TLP characteristics (**c**)

devices can provide for a 40–80 V domain a high holding current to roughly 20–30 % of the triggering voltage range.

In the alternative, the high holding voltage avalanche diode, the final clamp is assembled as a multi-finger diode array (Fig. 3.33a). The avalanche current may reach up to ~ 1 mA/um before the irreversible breakdown occurs. However, a rather high on-state resistance makes this device only useful as a reference component for the turning on of different SCR clamps (Fig. 3.33b).

A lateral avalanche diode (LAD) is usually designed by reusing the regions of the corresponding NLDMOS (Fig. 3.34a). The design transformation includes eliminating the source region, forming just a pAnode, while the positions of the active-to-active anode-cathode space nWell and the poly-RESURF are adjusted so that the on-state resistance per area is minimized. This device has the advantage that it is fast responding and simple. A major advantage of the HV lateral avalanche diode in comparison to a lateral PNP and even a SCR is the absence of a

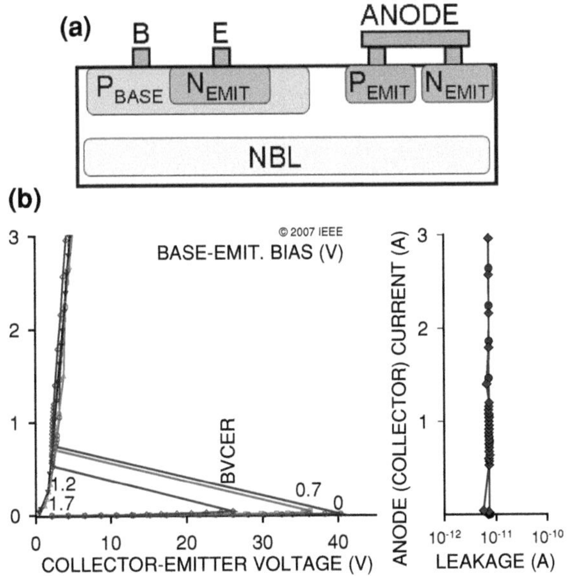

Fig. 3.32 Example of a 20 V BSCR device's simplified cross-section (**a**) and TLP characteristics at different voltages at the base (**b**)

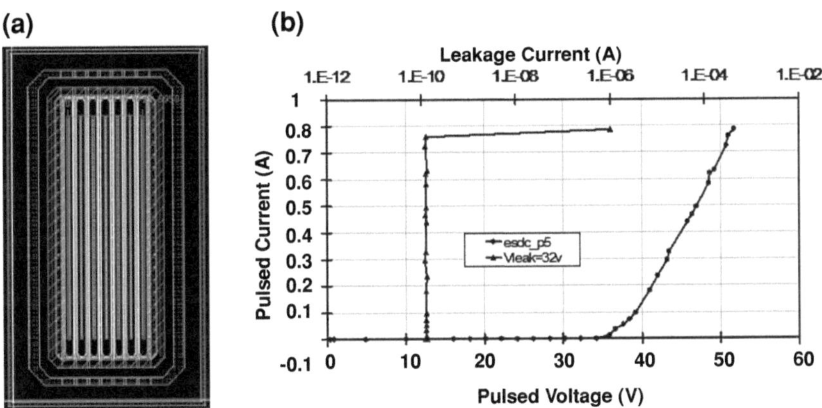

Fig. 3.33 Example of an avalanche diode device layout (**a**) and TLP I–V characteristics (**b**)

high-side p-region. This absence eliminates the potential problem of an SCR
current path forming in the layout when low-side n-regions are placed in the
vicinity of the clamp.

   Effective use of the features that provide reduced surface field (RESURF) effect
is applied for the on-state resistance optimization of the lateral avalanche diodes.
Alternatively, with some process compatibility, the avalanche diode's active

region length and resultant footprint can be more aggressively reduced using an alternative p-i-n diode architecture. A description for an example of such a HV ESD protection solution is presented below with numerical analysis and experimental validation results.

The p-i-n device utilizes the double-injection conductivity modulation effect to maintain a lower on-state resistance compared to the conventional HV avalanche diode architecture. Similar to the utilization of the other types of electrical conductivity modulation effects the resulting internal negative differential resistance (NDR) can be used to compensate the positive differential resistance of the structure contact regions.

Both the avalanche injection in bipolar n-p-n (or p-n-p) structures and the double injection in the thyristor's (SCR) p-n-p-n structures conductivity modulation mechanisms provide rather strong positive feedback. It results in a relatively low holding voltage in the high current mode (Chap. 1). In case of high voltage devices with the high holding voltage requirements close to a breakdown voltage level is practically hard to achieve. Thus any improvements to the on-state resistance per area of the relatively large footprint lateral avalanche diode are favorable. In spite of a partial conductivity modulation of the drift regions and a complex dependence of the electric field upon the breakdown current level, in general, the LAD is usually assumed to operate in an avalanche breakdown mode with the corresponding exponential dependence of the current upon the applied voltage thus maintaining the positive differential resistance across the entire range of operation regimes.

Also there is another physical conductivity modulation mechanism—double avalanche injection (Chap. 1). It can be observed in p-i-n diode structures with proper region parameters. Theoretically, this type of conductivity modulation can be used to improve the electrical characteristics of the HV avalanche diodes due an ideal uniform distribution of the electric field in the i-region. However, due to the operation regime with very high avalanche multiplication coefficients, an appropriate avalanche current density level is hard to achieve for practical design due to lack of scalability with structure width [2].

The physical principle of the double avalanche injection in p-i-n structures is based on the space charge neutralization in the middle of the i-region by the carriers generated due to impact ionization at the both n-i and p-i junctions simultaneously. Similarly to the avalanche injection phenomena (Chap. 1) the double avalanche injection phenomenon can be illustrated using 1-D numerical simulation by comparison of the electric field and carrier distributions for two regimes which correspond to the low current positive (I) and high current negative (II) differential resistance modes of the structure operation (Fig. 3.34).

The experimental evaluation of the HV p-i-n LAD design has been done through the comparison with the conventional lateral avalanche diodes in a 60 V 0.18 μm BCD process. The lateral blocking junction of the conventional device is reusing the reduced surface field (RESURF) features of the corresponding HV NLDMOS device ancestor. The initial device design is based on numerical analysis using the parameterized mixed-mode simulator DECIMM^{TM} [19]. The

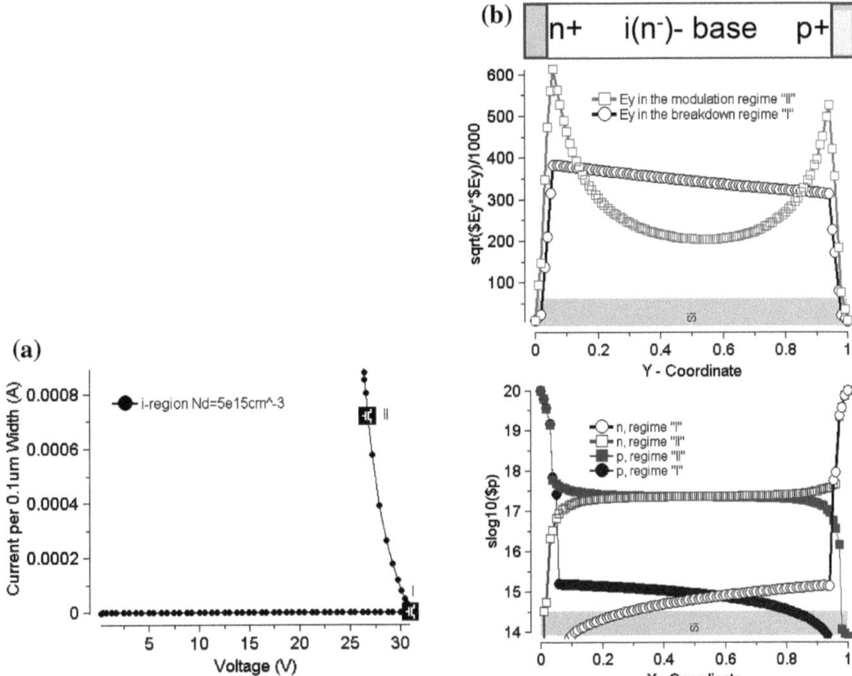

Fig. 3.34 Conductivity modulation in the p-i-n structure: the I–V characteristics (**a**) and comparison of distributions for the electric field magnitude, electron and hole density (**b**) for the avalanche in the breakdown state "I" and the avalanche injection state "II"

NLDMOS drain region was reused to form the n-Cathode, while the p-Anode was formed by a modification of the Source-Body region eliminating the Source n+ diffusion. Both the RESURF poly field plate and the deep p-RESURF region were retained from the original isolated NLDMOS device architecture. The parameters of the anode to cathode diffusion spacing's as well as poly-RESURF were adjusted for optimal breakdown and clamping voltage levels suitable for the targeted 30 V domain ESD protection (Fig. 3.35"A").

The on-state resistance in the avalanche breakdown mode is a function of the anode-cathode spacing and the n-drift region doping. Unlike LDMOS, LAD is not expected to deliver a functionality to support the unipolar current conduction from a channel region. Therefore to maintain the high breakdown voltage capability either the Pbody/Pwell anode implant region or even the cathode n-drift region can be eliminated from the device structure. The flexibility in the design changes can be applied to obtain new HV tolerant devices with pseudo i-region that can occupy either entire the anode-cathode spacing, (Fig. 3.35"C") or a partial space between p+-anode and n-drift (Fig. 3.35"D"). Both devices allow a significant scaling down of the anode-cathode spacing in comparison with the original conventional device (Fig. 3.35"A") while still meeting the high breakdown voltage requirement.

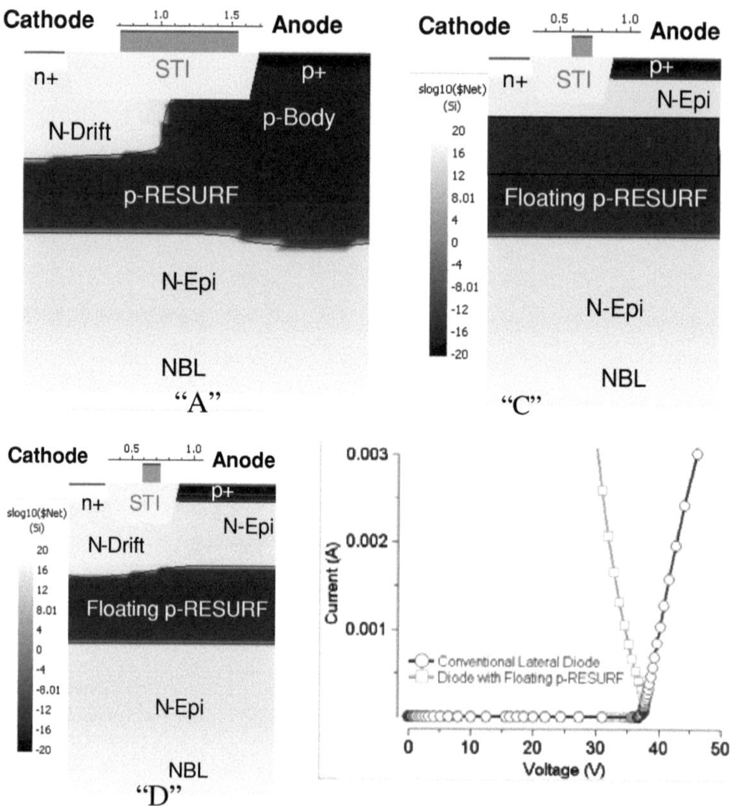

Fig. 3.35 Example of the conventional lateral "A" and pseudo p-i-n diodes without n-drift "C" and with n-drift "D" that correspond to the experimental structure (Table 3.3) and their simulated I–V characteristics

Certainly, to achieve an optimal performance both poly and p- RESURF regions must be changed in order to support a close to uniform electric field distribution in the anode-cathode spacing with the diluted doping level. Another key feature of the new design is the floating p-RESURF. It is formed automatically due to the elimination of the Pbody/Pwell region and is required to provide proper depletion of the pseudo i-region. The diodes are formed by p+-anode and n+-cathode contact diffusion regions and the lightly doped nEpi region representing the physical equivalent of the i-region. If the floating PRESURF region is completely removed, then relatively thick nEpi and NBL regions lead to the breakdown limitation at the p+-nEpi level ∼ 30 V only.

According to the results of numerical simulation of the I–V characteristics for the conventional (Fig. 3.35"A") and the pseudo p-i-n diode (Fig. 3.35"C") both devices are expected to meet the high voltage tolerance target above 35 V. However the negative differential resistance region is specific for the new pseudo

Fig. 3.36 Layout views with active, poly and contact layers for the same area conventional (**a**) and pseudo p-i-n diode (**b**)

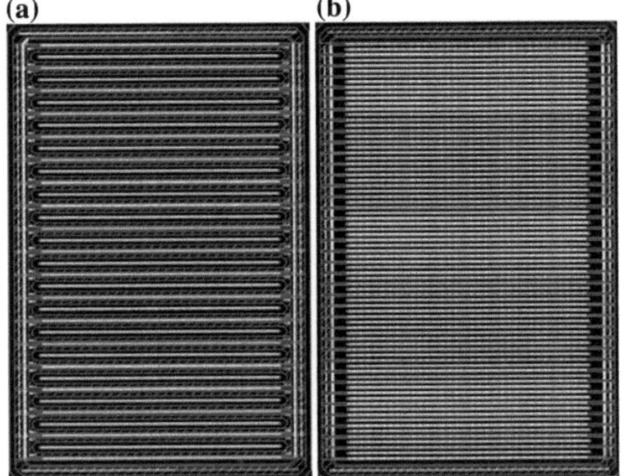

p-i-n LAD structure only. By using a cutline in the lateral direction it can be easily shown that the lateral electric field and carrier distributions at the NDR formation are based on the same double avalanche injection mechanism as in the simple elementary p-i-n structure (Fig. 3.34).

While the simulation results provide a high level of confidence for the breakdown electrical characteristics, the high current operation can be verified only experimentally. Since the new p-i-n device has a smaller anode-cathode active region spacing a fair comparison must take into account the area-factor too. To normalize the on-state resistance per unit area the LAD clamp diode arrays were experimentally compared with the same area, but with corresponding different number of fingers (Fig. 3.36).

A design of experiments with only four avalanche diode clamps is sufficient to explain the major regularities observed. Table 3.3 combines both the design parameters for these four representative diode clamps and the summary of the experimental results. The devices "A" and "B" represent the conventional design for the LAD clamps with n-drift region, the poly- and p-RESURF features similar to the simulated cross-section (Fig. 3.35"A"). Two new pseudo p-i-n diodes with floating p-RESURF are represented by the device "C" similar to the simulated cross-section (Fig. 3.35"C"). In this device the Ndrift region is completely eliminated. Finally, the device "D" is similar to the simulated cross-section (Fig. 3.35"D"). In the structures "C" and "D" Pbody/Pwell regions are blocked thus resulting in a formation of the floating pRESURF diffusion region feature (Table 3.3, DOE section)

The initial conventional design "A" has 26 fingers for the anode-cathode spacing $LAC = 1.7$ and 70 µm finger width, arranged within the fixed footprint. While the breakdown voltage of ~ 40 V meets the 30 V tolerance target, the clamping voltage for 2 kV HBM equivalent current is rather high at ~ 54 V. The anode-cathode spacing LAC is at the physical limit for this design. Reduction of

Table 3.3 Comparison of the simulation results for 65 V tolerant clamps with roughly estimated area factor normalized to the active clamp

Plot label	Clamp desctiption	DOE parameters							Experimental data								
		N fingers x width	LAC "A" to "C"	LPA Poly to "A"	LPC Poly to "C"	Lpwell over active	Lnw (nwell)	Pbody & Pwell	pos Vbr @ 1uA	Vcl @ 1.5A (V)	pos It2 (A)	pos Vt2 (V)	pos. HBM (kV)	pos. MM (kV)	neg. It2 (A)	neg. Vt2 (V)	neg. HBM (kV)	neg. MM (kV)
A	Conventional LAD	~26 × 70	1.7	0.2	0.5	0.35	1.15	YES	42	54.4	2.92	58.4	6.2	0.15	4.5	1.77	>8	0.1
B	Conventional LAD reduces	~38 × 70	1	0.22	0.22	0.35	0.6	YES	26	30.8	5.22	40.2	>8	0.15	4.8	1.94	>8	0.1
C	New FP-RESURF diode wNW&Poly	~38 × 70	1	0.22	0.22	–	–	NO	37		0.06	39.3			81	1.29		
D	New FP-RESURF diode no NW&Poly	~38 × 70	1	–	–	–	0.6	NO	38	41	1,79	41.8	4.3	0.25	13.7	1.17	>8	0.25

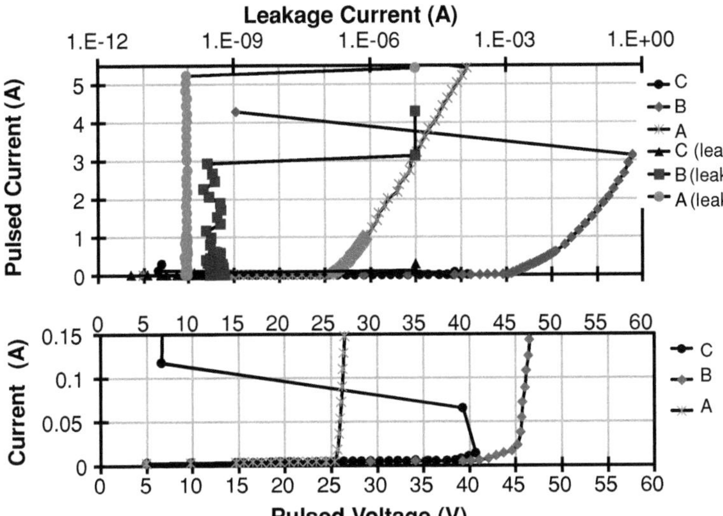

Fig. 3.37 Comparison of the TLP I–V characteristics for the structures "A", "B" and "C"

the *LAC* down to 1 µm results in a breakdown voltage reduction down to 26 V in the structure "B" (Fig. 3.37).

As it was expected from the simulation results, the high breakdown voltage target ∼37 V is indeed met in the pseudo p-i-n diodes "C" and D" with the floating p-RESURF architecture (Table 3.3, experimental data part). However, only the diode "D" provides a suitable high current capability (Fig. 3.38), while the diode "C" fails immediately after the snapback (Fig. 3.37). The diode "D" with completely blocked nDrift region provides a fairly good clamping voltage ∼42 V with an S-shape TLP I–V characteristic explained by the double avalanche injection phenomenon. The HBM and MM test results were found to be in correlation with the TLP I–V characteristics (Table 3.3) thus validating the usability of the new device.

Thus an important question to answer is why clamp "C" has a poor high current performance while clamp "D" with the partial n-drift region provides an adequate high current capability. The main hypothesis explaining the experimental results is based on the ballasting effect of the n-drift region. Indeed in the p-i-n structure (Fig. 3.34), as well as in the pseudo p-i-n lateral diode "C" (Fig. 3.35c), the current density is limited only at rather high level that provided by current saturation in the contact n+ and p+ regions only. In the high current NDR state due to the strong positive feedback between the current and the holding voltage, a very narrow current filament can form and rapidly achieve some critical amplitude for the device local burnout. At this current density the fast local overheating of the structure is realized practically in adiabatic conditions results and the instantaneous local damage of the device becomes imminent.

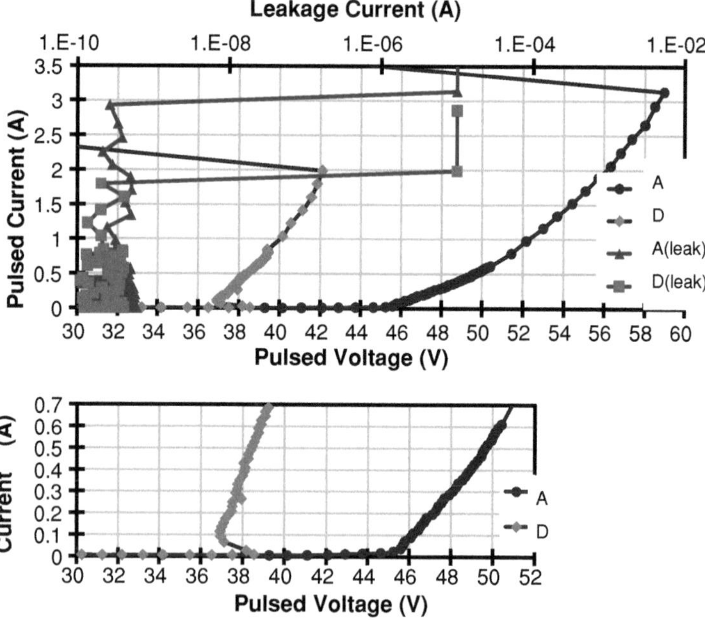

Fig. 3.38 Comparison of the TLP I–V characteristics for the conventional structure "A" and new pin-like structure "D"

The LAD device "D" incorporates the n-drift region which serves in a role of the local negative feedback component thus limiting the current filament amplitude on the local level due to the corresponding voltage drop.

To illustrate this nonlinear phenomenon from the 2-D simulation, the distributed p-i-n structure (Fig. 3.34) has been modified to include an additional n-region representing the n-drift part (Fig. 3.39). The final structure with 10 μm contact width limited by the oxide regions in X-direction was included in the mixed mode circuit (Fig. 3.39a) with a pulsed current source and simulated for different n-drift region length at a constant anode-cathode spacing of 1 μm. The presence of the numerical noise in the finite element model provides a physical equivalent of the real physical nonuniformity [4]. This enables the numerical solution for a spatial current instability in the form of the resulting current filaments [4].

During the transient process the initially uniform current distribution in the device becomes unstable after the first few nanoseconds which results in a spontaneous current filament state formation (Fig. 3.39b). The filament amplitude can be calculated as a function of the n-drift region length (Fig. 3.39c). Thus the critical role of the n-drift ballasting region in this pseudo p-i-n LAD is somewhat similar to the drain ballasting region in the snapback NMOS structures. This explains the difference in high current capability between devices "C" with no n-Drift ballasting and "D" with this feature.

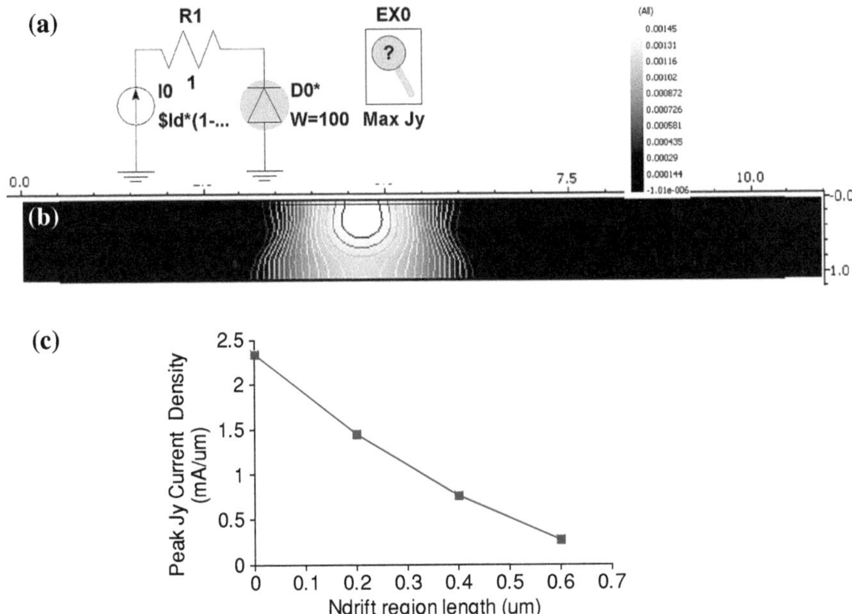

Fig. 3.39 Mixed mode circuit for current filament analysis (**a**), the example of the end of transient current distribution (**b**) and dependence of the peak current in the structure upon the n-drift region length (**c**)

Thus a new design for the high holding voltage lateral pseudo p-i-n avalanche diodes with floating p-RESURF region can be presented as one of the rare ways to improve the avalanche diode characteristics towards the system level high holding voltage solutions as it is shown above both by simulation and experimental results.

Due to the elimination of the non-self-aligned pWell implants a lower sensitivity of the new design to the mask alignment potentially might be achieved. This aspect will be discussed in more details in the last section of this chapter.

3.3.4 Lateral PNP ESD Devices

A more effective high holding voltage device alternative to the lateral avalanche diode is the lateral PNP (LPNP). Versions of this device were proposed for HV I/Os protection in smart power technology [67–70].

Redesign of a LPNP device (Fig. 3.40a) from the lateral avalanche diode consists of embedding the diode device with a pEmitter region while minimizing the nBase contact diffusion area at the layout level.

The avalanche diode's cathode and anode become the nBase and pCollector regions, respectively, while the HV blocking junction with double RESURF

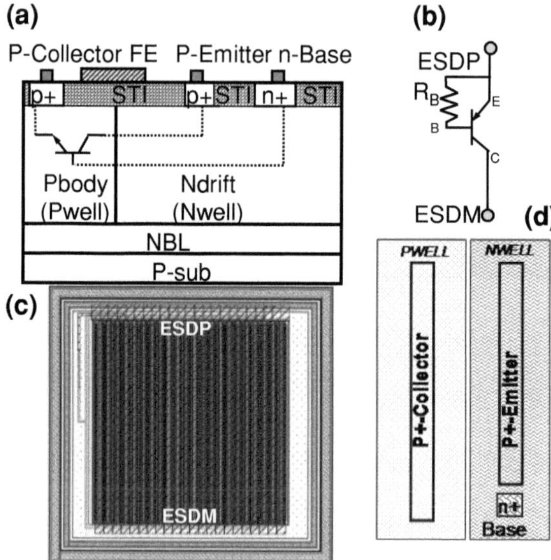

Fig. 3.40 Simplified cross-section (**a**), schematic for a lateral PNP (**b**), real clamp (**c**) and simplified (**d**) layout views

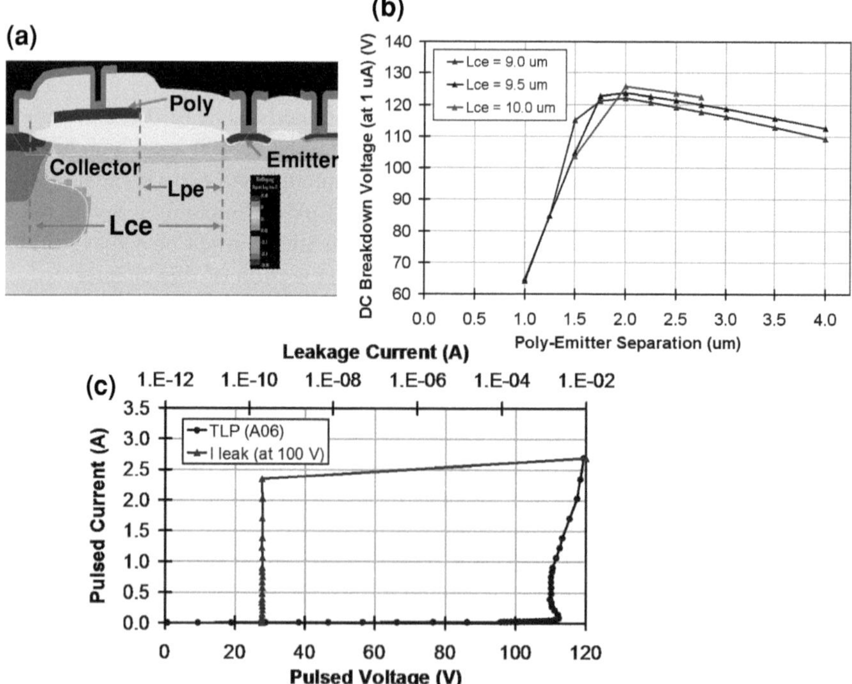

Fig. 3.41 Example of the breakdown voltage optimization: cross-section of FEM 100 V LPNP device (**a**), calculated base-collector breakdown voltage dependence upon the field-electrode position (**b**) and TLP I–V characteristics for the clamp with 4 mm total active region width (**c**)

architecture remains the same in both devices. The on-state resistance of the LPNP is substantially lower than that of the HV lateral avalanche diode, and limiting local nBase contact diffusions in the layout provides additional resistance reduction per unit area (Fig. 3.40d).

The LPNP adds both an internal gain and avalanche-injection conductivity modulation of the lengthy nBase region [67]. The latter provides a "weak" S-shape that counters the high on-state resistance, giving more vertical I–V characteristics (Fig. 3.41). The internal gain of the clamp can be increased using a base-emitter resistor; however this measure eliminates the efficient collector-base reverse path diode. The reverse current path is still provided by the internal lateral PNP bipolar structure with interchanged collector and emitter diffusions.

In the LPNP, the conductivity modulation is avalanche-injection of the lengthy nBase [70]. The negative differential resistance region is weak or completely hidden by the positive differential resistance of the avalanche breakdown and voltage drop on the saturation regions. Due to the difference in electron and hole carrier mobilities in Silicon, the carrier balance in the LPNP results in a higher holding voltage.

For the most optimal design of HV devices with a minimal active region length, the I–V characteristics require careful optimization using TCAD tools (Fig. 3.41).

For a stacked solution, a snapback clamp can be produced either by stacking the same components or diverse components with different holding voltages. When the same components are stacked, the number of clamps is simply a multiplier of the holding voltage of a single clamp, as long as no parasitic path is formed in the real layout. This approach is applied to dual-direction devices in the next section. The drawback in this case is the corresponding increase of the triggering voltage.

This problem can be addressed by combining a non-snapback clamp component with high positive differential resistance with a low holding voltage snapback component. The approach is illustrated based upon parameterized mixed mode simulation with DECIMM^{TM} tool [19] using circuit (Fig. 3.42a). The circuit presents a snapback clamp composed from the lateral avalanche diode with high width W = 5 mm (Fig. 3.42b) and small LVTSCR structure with W = 100 μm (Fig. 3.42c). The avalanche diode provides high voltage avalanche breakdown with high positive differential resistance (Fig. 3.42d). Therefore to fit the clamping voltage into the ESD protection window range a significant width scaling would be required. On the contrary LVTSCR delivers the snapback characteristics with high triggering ~ 10 V and low holding voltage ~ 2 V (Fig. 3.42d). The resultant characteristic of the stacked clamp (Fig. 3.42d) is the result of compensation of the negative and positive differential resistances toward the desired more vertical I–V characteristic.

3.3.5 HV Dual Direction Devices

HV versions of dual-direction devices are based on the same device and clamp design principles as those previously described for the LV dual-direction devices.

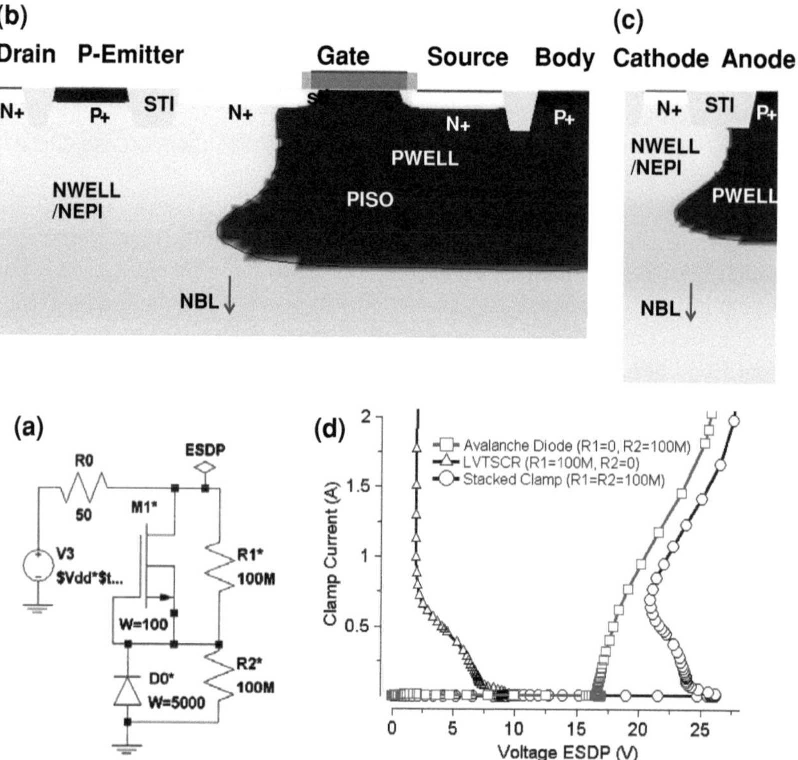

Fig. 3.42 Mixed mode circuit (**a**) to illustrate stacked HV clamp composed from LVTSCR (**b**) and lateral avalanche diode (**c**) devices and comparison of the I–V characteristics of the separate clamp components and the entire clamp (**d**)

The same approach can be used to compose a back-to-back stacked clamp from a HV nLDMOSSCR, Lateral PNP or Avalanche diode unidirectional clamp by merging and floating the nEpi regions, or just by a floating metal connection in the separate nEpi regions. In this type of stacked clamp design, major unexpected issues can arise due to the interaction of clamp components from imperfect isolation between the clamps in stack.

Nevertheless, a major space-saving solution for system ESD stress current levels is the DIAC device. In terms of HV system-level architecture, the DIAC is advantageous because it reuses chip space for the lengthy blocking junction region.

The HV dual blocking junction of HV DIAC can be formed by lateral nWell, pWell, pBody or nDrift regions arranged in both p-n-p (Fig. 3.43a) and n-p-n sequences (Fig. 3.43a). A major transformation step for both devices is back-to-back stacking of the initial NLDMOSSCR components. A DIAC with a p-n-p lateral blocking junction is created by merging the drain regions of two

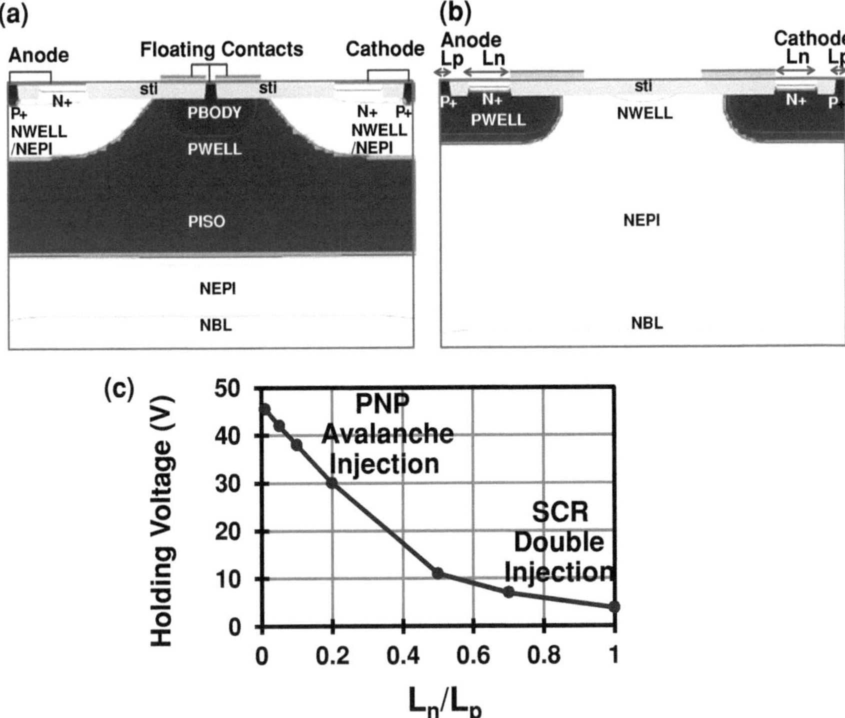

Fig. 3.43 Cross-sections for FEM HV DIAC devices with p-n-p (**a**) and n-p-n (**b**) dual blocking junctions and the dependence of the holding voltage upon the injector ration (**c**)

NLDMOSSCR clamps, reusing the common nDrift region to form depletion regions for both junctions in corresponding bias directions.

Theoretically, the body regions can be merged (Fig. 3.43b). In this case, the device requires two drift regions. However, a DIAC with an n-p-n blocking junction arrangement is usually impractical. Due to typically insufficiently high side body region isolation from the pSubstrate the architecture can be practically useful only in SOI processes. Therefore, DIAC architecture with a p-n-p dual blocking junction represents the practical device solution (Fig. 3.43a).

Similarly to LV DIACs, the HV DIAC combines pairs of nplus and pplus contact diffusion regions that act as SCR bases and emitters, depending on the current direction (Fig. 3.43a).

The challenging task of implementing both a high breakdown voltage and high current capability simultaneously in both current directions becomes even more complicated when the high holding voltage requirements are specified. This problem cannot be practically solved at the 2-D level of device cross-section design and requires a topological solution.

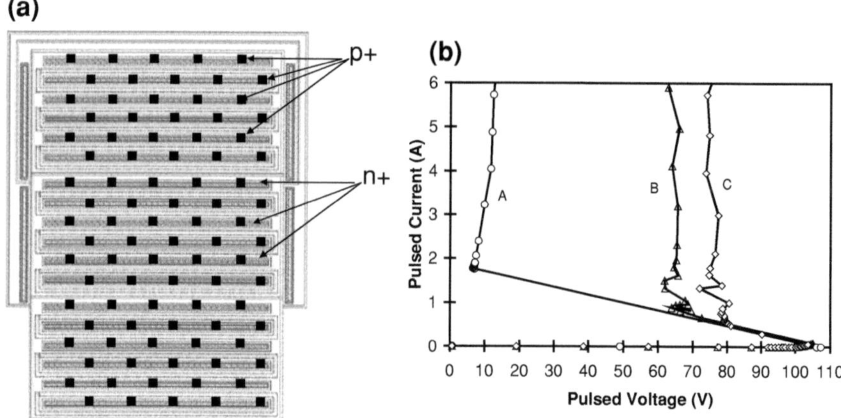

Fig. 3.44 Layout design example for a high holding voltage solution with a 60 V tolerance in the form of triple stacked DIAC cells with interdigitated injectors (**a**) and experimental TLP I–V characteristics (**b**) for: device "A"—no interdigitation with solid n+ and p+ diffusion stripes; device "B"—interdigitated device with 20 % of p+ area; and device "C"—interdigitated device with 10 % of p+ area

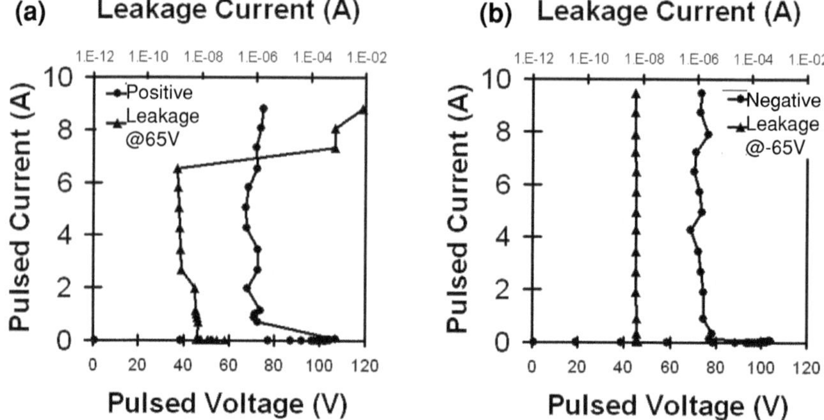

Fig. 3.45 TLP I–V characteristics of the dual direction device for positive (**a**) and negative (**b**) pad current

The layout-level solution is based upon the formation of interdigitated contact regions that balance the carrier injection and collection in the structure. Such a balance allows for an optimal electric field distribution in the structure, while still maintaining high ESD current capabilities (Fig. 3.44a). An optimal p+/n+ contact diffusions length ratio for the injector regions is usually in the range of 5–10 (Fig. 3.44b).

The TLP I–V characteristics of the DIAC cell in both directions can be obtained after careful design efforts toward balance in the injectors current and internal electric field distribution (Fig. 3.45).

3.4 ESD Cell Design Principles

One of the major challenges in HV system-level ESD cell design lies in achieving the proper pulsed type independent width scaling. Some devices in particular process technologies can provide a rather high current density on small structures. However, width- and array- scaling for system current level performance can be significantly non-linear.

Another unexpected side-effect is a miscorrelation between high measured TLP current, passing contact IEC pulse level and air gap pulse performance. The miscorrelation is accompanied by the so-called "windowing effect," when the designed cell is passing low and high ESD pulse levels, but failing in some intermediate range.

The major reasons for non-linear performance of the structure width scaling are related the effects of current crowding due to an improperly balanced layout, the multifinger turn-on effect or an undedicated "sneak" current path formed in a particular cell and product layout. The last class of phenomena is addressed in the next chapter. This section discusses the measures that can be applied to cell design.

3.4.1 Undesirable Multifinger Turn-On Effect

The layout of high current devices and clamps is a rather critical part of ESD clamp design. The topological degree of freedom is primarily used to achieve appropriate characteristics for the clamp, but can also significantly impact the ESD cell's high current performance. This section is focused on the device-level front-end ESD cell layout, while the following section brings some insight to cell metallization design.

Successful system-level ESD clamp implementation combines a broad range of aspects: device width scaling, lateral isolation, vertical isolation from the substrate, latch-up isolation by guard rings, clamp-level design for voltage reference, dynamic coupling as well as implementation of a more "intelligent" driver circuit.

The usual layout implementation of an active ESD device in clamps is a multifinger distributed array. Ensuring that all fingers turn on in ESD pulse conditions with current balance at the cell level is a challenge. HV devices especially require that 3-D design of array fingers, the metallization scheme, and the driver circuit connection are taken into account. From this perspective, when system-level width scaling is required, all fingers of the ESD device should be scaled within the same Epi pocket shared space, rather than by instantiation of several

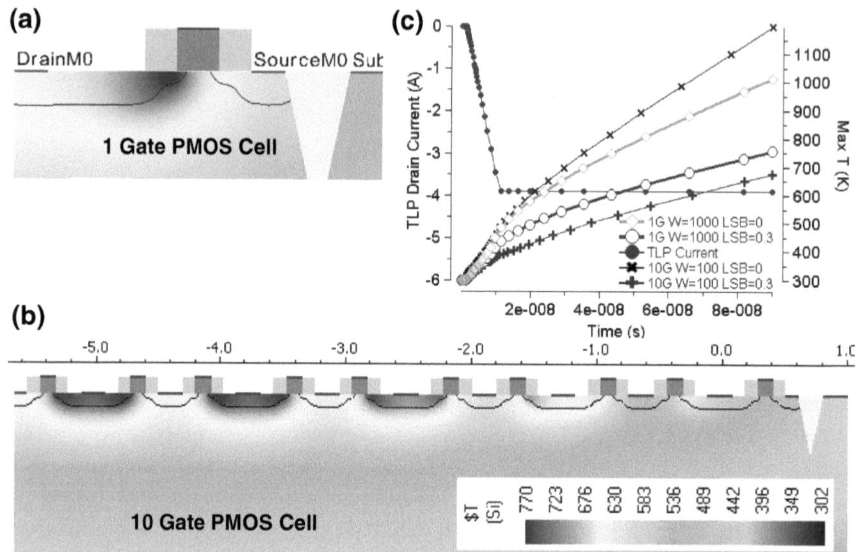

Fig. 3.46 Thermal coupled mixed-mode analysis for cell-level current redistribution in a 2 V 0.13 μm PMOS cell with a common n-Body diffusion for single finger (**a**) and a multifinger device (**b**), comparing the temperature to time for the same normalized width of both devices (**c**)

isolated devices in parallel. This design principle safeguards that all fingers in an ESD device array with a large total width will turn on.

In general, it is hard to expect an absolute symmetry in any ESD clamp device array. There are different connection paths from control electrodes to drivers, shared components in the entire array that require too much space to bring an identical arrangement to every finger, varying distance to the lateral isolation rings, and not even metal connections.

In many cases, a LV ESD device array is designed with a common well tie or epi connection for the active region, with multiple active region fingers. It is important to realize that in this case, the current distribution inside the ESD array is already non-uniform due to different effective resistances from the common region to each finger. For example, in a non-snapback PMOS cell (Fig. 3.46) with common nBody contact diffusion, the multifinger design results in a different cell capability at width scaling.

A simulation comparison was performed for the physical equivalents of a large footprint cell with an n-Body connection evenly placed at each active finger with a polygate (Fig. 3.46a) and a more compact cell with the 10-gate FEM cross-section (Fig. 3.46b). With same total width, normalized in the simulation by the scaling factor W, the cell with the common nBody (Fig. 3.46b) has a 20 % worse performance than the cell with an nBody added to each active finger (Fig. 3.46c). A small additional drain ballasting region improves the current capability by almost two times. This corresponds to experimental results [71].

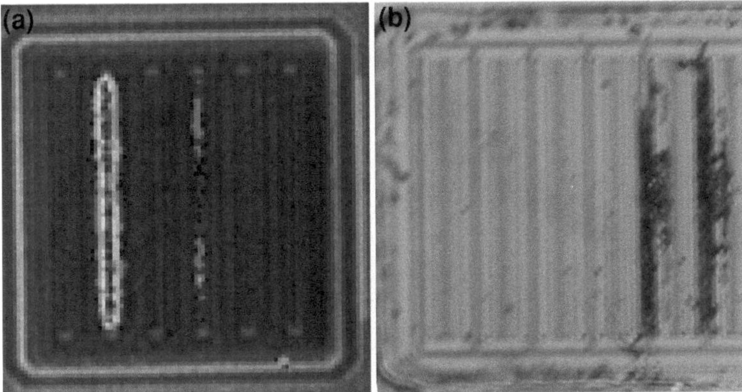

Fig. 3.47 Backside emission from an ESD cell at HMM pulse in the reversible operation conditions (**a**) with one mostly one finger turn-on and the view for cell burned-out cell with several fingers destroyed (**b**) (courtesy of Todd Mitchell)

In general, ESD cell design cannot be based on an assumption that turn-on of any single finger in the conductivity modulation mode inevitably or at least eventually results in some coupling or a mutual interaction effect, bringing all fingers in the array to a turn-on state. In other words, a turn-on domino effect is not necessarily always present in the ESD device array, especially in the case of HV snapback devices with a triggering voltage that significantly exceeds the holding voltage (Fig. 3.47). A variation of this effect is the inconsistent operation of the ESD cell for different pulse types and loads, and the already mentioned windowing effect.

On the contrary to the multifinger effect, a local turn-on within the same finger width (for example in the middle or at the end of the finger) into the high conductivity state practically guarantees that diffusion of the injected carriers will propagate laterally with current increase until the entire finger is in the on-state (Fig. 3.47).

In SCR devices, when coupling between the fingers is low, the spreading of the conductivity modulation current along a single finger results in the formation of a high current state with a low holding voltage. The following scenario depends on the triggering voltage V_{T1} and the critical voltage for maximum current V_{T2}. If V_{T2} exceeds V_{T1}, then additional fingers may turn on after current saturation in the initial finger, with the voltage dropping to this level. In this case $V_{T1}(t, I_{inj})$ is dependent on both the rise time and internal injection current. For HV SCR devices, in spite of a strong reduction in V_{T1} due to internal current injection from already turned on fingers, the resultant $V_{T1}(t, I_{inj})$ for off-state fingers will still be much higher than the holding voltage V_H.

This non-simultaneous finger turn-on may produce multiple S-shape regions that correspond to the on-state of different fingers [72]. Non-uniform turn-on in multi-finger NMOS and SCR was studied under TLP and HBM stress in [73, 74].

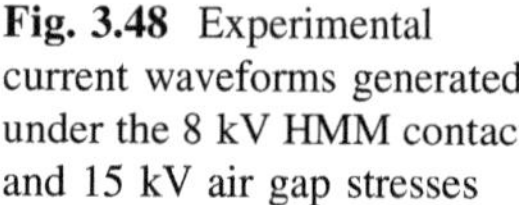

Fig. 3.48 Experimental current waveforms generated under the 8 kV HMM contact and 15 kV air gap stresses

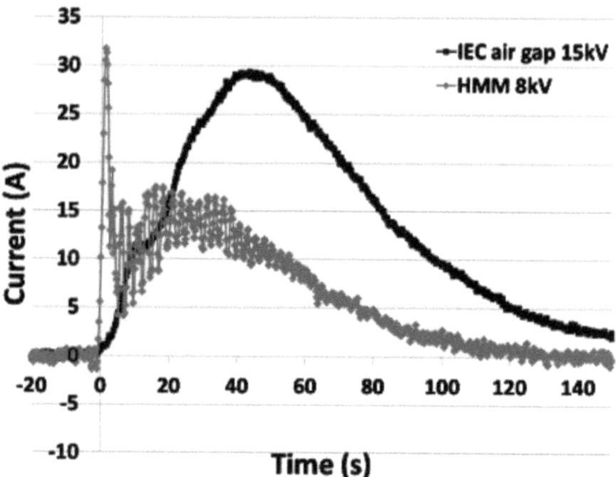

Typically, $V_{T2} < V_{T1}$ for HV SCR devices, therefore burnout of the first finger can be expected before the remaining fingers in the ESD device array are engaged. The same level of speculation is applicable to when two isolated ESD snapback clamps are used to increase the total clamping current.

On the other hand, if there is at least local or partial turn-on of each finger, the following current growth has an even array turn-on operation. It can be assumed that this scenario is realized at the experimentally observed miscorrelation between the stress results for the contact and air gap system-level pulses due to different waveforms of the pulses themselves.

In case of contact IEC 61000-4-2 stress, the first fast short peak of the pulse (Fig. 3.48) ensures turn-on of all the fingers that then evenly re-distributes the current density during the rest of the pulse, providing the highest current capability. In the opposite case with a slow air gap ESD pulse (Fig. 3.48), the finger that turns on first clamps the voltage to a low level and prevents the remaining fingers from engaging, thus limiting the current capability of the entire array down to the performance of a single finger.

This "fundamental" effect in system-level cell design was studied for multi-finger HV NLDMOSSCR devices in [75], using experimental and numerical simulation analysis with tool DECIMM [19]. Indeed, according to experimental results the device has demonstrated likely the uniform triggering with expected high current performance for 50 Ω TLP and HMM IEC contact pulses (Table 3.4). On the contrary, the air gap IEC pulse passing level did not correlate with these results, pointing to pulse-type-dependent operation (Table 3.4).

The effect of sensitivity to ESD pulse type due to non-uniform finger triggering was reproduced using a numerical simulation with fast contact and slow air gap IEC pulse stresses. For DECIMM mixed-mode analysis, a parameterized 4-finger HV NLDMOS-SCR (Fig. 3.49a, b) was studied in the mixed-mode circuits with corresponding sub-circuits for the contact (HMM) (Fig. 3.49c) and IEC air gap (Fig. 3.49d) pulses. To introduce a factor physically equivalent to cell asymmetry,

Table 3.4 Experimental results of SCRs under TLP/HMM/IEC air gap

Cell name	Vtrig (V)	Vh (V)	It2 (A)	HMM (kV)	Gun air (kV)
Cell A	95.6	6	>25.2	12	3.5
Cell B	89.3	4.8	>26.4	11	4.5
Cell C	80.5	4.8	>26.4	12	5.5
Cell D	69.4	4.7	>26.4	12	4.5

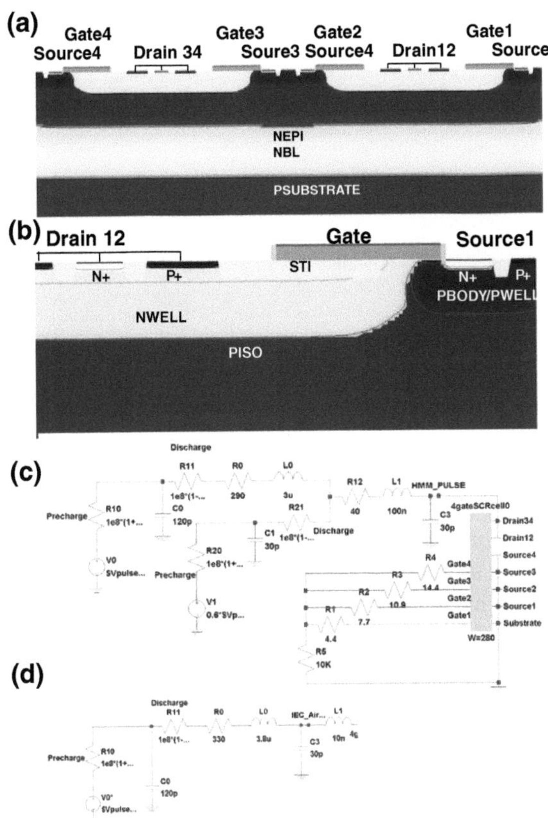

Fig. 3.49 Cross-sections of four-finger SCR (**a**) with zoom in the first finger (**b**) and mixed-mode circuits for HMM (**c**) pulse simulations with connected device and for air-gap pulse source (d)

optional separate gate-to-ground R_{GS} resistors were included in the circuit (Fig. 3.49c, d).

However, the originally expected effect was already observed with same values of the gate resistors. An example of the simulator's output for source terminal current waveforms with non-uniform multifinger turn-on and a slow air gap pulse

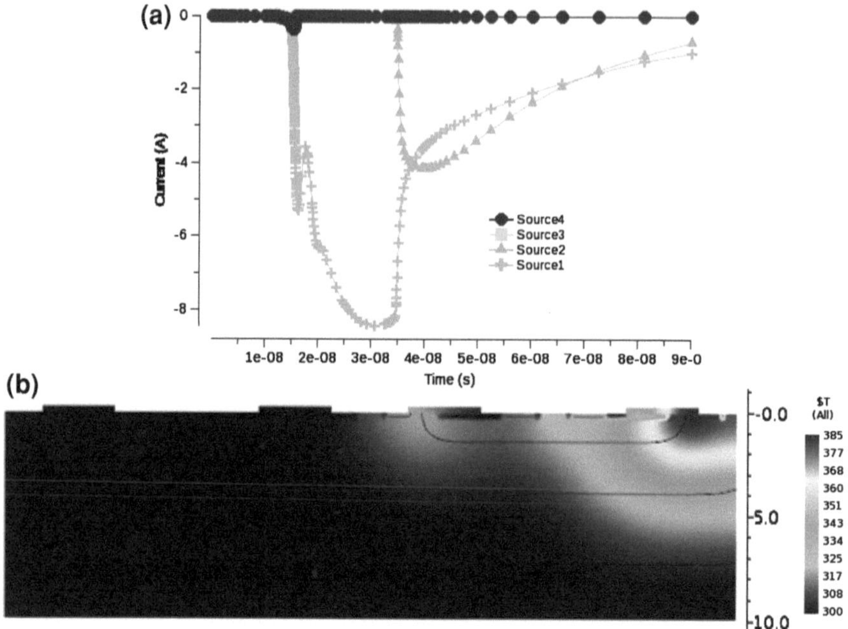

Fig. 3.50 Four-finger SCR transient source current waveforms under 4 kV IEC air gap stress
(**a**) and temperature distribution under 4 kV IEC air gap stress at 90 ns from the beginning of the
pulse (**b**)

(Fig. 3.50) demonstrates that on-state is triggered only in Finger1 and Finger2. Finger2 turns on 20 ns later than Finger1 with corresponding current sharing between the two to achieve the same level at the end of the pulse. The temperature distribution under 4 kV IEC air gap stress at 90 ns (Fig. 3.50b) indicates that the peak is at the Finger1 source.

The transient source current waveforms under 12 kV HMM stress (Fig. 3.51), on the contrary, show simultaneous turn-on of all the fingers with an even source current distribution. A similar current distribution is seen for the IEC contact pulse.

This simple numerical experiment demonstrates the multi-finger turn-on effect and the windowing effect as a result of non-even turn-on of the fingers at a slow IEC air gap pulse. When the pulse amplitude is small enough for only one finger to sustain the ESD current, the device operates reversibly. If the pulse is fast, as in the case of a 50 Ω TLP, or has a high amplitude, first peak to "ignite" the conductivity modulation in all active regions of the device turns on all the fingers and can pass a relatively high current level.

However, for some intermediate pulse range where only one finger is in the on-state, the critical power for burnout is reached under passive conditions on the rest of the fingers. This understanding is in correlation with experimental results (Table 3.4) and additional experiments demonstrating the windowing effect.

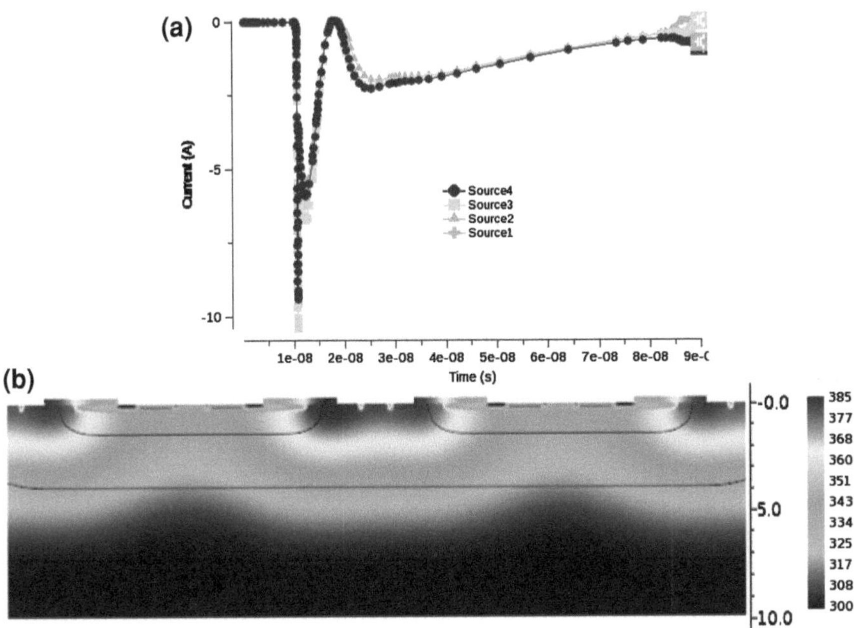

Fig. 3.51 Four-finger SCR transient source current stress (**a**) and temperature distribution (**b**) under 12 kV HMM stress

3.4.2 Poly Ballasting to Overcome Multi Finger Turn-On Effect

Proper design of back-end metallization routing, well tap connections and symmetrical gate control electrode connections is important for avoiding the non-simultaneous multifinger turn-on effect as well as contact enclosure in the diffusion region. Excessive power generation can cause an elevated local temperature at the contact, which may result in irreversible processes in the metallurgical contact structure.

Two types of cell-level design measures can be applied to the device and cell to reduce this undesired effect. The first type involves current ballasting by metallization, poly resistors, or diffusion regions.

Experimental results for the poly-ballasting approach applied to HV DIAC cells are shown in Fig. 3.52. An additional distributed poly-resistor region has been introduced between the cell injector regions and the connection between the pad and ground (Fig. 3.52a). Although the cell has demonstrated better finger scaling, the system-level performance was lower due to the backend limitation because less vias and contacts were used.

However, an interesting side-effect was the favorable increase of the holding voltage (Fig. 3.52b) in comparison to a cell with a conventional metal connection.

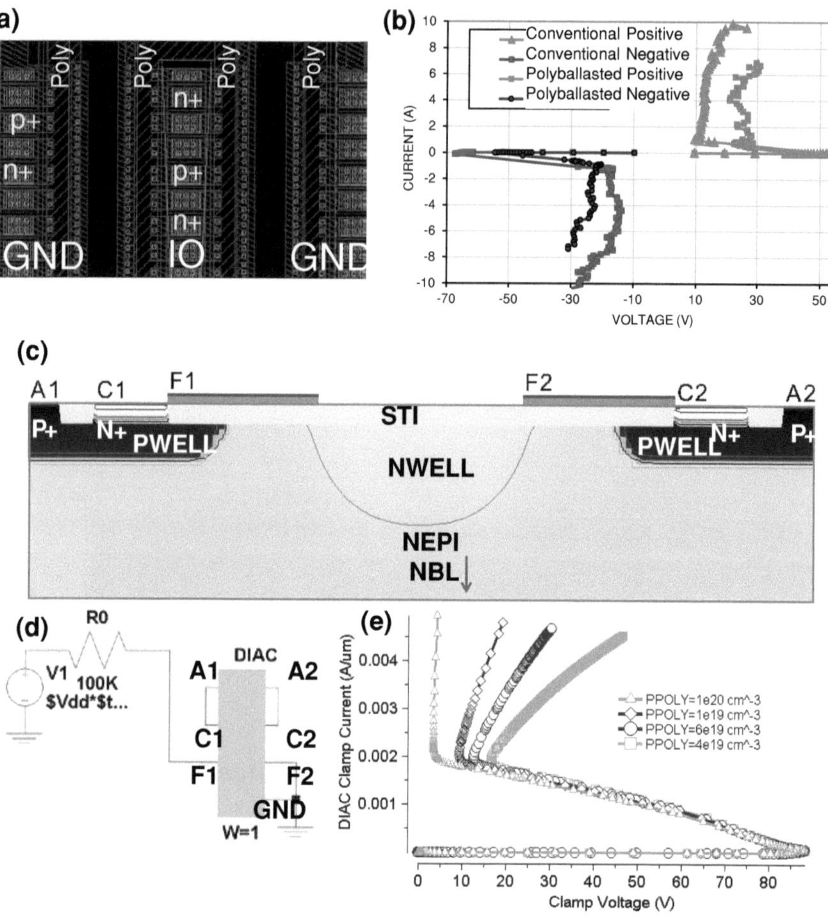

Fig. 3.52 Layout view of poly-ballasted DIAC cell with interdigitated emitters (**a**) and comparison of the experimental TLP results for conventional and poly-ballasted DIAC cells (**b**), simulated FEM DIAC device (**c**), mixed-mode circuit for the curve trace (**d**), and calculated I–V characteristics for elevated poly region doping levels (**e**)

The effect of the higher holding voltage is understood using numerical simulation analysis (Fig. 3.52c–e); it is the result of the additional voltage drop on the distributed polyresistor (Fig. 3.52e).

The miscorrelation of dual-direction DIAC devices in high current operation with different system level pulsed can be both experimentally observed and explained by a non-simultaneous turn on of the DIAC device fingers at slow negative air gap IEC standard ESD pulses versus a high performance for the contact IEC standard pulse, HMM or TLP or contact IEC standard ESD discharge.

Similarly to other examples of this Chapter, the poly ballasting effect can be studied using numerical simulations which generate relevant information about

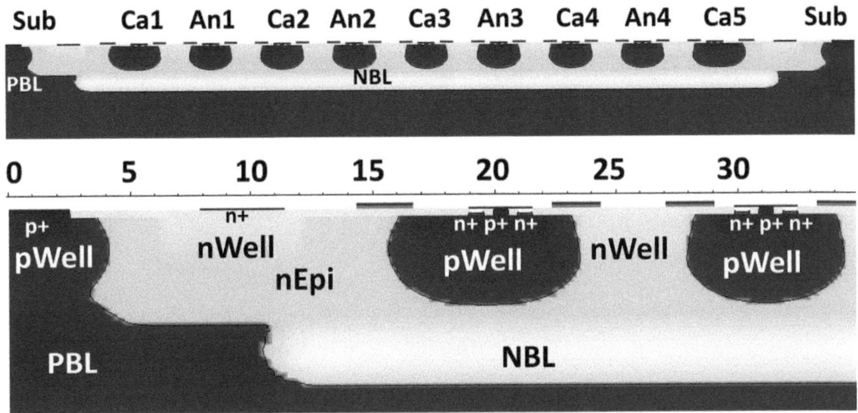

Fig. 3.53 Full cross-sections of four-finger DIAC (**a**) with zoom of the first finger (**b**)

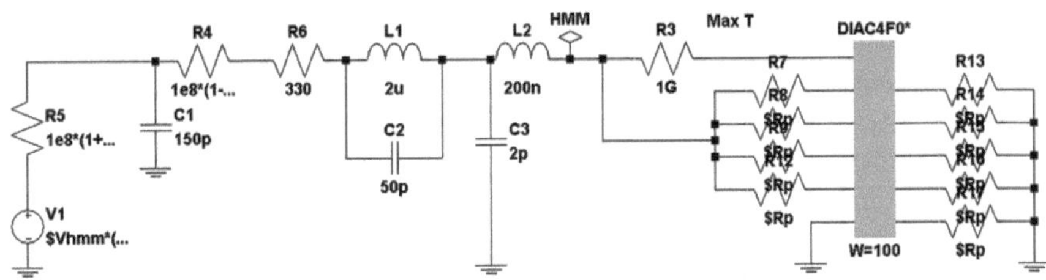

Fig. 3.54 Mixed-mode circuits for HMM pulse simulations with connected device and ideal resistors representing the poly-ballasting features

physical devices operation, followed by the experimental validation of two new proposed innovative designs for the DIAC devices.

The device used in this section for the demonstration of the poly- ballasting effect is the dual blocking junction of HV DIAC with the full cross-section of a 4-finger array (Fig. 3.53a). Each active finger is formed by lateral nWell, pWell, pBody or nDrift regions arranged in an n-p-n sequence. The pWell regions include the nplus and pplus contact diffusion regions that have an interchangeable role of SCR bases and emitters, depending on the current direction (Fig. 3.53b).

To physically represent the effect of poly-ballasting resistors an additional ideal resistors can be included in the mixed mode circuit for the device with the HMM pulse source (Fig. 3.54). In this circuit the parameter for all Anode and Cathode resistor are the same.

In spite of an insignificant difference in the clamping voltage waveforms (Fig. 3.55), according to transient analysis the structure operation indeed greatly depends on the value parameter Rp of the polyresistor parameter physically represented in the circuit by the equal ideal resistors $R7$-$R17$ (Fig. 3.54). For example in case of $Rp = 0.1\ \Omega$, that physically represents the absence of the poly

Fig. 3.55 Results of mixed-mode simulation for the physical equivalent DIAC device cross-section for the 4-finger cell. Waveforms for the clamping voltage and peak temperature for 8 kV HMM ESD pulse for different ballasting resistor values Rp

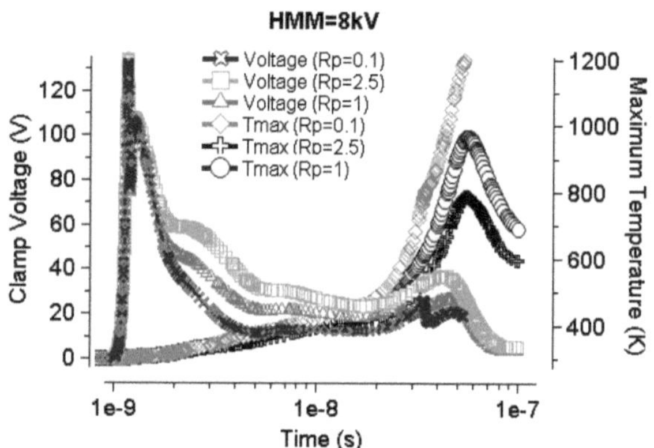

ballasting, the peak temperature during the transient exceeds 1190 K. This certainly represents the device irreversible failure at the given 8 kV HMM pulse. However, if the Rp parameter set at 2.5 Ω, that physically represents the poly ballasting resistors, the peak remains below 600 K during the entire transient. Thus likely represents the passing of the stress level (Fig. 3.51).

A more detailed insight into the reason for the difference in peak temperature reveals the effect of non-uniform finger turn-on. According to the peak temperature distribution (Fig. 3.56) the DIAC clamp with $Rp = 2.5$ Ω has a uniformly distributed temperature along all active finger regions, while the clamp with $Rp = 0.1$ Ω has a significant overheating of the middle fingers after a time of 52 ns.

These results can be complemented by the simulator output for the cathode current of the clamp device (Fig. 3.57). An uniform current component distribution for the $Rp = 2.5$ Ω case is observed (Fig. 3.57a), taking into account that Cathode 1 and 5 must supply only half of the other cathodes current to maintain the same current level for all 4 anodes. On the contrary, for the device with $Rp = 0.1$ Ω, that physically represents the absence of the poly ballasting, the cathode current components through the Cathode 1 and 5 are practically equal to zero.

3.4.3 Overcoming Multi Finger Turn-On by Proper Cell Layout Engineering

In HV DIACs and SCRs the active regions of the adjacent finger are substantially separated and therefore face a rather low coupling effect. As a result in relatively low clamping voltage conditions the injection in one finger may not create a domino-effect to turn-on the remaining fingers prior to this finger's burn-out.

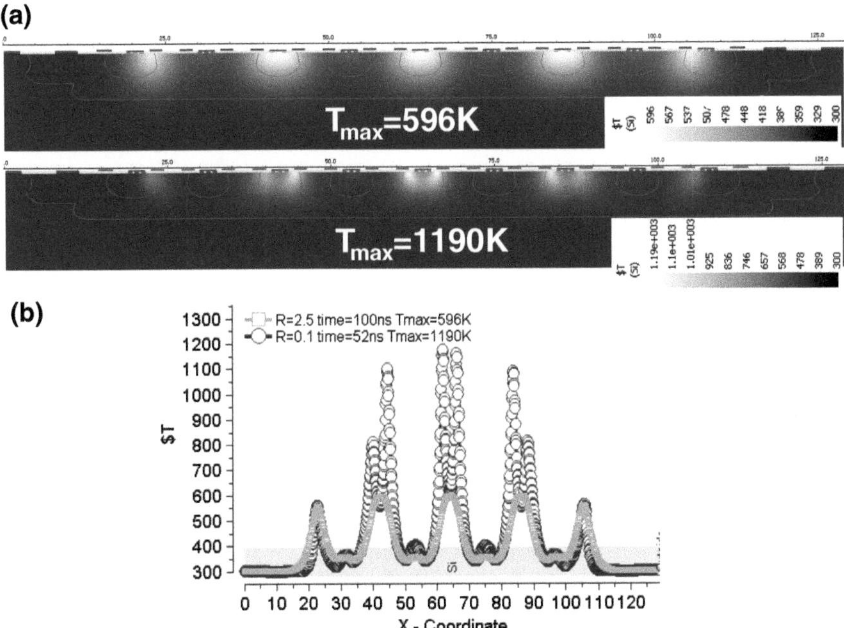

Fig. 3.56 Results of mixed-mode simulation for the physical equivalent DIAC device cross-section of a 4-finger cell. Comparison of the temperature depth profiles for the clamp with $Rp = 2.5\ \Omega$ and $Rp = 0.1\ \Omega$ for 8 kV HMM ESD pulse (**a**) and plotted temperature distribution for a lateral cutline near the device surface (**b**)

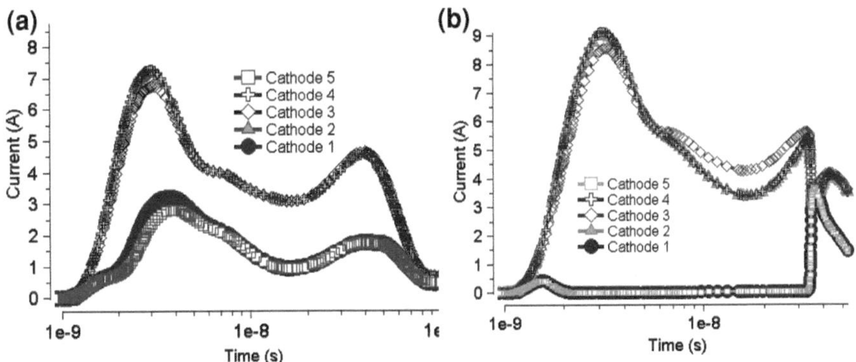

Fig. 3.57 Results of a mixed-mode simulation for the physical equivalent DIAC device cross-section of a 4-finger cell. Comparison of the temperature depth profiles for the clamps with $Rp = 2.5\ \Omega$ (**a**) and $Rp = 0.1\ \Omega$ (**b**) for 8 kV HMM stress for the cathode current components

At the same time the spreading of the conductivity modulation current along a single finger remains consistent due to a direct lateral injection. C-shape (or horseshoe) and race track device layout architecture or just very long fingers are proven

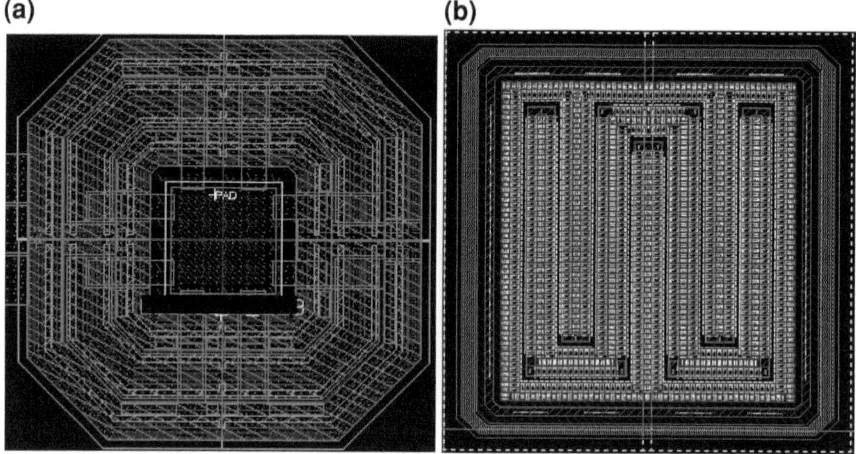

Fig. 3.58 Topologies of the race track (**a**) and snake cell (**b**) layout to eliminate multiple fingers

in many industrial designs. Thus, the alternative to the ballasting measures is the cell layout that aims to "eliminate" the multiple fingers themselves using different layout topology. Certainly, a single long finger will provide such a solution. Therefore, C-shape, race track (Fig. 3.58a), or "snake" (Fig. 3.58b) device architecture is often used to provide the physical equivalent of a single finger. A more consistent performance of these devices is an advantage. However, unlike rectangular cells, these devices are both hard to scale and to provide properly balanced current connection by metallization routing.

3.4.4 Metallization Limitations and Optimization

For simplification, an empirical ratio between the critical ESD current metallization burnout and the guaranteed electromigration current for long-term reliability of the given interconnected layers is often used to estimate cell metallization limits. The long-term reliability parameters are usually listed in the process rules. For example, the "20x" or "40x" rule to multiply the specified electromigration limit parameter is often used as an empirical correlation factor. For ESD cell layout design, the expected current density must be below the critical current level to realize the full potential of the active device and to avoid physical limitation of cell performance at the system level by the backend. The correlation factors for the metallization and via layers can be directly measured on test structures for given standard ESD pulses and TLP current.

Historically, ESD robustness was mostly studied on the component level for packaged-level ESD pulses—HBM, MM, and CDM [76–80]—rather than system-level stresses. To bridge the gap in understanding of robustness of BEOL metal

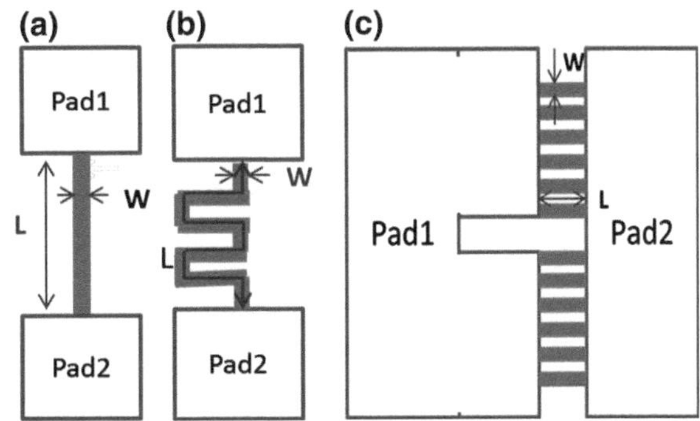

Fig. 3.59 Simplified layout views for **a** straight line, **b** meander, and **c** bridging interconnected test structure topologies, where W is the line width and L is the line length

lines under IEC and HMM pulses, a correlation between IEC/HMM and TLP pulses was studied in [81]. The study was taking into account the effect of metal lines design topology. It was shown that the main physical mechanism of metal line failure is Joule heating, which is determined by the total energy of the applied ESD pulse.

The study was done for system level ESD IEC 61000-4-2 standard pulses applied to copper metallization test structures fabricated in standard CMOS process technology. The basic topologies used in this experiment are the straight metal line, the meander metal line, and the bridging metal lines (Fig. 3.59), with varied width and length parameters.

The step-by-step pulse amplitude increase was applied to the test structures until an open circuit failure. The collected waveforms from wafer level HBM and HMM results was used to compare the pulse energy (Fig. 3.60).

The accumulated pre-failure energies integrated from the analytical pre-failure waveforms are compared with the experimental data (Table 3.5). With the exception of the results for IEC 61000-4-2 pulse stress, the analytical pre-failure energies for the pulses are in agreement with the experimental results. Under IEC stress, the miscorrelation between the experimental and analytical results is likely due to the inductive load, according to IEC 61000-4-2 standard.

Mixed-mode DECIMM [19] simulations were carried out using Si material with altered parameters to physically approximate metallization properties. Since the pre-failure energy is saturated at 90 ns, the heat generated at 90 ns was adopted as the failure criteria under a peak temperature of 1358 K—the copper melting temperature. The simulated and measured pre-failure levels of straight (Fig. 3.61), meander, and bridging metal lines match well.

The temperature distributions of these three topologies under a 1 kV HMM stress were simulated and compared (Fig. 3.62), demonstrating that the lowest peak temperature was in the bridging metal line due to less specific self-heating (Fig. 3.62b), and the highest peak temperature was in the meander metal line due to high current crowding in the corners (Fig. 3.62c).

The average correlation factors and their standard deviations from the pre-failure levels (Table 3.6) were obtained for the experimental data for these basic

Fig. 3.60 Straight metal line pre-failure current waveforms (**a**) measured for TLP, HBM, IEC and HMM pulses, and calculated pre-failure energy waveforms (**b**)

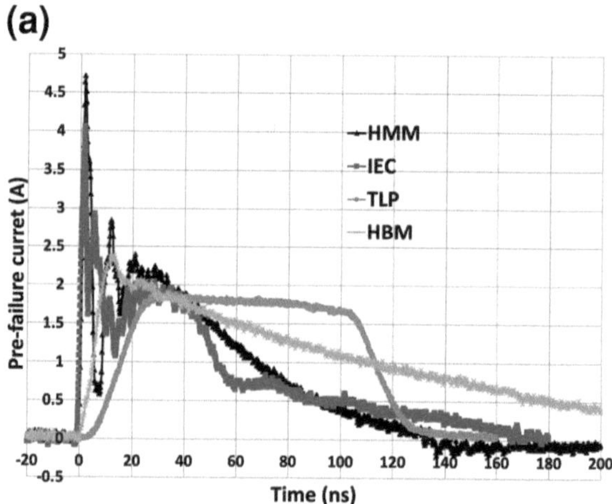

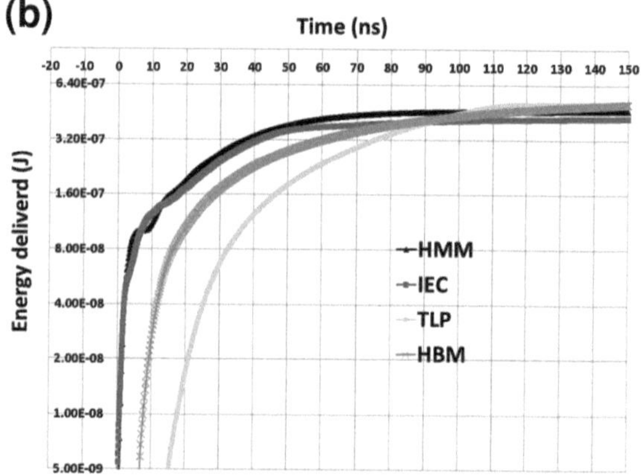

Table 3.5 Straight metal line experimental and analytical pre-failure energies under TLP, HBM, HMM and IEC pulses

	TLP (1.6 A)	HBM (2.87 kV)	HMM (1.0 kV)	IEC (1.3 kV)
Experimental pre-failure energy (J)	4.4E-7	4.3E-7	4.6E-7	4.1E-7
Analytical pre-failure energy (J)	4.0E-7	4.7E-7	4.9E-7	8.4E-7

metal lines under TLP, HBM, HMM and IEC stresses. From these data, we see that the ratio of system-level pulse to TLP current is in the range of 0.6–0.8 kV/A.

Based upon established regularities, a more complex cell layout can be analyzed using numerical simulation of more practical cases. An example of simulated current distribution in a cell with the same parameters but different contact boundaries is presented in Fig. 3.63. In a cell with contact placement along the entire horizontal cell boundary, the distribution is balanced between 4 array fingers

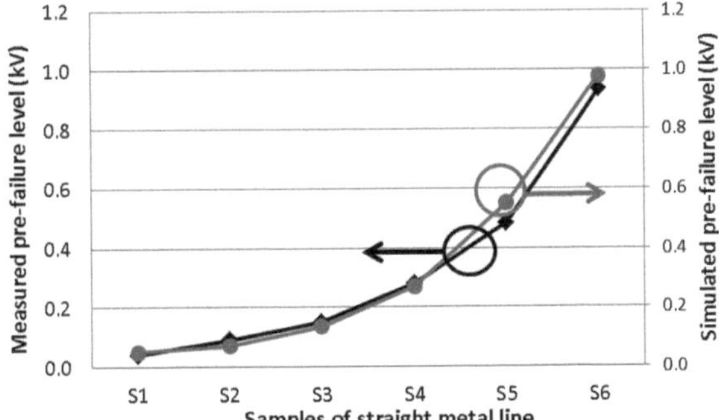

Fig. 3.61 Comparison of simulated and measured results of the straight metal lines at 90 ns

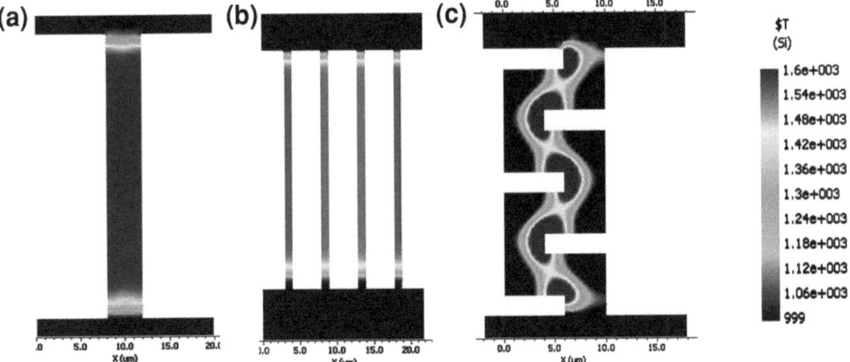

Fig. 3.62 Simulated temperature distribution of straight (**a**), bridging (**b**) and meander (**c**) metal lines at a width of 4 μm under 1 kV HMM stress

Table 3.6 Correlation factors for pre-failure voltage versus pre-failure current for HMM versus TLP (VHMM/ITLP), IEC versus TLP (VIEC/ITLP) and HBM versus TLP (VHBM/ITLP) for straight, bridging, and meander metal lines

Metal line type	V_{HMM}/I_{TLP} (kV/A)	V_{IEC}/I_{TLP} (kV/A)	V_{HBM}/I_{TLP} (kV/A)	V_{MM}/I_{TLP} (kV/A)
Straight	**0.65±0.03**	**0.80±0.04**	**1.90±0.14**	**0.13±0.02**
Bridging	**0.63±0.03**	**0.81±0.03**	**N/A**	**0.10±0.01**
Meander	**0.63±0.04**	**0.81±0.06**	**1.88±0.25**	**0.11±0.02**

(Fig. 3.63a), while connecting the cell only on the left side (Fig. 3.63b) creates a significant imbalance of the current distribution, with overstress of the left finger to the same critical temperature at a much lower current through the cell.

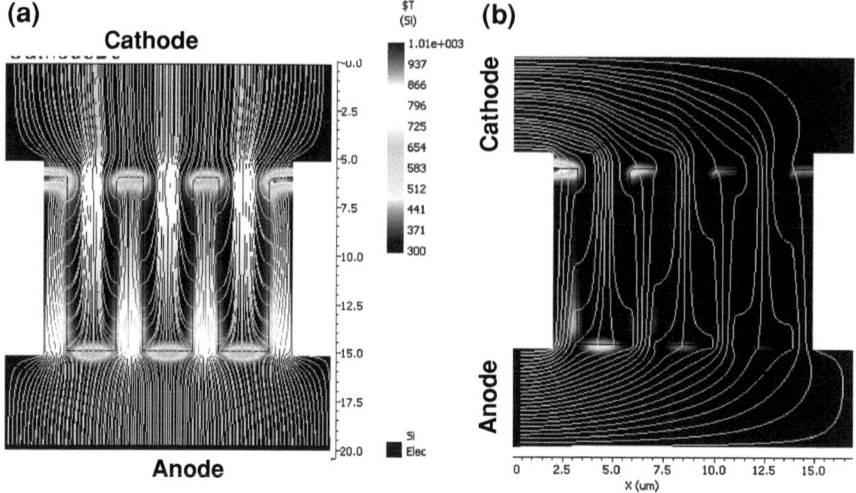

Fig. 3.63 Example of DECIMM cell-level analysis using a physical representation of the current through the cell metallization—with a connected diode demonstrating the temperature distribution and current flow lines—in a 4-finger n-p-n diode cell with *top* and *bottom* (**a**) and *left* (**b**) contact placement

3.5 Process Capability Index for ESD Devices

3.5.1 Understanding Process Capability Index for ESD Devices

The final section of this chapter brings a new angle of view on ESD device design. It is done from the perspective of the process variation across the production lots and wafers. In real integrated process technology, finite mask alignment tolerances, variation in etching, diffusion time and temperature, implantation dose, deposition and chemical-mechanical polishing (CMP) result in variation of the devices fabricated on wafer and between different production lots. The standard deviation for each parameter is known at least with some accuracy and can be measured using corresponding test structures. For example, mask misalignment tolerance, dose of ion implantation and gate oxide growth are the well-characterized parameters guaranteed by the used tools. The process variation directly affects the particular process yield and thus is thoroughly controlled and monitored in the manufacturing facilities. Essentially the control in semiconductor process technology is similar to the historical quality control methodologies [82].

A similar problem of process variation can be applied to the confidence of the ESD protection window, targeting also the use of a non-self-aligned ESD device architecture. For example in a local clamp the HV ESD device triggering into snapback often relies on the internal blocking junction breakdown or on the external reference clamp component breakdown voltage. Both components are subject to the process variation. In case of component level protection requirements the voltage tolerance of the ESD solutions is not critical as long as it is not impacting the circuit functionality. This is because the selected parts need to pass only the ESD stress levels during qualification. Also the qualification is done on a very limited amount of samples usually from a single lot. Nevertheless, it is usually assumed that the ESD performance during the remaining lifetime of the process and semiconductor product can be guaranteed. An argument in favor of this approach, perhaps, can be made by taking into account the fact that the passing qualification level for the components (typically 2 kV HBM) usually significantly exceeds the real possible events in an ESD controlled environment (<500 V).

A way more complex scenario can be envisioned for the system level ESD protection requirements. Assuming that indeed the part can experience multiple real live ESD stress events during the entire product life cycle, the ESD clamps are still supposed to provide adequate clamping transient characteristics across all wafers and production lots. Thus, theoretically, the system level ESD performance is expected to meet the specification level taking into account the process variation and ESD related yield parameters. Thus, if an ESD clamp design has been verified on the test chip using just a few samples for the pulsed testing, it hardly can be expected that it will automatically provide the repeatable I–V characteristics across many production wafer lots and across the wafers.

The fact that different characteristics of the semiconductor devices in integrated process technologies are not identical is very well understood in the semiconductor industry. The corresponding integrated semiconductor process parameters are continuously monitored and evaluated using the statistical test results from the scribe line test pattern (SLTP) of each wafer monitoring to verify the compliance of the device parameters within the specified limits of the process.

Repeatability of the device characteristics is usually reflected by the process capability index Cpk that represents a statistical measure of the given process reproducibility. A quick start introduction for the topic can be found in [83, 84]. Cpk is defined as the ability of a process to produce output within certain specification limits. The concept of process capability index gains meaning only for processes with a statistical control in place. It reflects the level of process parameter variation relatively to the adopted specification limits, rather than the variation of the parameter itself represented for example by standard deviation.

In general the specification for semiconductor device parameters can include both upper (USL) and lower specification limits (LSL). These limits are compared to the measured statistical mean μ value taking into account the standard deviation σ. Often only one of the specification limits can be important for a given product or a device at a time. For example only LSL parameter can be important for the breakdown voltage of the standard devices, while only USL parameter can be important for the leakage current at operation voltage. Thus not important limits can be always specified sufficiently away from the mean value. In this case the process capability of the parameter is determined by the critical single sided specification limit. For the case of critical LSL rating the process capability index can be calculated as a difference between mean and lower specification limit normalized to the tripled standard deviation value:

$$C_{p,lower} = \frac{\mu - LSL}{3\sigma} \tag{3.1}$$

Often, however, both *LSL* and *USL* parameters are critical, for example, for the threshold voltage of the standard device or for the breakdown and triggering characteristics of the ESD devices. In this case *Cpk* can be logically defined by a minimum value between the corresponding upper and lower process capability indexes:

$$C_{pk} = \min\left[\frac{USL - \mu}{3\sigma}, \frac{\mu - LSL}{3\sigma}\right]. \tag{3.2}$$

The *Cpk* indexes are calculated separately for each particularly monitored individual device or process parameter. This approach is covering the majority of the supported integrated device parameters in the integrated process technology. The exception is usually given for so-called free or category II devices, not mentioned in the process specification. Unfortunately the majority of ESD devices are usually assigned to such category. As a result they usually are not even placed on the SLTP due to lack of space which is taken by the standard devices. This fact contributes to the problem statement discussed below.

In case of a normal distribution the process yield can be calculated by the area under the probability density function $F(\sigma)$):

$$F(\sigma) = \frac{1}{\sqrt{2\pi}} \int_{-\sigma}^{\sigma} e^{\frac{-t^2}{2}} dt. \tag{3.3}$$

Respectively, the remaining set of samples beyond the desired specification limits represents the process fallout in form of the defective part quantities that are not meeting the specification limits. Apparently, the fallout rate depends on both how close the specification limits are set to the mean value and the standard deviation of the parameter distribution across the pool of measured samples. The major figure of merit to represent the process fallout is called either "defects per million opportunities" (DPMO) or defective part per million (DPPM).

For an established robust semiconductor process the critical parameters should have a *Cpk* of at least $1.33 = (4\sigma/3\sigma)$ logically representing an acceptable 99.7 % process yield per given parameter for non-critical parameters. For the critical parameters the target is $Cpk \sim 1.67 = (5\sigma/3\sigma)$. The value $Cpk \sim 1.67$ represents a fallout of only one part per million (PPM) (Table 3.7).

At the same time some automotive and high reliability products may require the semiconductor process quality level with so-called 0 DPPM. This requirement simply represents at least six sigma margin and corresponding $Cpk = 2 = (6\sigma/3\sigma)$ according to (3.2) and Table 3.7. Certainly, from physical point of view the real calculated fallout is expected to be below 0.002, rather than actual zero.

Table 3.7 Relationship to measures of process fallout

C_{pk}	Sigma level (σ)	Area under the probability density function $F(\sigma)$	Process yield (%)	Process fallout (DPPM)
1.00	3	0.9973002039	99.73	2700
1.33	4	0.9999366575	99.99	63
1.67	5	0.9999994267	99.9999	1
2.00	6	0.9999999980	99.9999998	0.002

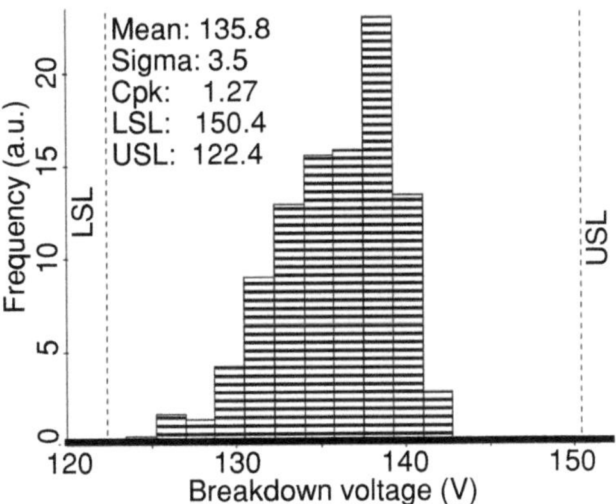

Fig. 3.64 Illustration of the experimental variation of the breakdown voltage across multiple wafer lots for Nepi to P-Substrate and calculated *Cpk* for adopted specification limits

An illustration of the experimental variation of the breakdown voltage across multiple wafer lots for a 100 V BCD process and calculated *Cpk* for the process limits is presented in Fig. 3.64. The breakdown voltage reflects a nEpi pocket with n-buried layer (NBL) to p-Substrate region isolation with p-buried layer (PBL). The breakdown voltage is determined by the spacing between the in-depth NBL and PBL layers and thus is sensitive to the alignment of the NBL and PBL mask layers. These layers are implanted with rather high energy before nEpi is grown. That requires a rather thick photoresist layer with a resulting large misalignment followed by the significant diffusion during nEpi growth and other process anneal cycles. As a result, the standard deviation of the breakdown voltage is rather high. At the same time a possible correction of the breakdown voltage with higher NBL-PBL spacing towards the breakdown voltage saturation both consumes higher isolation space on chip and as a side-effect an undesirable lower latch-up isolation.

Monitoring *Cpk* and improving yield of the semiconductor process are supported by many standard operation procedures both in semiconductor fabrication facilities and process development organization. Respectively, historically the topic was broadly addressed both experimentally, by modeling of the process variations as well as by the development of statistical methods for the estimation of the effects impacting the device-circuit operation [85, 86]. For example the effect of the process variation impact on the integrated timing references (ring oscillators) and other analog circuits were studied. However, only a few studies can be found regarding similar effects in ESD protection structures [87, 88], while some efforts were mainly targeting the statistical effects for endurance [89].

As it was mentioned in the introductory Chap. 1, design of ESD protection devices is often based on approach "free" from additional implants and masks. Thus, often, a non-self-aligned ESD device architecture is the only choice to target the specified ESD protection window. Typically, a very limited amount of the test structures is evaluated due to the rather time consuming and costly pulsed measurements. Such methodology to guarantee ESD device parameters barely looks like an adequate approach in comparison with the automated electrical testing from scribe line test patterns adopted for the standard devices. Thus, the statistical content of ESD device pulsed testing involved in the development hardly can represent the desired adequate degree of confidence for high reliability components, if variations across the wafers and future lot-to-lot are truly taken into account. Meantime, the work on process yield is usually continued for a long period of time after the process release for production, while the corresponding deviation of ESD device parameters often remains unmonitored.

This problem, perhaps, was not so critical for the older generations of semi-conductor process technologies. Such processes were typically integrated to provide a large margin for high voltage devices, while the low voltage domains were mostly relying on the non-process sensitive active clamp protection approach. In parallel to this, one of the major assumptions to increase the confidence was the design of ESD devices by applying a transformation to the standard devices. This methodology was explained in the above sections of this chapter. In this case the expectation is that the ESD device characteristics, at least for non-ESD operation, will follow the same trend in change of electrical characteristics as the standard devices will. For example the breakdown and triggering voltages of the ESD device will follow the breakdown and absolute maximum rating voltage of the standard devices.

The implementation of new generation analog processes with power optimized devices results in a significant shrink of the relative ESD protection window. IC components for automotive, medical and other six sigma quality high reliability applications are generating a demand for 0 DPPM fallout. This automatically translates itself into a new trend, when *Cpk* parameters of ESD devices must not interfere with the protected power device parameters. Moreover, in case of system level, they should also provide the repeatable ESD transient characteristics. Thus, to avoid both a yield and reliability impact the issue of the process capability figures of merit for ESD devices hardly can be neglected any longer for some applications.

Indeed, an overall understanding of the impact of the "natural" process parameters variation on the ESD device breakdown voltage and high current characteristics presents a primary practical interest towards a level of confidence. In spite of that ESD IP is often released in the process design kits based on rather limited data. This subject is quite rarely addressed in regular way. For example due to low productivity of the TLP testing a regular monitoring of pulsed characteristics of ESD devices is rather difficult. Only over recent years some automated wafer level testers were developed by vendors to address the increasing demand. Therefore, a progress of revealing any impact of the process variability on ESD devices can only be made by using numerical simulation. This is demonstrated in the next section.

3.5.2 Cpk Simulation for the Avalanche Diodes Breakdown

Typically the numerical analysis of the devices is defined manually. This is although some limited sensitivity data can be obtained for the so-called corner parameters. This approach hardly can output sufficient data for a statistical variation analysis taking into account multiple variations of possible parameter combinations.

To demonstrate the major aspects of the ESD devices across the wafer and production lots variation a new simulation approach has been intentionally developed and applied [90]. It is based on parameterized mixed-mode device-circuit analysis with an automatically generated input based on an additional casual parameter variation. Such automatically generated input is achieved by applying the Monte Carlo method to each independent parameter within a defined desired standard deviation. Within the capability of the parameterized device template any process sensitive device implant, region or even physical material parameter can be varied. To enable the statistical simulation of the device characteristics, taking into account both the process and mask alignment variation, the simulation tool DECIMM^{TM} [19] is upgraded with a unique capability in form of a so-called "DoE" tool interface. With the parameterized device definition approach supported by this unique tool, the automated generation of the parameter table for the simulation runs based on the Monte Carlo algorithm is done.

The methodology is applied for two ESD design specific examples to analyze the breakdown voltage of HV avalanche diodes. This is presented in this section. The simulation analysis for the *Cpk* of the triggering voltage of the NLDMOS-SCR is presented in the following section. The first example brings the comparison of the conventional double RESURF avalanche diode (Fig. 3.65a) with a p-i-n diode (Fig. 3.65b). The operation principles of the avalanche diode ESD devices are similar and have been described in Sect. 3.3.3.

The parameterized template of the device has been updated with additional geometrical parameters to physically represent the "natural" deviation of the critical device parameters in the simulation analysis. These parameters have been used to represent the casual diffused peak doping profiles variation and the mask misalignment assuming normal distribution of the initially set mean value of the parameter with a given standard deviation. The standard deviations were defined according to certain controlled limits of the real process variation.

The selected parameters physically represent mask misalignments and peak doping variation in form of small increments added to the corresponding template

Fig. 3.65 Device cross-sections with net doping profiles for conventional double RESURF (**a**) and pseudo p-i-n (**b**) diodes used in the statistical simulation analysis presented in this section

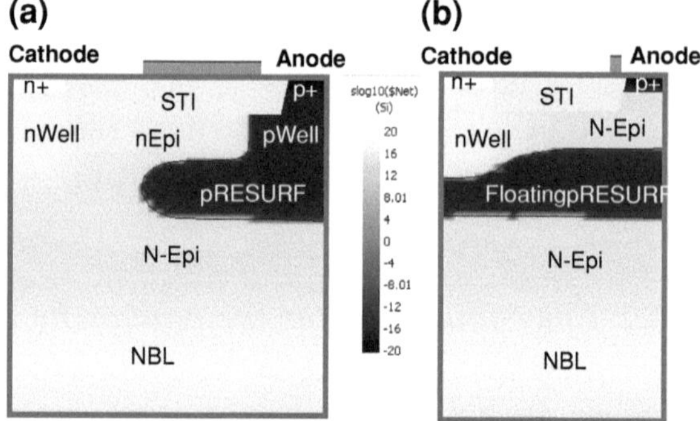

Table 3.8 Input parameters and defined standard deviation that has been used for automated runs Monte-Carlo generation

Physical process parameter	Variable	Sigma level (σ)
nWell peak doping scaling coefficient	Knw	1 %
pWell peak doping scaling coefficient	Kpw	1 %
pRESURF peak doping scaling coefficient	Krs	1 %
nEpi peak doping variation scaling coefficient	Kep	1 %
Shallow trench isolatiin (STI) depth variation	$ysti$	5 nm
Active region (STI position) misalignment	$xsti$	15 nm
Cathode nWell mask position misalignment	$xnwell$	15 nm
Anode pWell mask position misalignment	$xpwell$	15 nm
Anode pRESURF mask position misalignment	$Xprsf$	15 nm

parameter. For non-self-aligned diffused well profiles the scaling coefficient for the peak doping for nWell, pWell, pRESURF, nEpi were added as Knw, Kpw, Kpr, Kep, respectively, and originally set to 1. In the parameterized DoE runs generation the variation of the peak doping has been set at 1 % that corresponds to real process technology variation. The energy of the implant is controlled rather well and thus the profile peak depth variation can be neglected. Small increments to the nWell, pWell, pRESURF lateral diffused profile position and active region were added too, including the corresponding parameters $xnwell$, $xpwell$, $xprsf$ and $xsti$. These parameters were originally set to zero. For generation of the mixed-mode runs the standard deviation of these parameters was selected at 15 nm. Finally $ysti$ increment was introduced as an additive to the trench depth device parameter in order to physically represent process variation of shallow trench isolation etch and oxidation process with the standard deviation of 0.005 nm (Table 3.8).

The simulation runs were automatically generated in the user interface table for multiple transient simulations to calculate the I–V characteristics for the breakdown voltage. Another automated feature allows to stop the simulation if the level of the breakdown current is achieved. It extracts the numerical value of the breakdown voltage in the run tables.

(a)

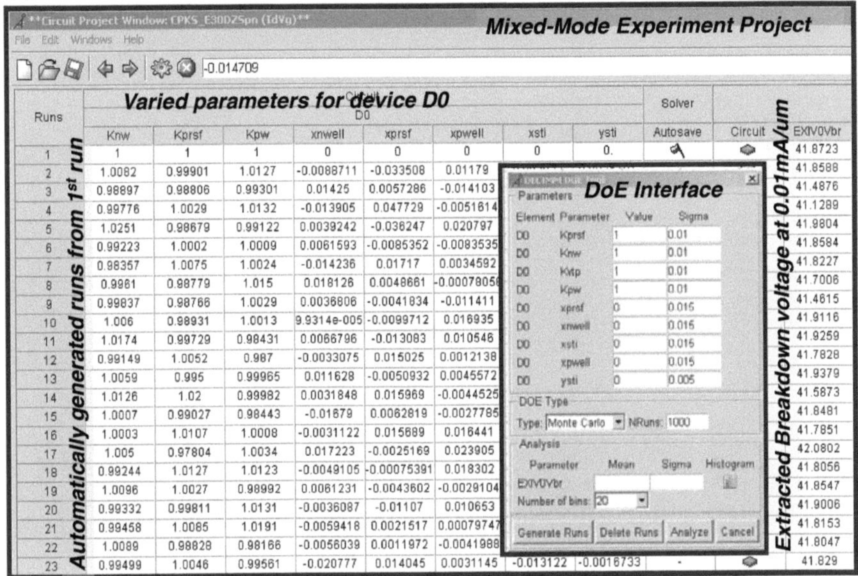

Runs	Knw	Kprsf	Kpw	xnwell	xprsf	xpwell	xsti	ysti	Autosave	Circuit	EXIV0Vbr
1	1	1	1	0	0	0	0	0.			41.8723
2	1.0082	0.99901	1.0127	-0.0088711	-0.033508	0.01179					41.8588
3	0.98897	0.98806	0.99301	0.01425	0.0057286	-0.014103					41.4876
4	0.99776	1.0029	1.0132	-0.013905	0.047729	-0.0051614					41.1289
5	1.0251	0.98679	0.99122	0.0039242	-0.036247	0.020797					41.9804
6	0.99223	1.0002	1.0009	0.0061593	-0.0085352	-0.0083535					41.8584
7	0.98357	1.0075	1.0024	-0.014236	0.01717	0.0034592					41.8227
8	0.9961	0.98779	1.015	0.018126	0.0048661	-0.00078056					41.7006
9	0.99837	0.98766	1.0029	0.0036806	-0.0041834	-0.011411					41.4615
10	1.006	0.98931	1.0013	9.9314e-005	-0.0099712	0.016935					41.9116
11	1.0174	0.99729	0.98431	0.0066796	-0.013083	0.010546					41.9259
12	0.99149	1.0052	0.987	-0.0033075	0.015025	0.0012138					41.7828
13	1.0059	0.995	0.99965	0.011628	-0.0050932	0.0045572					41.9379
14	1.0126	1.02	0.99982	0.0031848	0.015969	-0.0044525					41.5873
15	1.0007	0.99027	0.98443	-0.01679	0.0062819	-0.0027785					41.8481
16	1.0003	1.0107	1.0008	-0.0031122	0.015689	0.016441					41.7851
17	1.005	0.97804	1.0034	0.017223	-0.0025169	0.023905					42.0802
18	0.99244	1.0127	1.0123	-0.0049105	-0.00075391	0.018302					41.8056
19	1.0096	1.0027	0.98992	0.0061231	-0.0043602	-0.0029104					41.8547
20	0.99332	0.99811	1.0131	-0.0036087	-0.01107	0.010653					41.9006
21	0.99458	1.0085	1.0191	-0.0059418	0.0021517	0.00079747					41.8153
22	1.0089	0.98828	0.98166	-0.0056039	0.0011972	-0.0041988					41.8047
23	0.99499	1.0046	0.99561	-0.020777	0.014045	0.0031145	-0.013122	-0.0016733	-		41.829

Automatically generated runs from 1st run

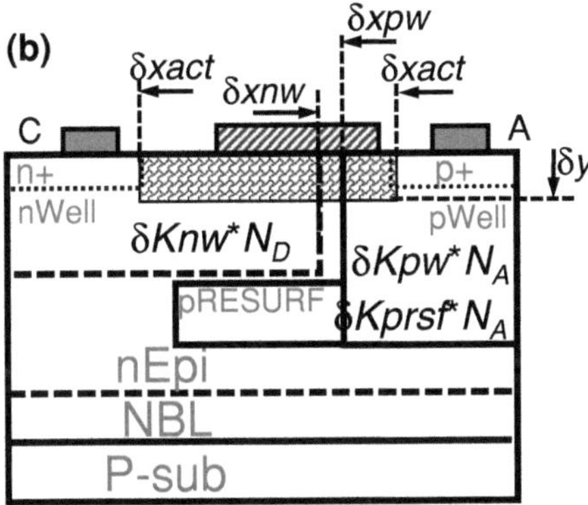

Fig. 3.66 Example DECIMM^{TM} DOE tool interface and portion of the generated table for 500 runs numerical experiment with the extracted breakdown voltage after simulation done (**a**) and simplified conventional RESURF avalanche diode cross-section with illustration the varied parameters (**b**)

Using the integrated DOE tool interface [19] (Fig. 3.66a) over 500 numerical experiments have been generated (Fig. 3.66a) and solved to simulate and compare the statistical distribution of the resultant breakdown voltage for both diodes (Fig. 3.65). The varied parameters are graphically illustrated in Fig. 3.66b. Based on these statistical results the physical process capability indexes were calculated using the simulated mean and standard deviation outcome.

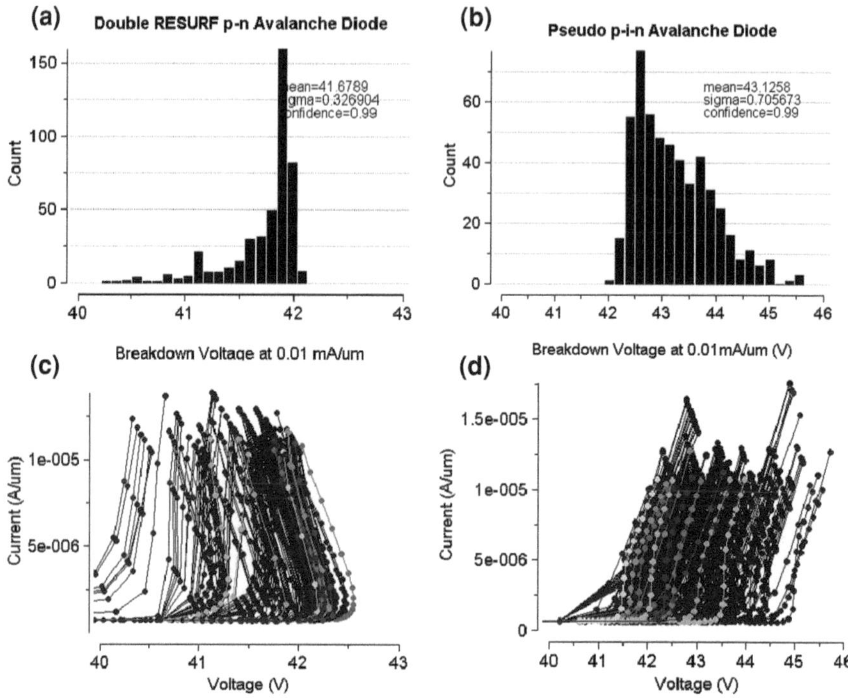

Fig. 3.67 Frequency plots for the calculated breakdown voltage for the lateral double RESURF (**a**) and pseudo p-i-n (**b**) avalanche diodes obtained for the statistical variation of the peak doping, STI depth variation and mask misalignment (Table 3.8) and plotted corresponding breakdown I–V characteristics for both diodes (**c, d**)

The comparison of the conventional RESURF avalanche diode (Fig. 3.67a) with the pin diode (Fig. 3.67b) is accomplished in order to evaluate the effect of the alternative device architecture. According to ~ 500 simulation runs with randomly generated offset values the distributions were close to normal with a mean breakdown voltage value for the double RESURF avalanche diode $\mu^{Vbr} = 41.7$ V and the standard deviation $\sigma^{Vbr} = 0.33$ V (Fig. 3.67a). For the alternative pseudo p-i-n avalanche diode (Fig. 3.67b) the generated values are $\mu^{Vbr} = 43.1$ V and $\sigma^{Vbr} = 0.71$ V. The views of simultaneously plotted transient I–V characteristics for the ~ 500 runs for both diodes are presented in Fig. 3.67c, d, respectively.

The intuitive expectation that the pseudo p-i-n avalanche diode might have a smaller standard deviation it is not true according to the simulation data (Fig. 3.67a, b). In spite of a smaller number of non-self-aligned diffused implant parameters, it appears that the sensitivity of this device to the process variation is actually higher (Table 3.9). Thus the sensitivity of the device to the process variation is determined by the device architecture itself. Moreover, according to the results (Fig. 3.67a, b) the statistical distribution of the breakdown voltage is

Table 3.9 Calculated from the ~ 500 simulation runs (Fig. 3.66) required specification limits for the avalanche diodes (Fig. 3.65) to meet different *Cpk* rating requirements for the breakdown voltage parameter

Desired *Cpk* for the breakdown voltage	Corresponding margins	Specification limits for double RESURF p-n diode with $\mu^{Vbr} = 41.7$ V and $\sigma^{Vbr} = 0.33$ V (V)		Specification limits for pseudo p-i-n diode with $\mu^{Vbr} = 43.1$ V and $\sigma^{Vbr} = 0.71$ V (V)	
		LSL	USL	LSL	USL
1.33	4 σ	40.4	43.0	40.2	45.9
1.67	5 σ	40.0	43.3	39.6	46.7
2.00	6 σ	39.7	43.6	38.9	47.4

Table 3.10 Calculated contribution of the different parameters variation into standard deviation of the breakdown voltage characteristics of the double RESURF p-n diode from ~ 100 to 200 simulation runs

Physical process parameter varied	Mean μ^{Vbr} (V)	Standard deviation σ^{Vbr} (V)	Required LSL for Cpk = 1.67	Required USL for Cpk = 1.67
All: *Knw Kpw Krs Kep ysti xsti xnwell xpwell Xprsf*	41.7	0.33	40	43.3
Only peak doping: *Knw Kpw Krs Kep*	41.82	0.116	41.2	42.4
Only mask misalignment: *xsti xnwell xpwell Xprsf*	41.81	0.167	41	42.6
Only STI depth variation: *ysti*	41.86	0.031	41.7	46.7

asymmetrical in a different way for the compared devices. The tail of the distribution is propagated towards a LSL for the double RESURF p-n avalanche diode, while the pseudo p-i-n diode has the tail towards an USL.

A reverse problem can be derived from the simulation results too. According to the simulation analysis the LSL and USL limits can be determined depending on the desired *Cpk* target. In this case for the simulated isothermal room temperature condition $Cpk = 1.33$ will require LSL and USL for p-n RESURF diode as 40.4 and 43 V, respectively. To meet $Cpk = 2$ the corresponding limits expansion is required to 39.7 and 43.6 V, respectively. Unlike the limitation of the experimental data the simulation results allow a quick estimate of the contribution of different factors into the standard deviation of the process parameter. For example for the p-n double RESURF diode the major contributing factor into standard deviation is well misalignment (Table 3.10).

Unlike a rather costly experimental analysis the simulation results can be easily applied to evaluate for example the impact of mask misalignment for the particular device, physically representing different tools. The comparison of the impact on the standard deviation for different well only alignment tolerances is summarized

Table 3.11 Required LSL and USL for the double RESURF avalanche diode at Cpk = 1.67 for different mask misalignment parameter for *xsti xnwell xpwell Xprsf*

Mask misalignment (nm)	Mean μ^{Vbr} (V)	Standard deviation σ^{Vbr} (V)	Required LSL for Cpk = 1.67	Required USL for Cpk = 1.67
10	41.8	0.1	41.3	42.3
15	41.8	0.167	41.0	42.6
30	41.7	0.44	39.5	43.9
60	41.4	0.99	36.5	46.4

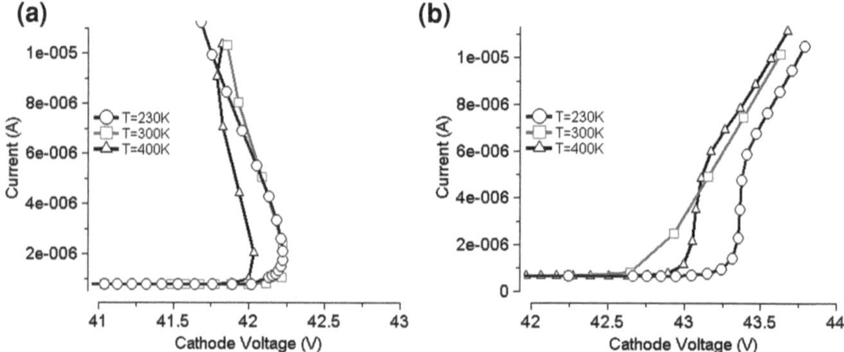

Fig. 3.68 I–V characteristics for both diodes for different temperatures for the lateral double RESURF (**a**) and pseudo p-i-n (**b**) avalanche diodes

in Table 3.11, with corresponding LSL and USL rating calculated to meet *Cpk* = 1.67.

The above simulation data reflect the constant room temperature effect. While for low breakdown current level the solution of the heat equation still can be avoided, taking into account the operation temperature range from −40 to 125 °C will require an addition of these limits. However even in this case it is not a simple effect due to the temperature coefficient of the avalanche breakdown process. Indeed, according to I–V characteristics (Fig. 3.68) the temperature dependence is observed.

However, a more complex dependence related to the effects in particular avalanche diode architecture with a rather complex cancelling out each other breakdown, RESURF and conductivity modulation effects is in place. As a result both mean and standard deviation parameters for both diodes are practically the same in the operation temperature range and not impacting the LSL and USL (Table 3.12).

Table 3.12 Results of simulation analysis from ∼300 runs for double RESURF p-n and pseudo p-i-n diodes for the parameters variation (Table 3.8) and different structure temperatures

LAD	Temperature	Mean μ^{Vbr} (V)	Standard deviation σ^{Vbr} (V)	Required LSL for Cpk = 1.67	Required USL for Cpk = 1.67
Double RESURF	−40 °C	41.55	0.33	39.9	43.2
p-i-n diode	**Room**	**41.7**	**0.33**	**40**	**43.3**
	+125 °C	41.60	0.31	40	43.2
Pseudo p-i-n diode	−40 °C	43.17	0.71	39.63	46.72
	Room (300 K)	**43.1**	**0.71**	**39.60**	**46.65**
	+125 °C	43.2	0.67	39.85	46.53

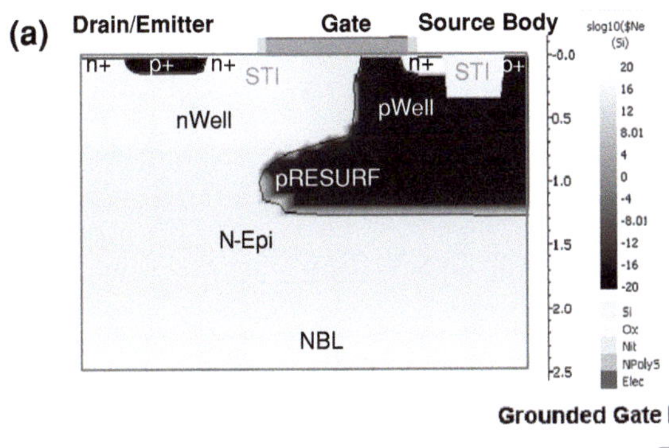

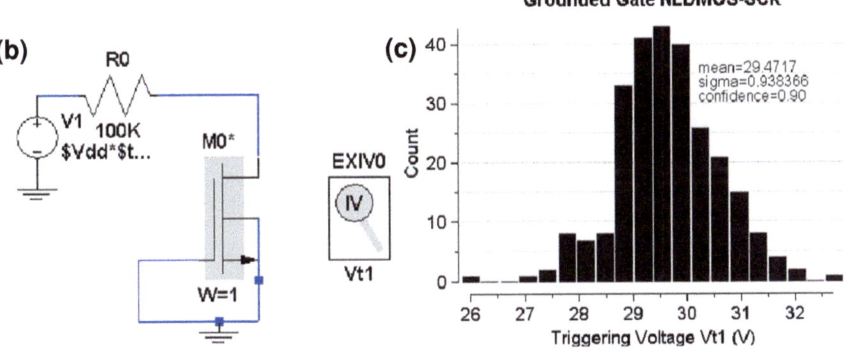

Fig. 3.69 Cross-section of the NLDMOS-SCR (**a**), mixed mode circuit with grounded gate (**b**) and simulation results from ∼250 runs in form of frequency plot (**c**) for the calculated breakdown voltage for the Monte Carlo generated statistical variation of the peak doping, STI depth and mask misalignment variations as defined in Table 3.8

3.5.3 Cpk Analysis for the NLDMOS-SCR Clamp

The second example is even more directly related to the pulsed characteristics of the snapback ESD devices. This example is created to study the statistical variation of a 20 V NLDMOS-SCR ESD device (Fig. 3.69a). In this case the same parameter like in Table 3.6, reticle misalignment variations, doping peak concentrations and trench depth variation, were included in the parameterized numerical analysis according to the methodology described in the previous section.

The major goal of the simulation is simulate the mean and standard deviation for triggering voltage $Vt1$ of this HV protection clamp in the grounded gate circuit, rather than the breakdown voltage. According the simulation result analysis with the parameters variation (Table 3.6) the triggering voltage of the grounded gate NLDMOS-SCR clamp (Fig. 3.69b) is rather sensitive to process parameter variations. A rather high standard deviation $\sigma^{Vt1} = 0.94$ V was calculated under the mean value $\mu^{Vt1} = 29.47$ V. Respectively, for this device and the grounded gate clamp to guarantee process capability with $Cpk \sim 1.33$ the LSL and USL should be set as:

$$LSL^{1.33} = \mu^{Vt1} - 4\sigma^{Vt1} = 25.7 \text{ V and } USL^{1.33} = \mu^{Vt1} + 4\sigma^{Vt1} = 33.2 \text{ V}.$$

If six sigma high reliability 0 DPPM $Cpk = 2$ is specified then it would require a further expansion of the specification limits to:

$$LSL^{2.00} = \mu^{Vt1} - 6\sigma^{Vt1} = 23.8 \text{ V}; \ USL^{2.00} = \mu^{Vt1} + 6\sigma^{Vt1} = 35.1 \text{ V}.$$

In the last case the ESD protection solution becomes hardly applicable to the high reliability products, if an aggressively power optimized process with a small ESD protection window is used.

An alternative known ESD clamp solution can be proposed based on the understanding of the on chip device design physics. A clamp with a high side avalanche diode reference can be used (Fig. 3.70a). The clamps turn on is based on the dependence of the NLMOS-SCR triggering voltage on the gate bias. The bias is provided by the voltage drop on the gate resistor from the lower voltage breakdown current of the high side reference avalanche diode (Fig. 3.70b). As a result the critical condition for current instability is achieved at much lower multiplication coefficients by multiplying the channel current of the device rather than the dark current of the p-n junction. It is logical to expect that in these conditions the NLDMOS-SCR triggering is less sensitive to the misalignment of the NLDMOS device parameters.

To demonstrate this effect the avalanche diode referenced clamp (Fig. 3.70a) has been "assembled" in mixed-mode simulation example using the NLDMOS-SCR device (Fig. 3.69a) and the finite-element model of the lateral avalanche diode (Fig. 3.70b) with the adjusted parameters for ~ 21 V. The comparison of the I–V characteristics for the separate NLDMOS-SCR and Avalanche diode Clamp components and the final clamps is provided by Fig. 3.70c.

For even better accuracy of comparison the Cpk simulation for the triggering voltage the clamp (Fig. 3.70a) has been accomplished by re-using the exact previously generated DoE for the grounded gate clamp (Fig. 3.69). Although the parameters of the avalanche diodes can be varied simultaneously with the NLDMOS-SCR too, for a more conclusive comparison the avalanche diode has been

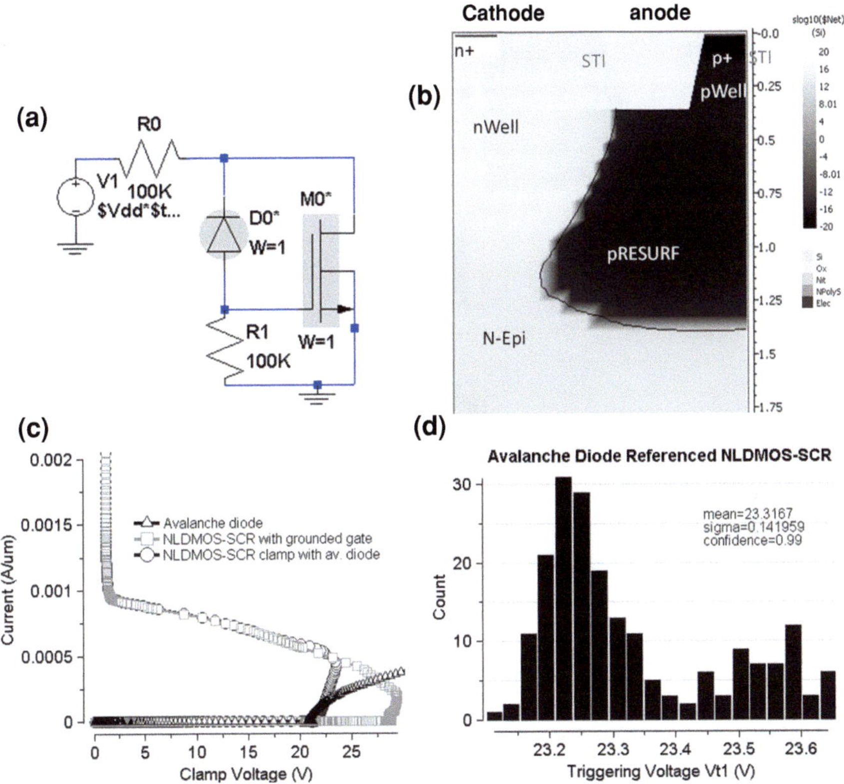

Fig. 3.70 Mixed mode circuit with avalanche diode referenced NLDMOS-SCR clamp (**a**), the avalanche diode cross-section (**b**), I–V characteristics of the clamps and the avalanche diode (**c**); simulation results in form of frequency plot (**d**) for the calculated breakdown voltage ~ 200 runs for the parameters generated as defined in Table 3.8

included in the circuit without any parameter variation. As a result a "purified" effect can be observed by simulation indicating for this clamp the expected significantly lower standard deviation of only $\sigma^{Vt1} = 0.14$ V under the mean triggering voltage $\mu^{Vt1} \sim 23.3$ V. Similarly to the above these results can be translated into the required LSL and USL ratings adequate to guarantee *Cpk* 1.33 and 2 as:

$$LSL^{1.33} = 22.8 \text{ V and } USL^{1.33} = 24 \text{ V}; LSL^{2.0} = 22.5 \text{ V and } USL^{2.0} = 24.2 \text{ V}.$$

The misalignment in the avalanche diode will result in the clamp replicating the standard deviation of the reference device that is still much smaller than the NLDMOS-SCR clamp based on the internal blocking junction.

To summarize the above material, the fallout of the ESD protection due to process variation can be demonstrated for case the 20 V open drain circuit with NLDMOS device (Fig. 3.71a) protected by NLDMOS-SCR clamp (Fig. 3.69a).

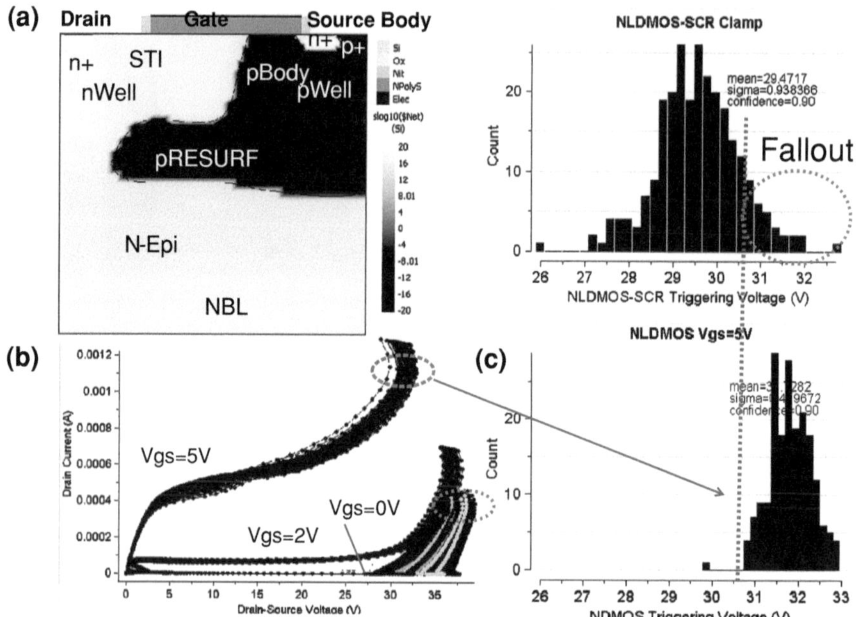

Fig. 3.71 Cross-section (**a**) and drain-source output I–V characteristics for $Vgs = 0$, 2 and 5 V (**b**) of the 20 V NLDMOS device from ~ 200 simulation runs, the frequency plot for the triggering drain-source voltage at the gate bias $Vgs = 5$ V compared with the frequency plot for NLMOS-SCR (**c**) to demonstrate the fallout portion of the open drain circuit

This protection approach assumes that the triggering voltage of the NLDMOS-SCR clamp will remain lower than the triggering voltage of the NLDMOS device at any transient operation conditions of the real circuit.

As a next step, the same methodology for the *Cpk* analysis is applied to the NLDMOS device (Fig. 3.71a). It returns the corresponding statistical variation of the drain-source triggering voltages at given gate bias values (Fig. 3.71b). This variation of the triggering voltage for given gate bias level can be translated into the variation of the pulsed SOA of the NLDMOS device.

The fallout scenario depends on the NLDMOS gate coupling. If the gate coupling is low then the NLSMOS SOA variation in a real circuit can be assumed similar to the triggering voltage variation at the gate bias $Vgs = 0$. In this case NLDMOS has the mean value $\mu^{Vgs0} = 37.6$ V under the standard deviation $\sigma^{Vgs0} = 0.68$ V. Comparing these numbers with the corresponding NLDMOS-SCR $\mu = 29.47$ V and $\sigma = 0.93$ V it can be concluded that the high yield protection can be indeed realized with 5-sigma $Cpk = 1.67$. However, if the gate coupling is high, the protection solution becomes unreliable. For example in case of the transient gate coupling during ESD stress up to $Vgs = 5$ V the triggering voltage of the device has the mean value $\mu^{Vgs5} = 31.72$ V and the standard $\sigma^{Vgs5} = 0.59$ V (Fig. 3.71c). By comparison of the distributions for both devices

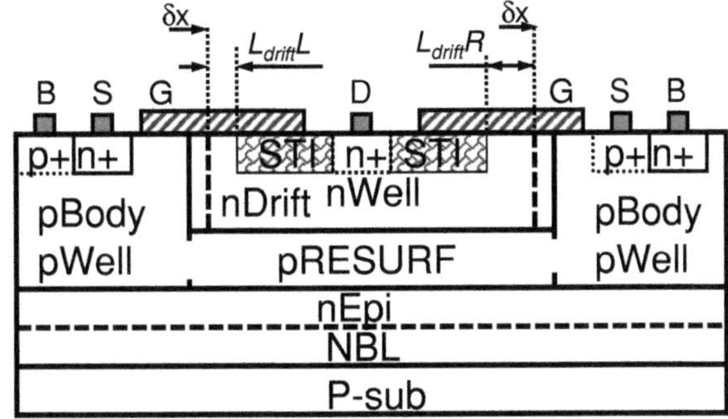

Fig. 3.72 Illustration of the nDrift mask misalignment effect in the symmetrical double finger array layout

the yield fallout corresponds to hardly acceptable 2-sigma $Cpk = 0.67$. Thus a reliable solution requires a different type of clamp. For example, as it was mentioned above, the NLDMOS-SCR clamp with high side avalanche diode can be used similarly to the design (Fig. 3.70), but with the triggering voltage adjusted into the desired voltage range for 20 V domain protection.

The scenario (Fig. 3.71) is brought under assumption of independent variation of the NLDMOS and NLDMOS-SCR parameters. Thus it represents the worst case for example for co-packaged devices in SiP. However, since both on-the same chip the NLDMOS-SCR clamp and the NLDMOS device will share the same physical location on the same wafer and the same lot, it can be assumed that both devices can be designed to have a similar variation of the particular parameters in order to reduce the reliability impact. For example increase of the triggering voltage of NLDMOS might be accompanied by a similar increase of the NLDMOS-SCR triggering voltage due a similar blocking junction design. This case can be further evaluated using numerical simulation in order to both verify and quantify these expectations. It can be done by creating a consolidated template that will combine both devices under a synchronized variation of the same process sensitive parameters. The mixed-mode analysis can be done to determine the corresponding correlation effects monitoring the current through NLDMOS device.

Another effect that improves the above worst case scenario is the consideration of the multifinger array layout with finger symmetry (Fig. 3.72). In this case mask misalignment δx, effectively impacts the half of the device in an opposite way. For example in case of misalignment of the nDdrift mask in of NLDMOS the smaller Poly-nDrift overlap *LdriftL* is formed for all left sides of the fingers with corresponding elevated the breakdown and triggering voltage. At the same time the bigger overlap *LdriftR* is realized on the right side of the finger resulting in the corresponding breakdown and triggering voltage decrease (Fig. 3.72). As a result of counter mismatch the average breakdown and triggering voltage will remain at a lower level thus likely improving the process variation sensitivity.

Similarly to the methodology used in the above numerical experiments, the possible self-canceling-out effect can be quantified by the numerical simulation too. The parameterized devices with full pRESURF layer have been used to

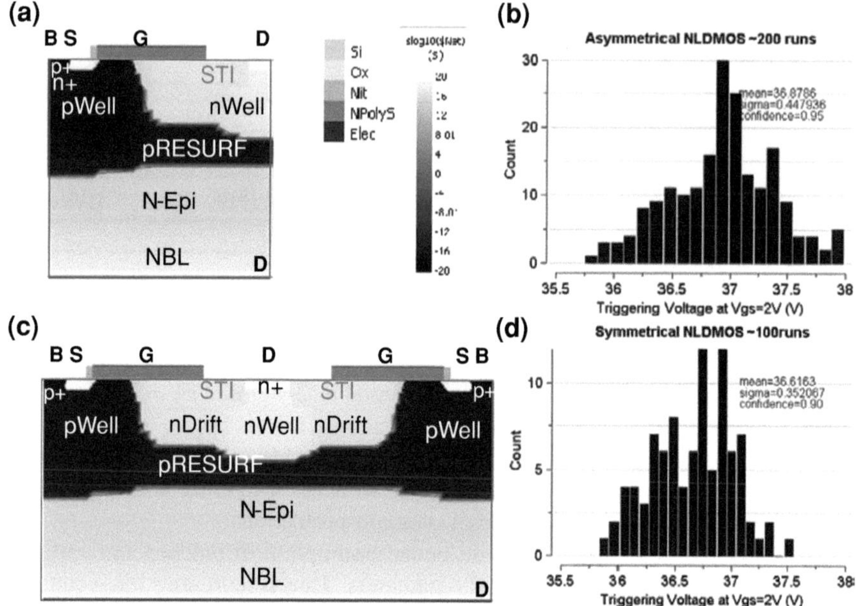

Fig. 3.73 Cross-sections of the asymmetrical (**a**) and symmetrical (**b**) NLDMOS device and the corresponding frequency plots (**c**, **d**) for the triggering drain-source voltage at the gate bias $Vgs = 2$ V both devices under width scaling factor for asymmetrical device $W = 2$

represent the asymmetrical (Fig. 3.73a) and symmetrical (Fig. 3.73b) NLDMOS device layout.

According to the simulation results an approximately 20 % lower standard deviation ~ 0.35 for the symmetrical layout is obtained in comparison with the standard deviation ~ 0.42 for the device (Fig. 3.73a) that physically represents the asymmetrical layout.

3.6 Summary

For analog circuits with system-level pins and wafer-level packaging, an optimal ESD protection network for the pins is based on local clamp protection. The embedded pulsed power network combines local clamps optimized to the system's current levels with appropriate electrical characteristics and metallization routing capable of withstanding these currents. In such design, the ESD clamps are treated simply as analog circuit building blocks released in the form of ESD libraries in PDK. System-level design complexity is determined by a greater level of inter-action of the ESD current path with the internal circuit blocks.

When the holding voltage is not required to be above power supply levels, the most effective local ESD device is an SCR. The operating principle of on-chip system-level MOS-based SCRs is based on a three-stage domino effect that consists of initial channel formation, avalanche multiplication of the injection of displacement currents, and avalanche-injection current instability in the primary embedded BJT structure to a certain current level, followed by the turn-on of the secondary complementary BJT after the high current provides internal emitter junction biasing. The positive feedback between the embedded BJT structures results in an SCR-type double injection conductivity modulation. In this mode, the impact ionization level at the blocking junction is suppressed and replaced by hole injection from the pEmitter.

There are several known feasible low voltage SCR-based solutions which overcome the problem of low holding voltage. The contrasting design of HV SCR clamps with increased holding voltages presents a bigger challenge. It can be accomplished in the few ways by limiting emitter injection with excessive isolation, a poly-ballasting voltage drop or the stacked clamps approach. When the high holding voltage requirements are critical, the alternatives to SCR devices include lateral avalanche diodes and lateral PNP ESD devices.

Among the critical design targets is reduction of sensitivity to mask misalignment in volatile device regions. A basic derived design approach for on-chip system-level ESD devices is applying certain transformation steps to modify standard devices already supported in the given process technology. The transformation simultaneously combines optimal conductivity modulation conditions and device-level negative feedback in the ESD device, suppressing uncontrollable spatial current instabilities and filamentation below the critical current density, which leads to irreversible phenomena. Negative feedback during the high current stage in ESD structures can be induced in several ways, including suppressing the impact ionization rate by substituting the avalanche hole current component with junction injection, embedding ballasting regions and using the substrate effect.

A substantial experimental verification and validation of the final solutions on the test chip is required for successful ESD development. For effective experimentation, a physical design requires in-depth TCAD-based mixed-mode analysis of the ESD devices, clamp circuits and analog peripheral circuit blocks. The most realistic way to setup and run the numerical experiments is through parameterized TCAD analysis using the DECIMM^{TM} tool [19].

Successful system-level ESD clamp implementation combines a broad range of aspects: device width scaling; lateral isolation and vertical isolation from the substrate; latch-up isolation by guard rings; clamp control; driver and enable circuits for voltage reference and dynamic coupling; and proper use of the multistage protection approach.

The electrical parameters of a clamp are the current and voltage for triggering into a high-current state, the minimum and high-current holding voltage, the transient response (turn-on time, dynamic voltage waveform, minimum turn-off current, and recovery time), DC voltage tolerance in long-term reliability operation conditions and leakage, breakdown voltage, parasitic capacitance, reverse current clamping, and some other parameters more specific to a particular application. The major layout parameters are dimensions of the clamp footprint as well as required mask count for active and metallization layers.

In addition to active high-current ESD devices, the clamp generally requires components that target these specified parameters. The most typical clamp components include:

(i)   Triggering circuit for reference voltage or dynamic coupling, which targets the turn-on inside the ESD protection window voltage range;
(ii)   Holding voltage control sub-circuit components;
(iii) High-current ballasting, current crowding prevention, measures for turn-on of all fingers
(iv)   Reverse current path diode
(v)   Optional driver to shutdown or enable functions realized depending on the internal circuit state.

The conventional layout of active ESD devices in the clamps is a multifinger distributed array. The cell-level current balance that ensures all fingers turn on in ESD pulse conditions is challenging to obtain in HV devices and requires taking into account the 3-D design of the array finger, the array metallization scheme, and the control circuit connection. From this perspective, when system-level width scaling is required, all fingers of the ESD device should be scaled within the same shared epi space rather than the instantiation of the several identical clamps. Not following this principle can create a passing current level issue due to multifinger turn-on, or can cause a very non-linear width scaling and including miscorrelation between contact and air gap standard IEC pulses.

In wafer-level packaging (with micro-SMD surface mounting devices) design, chip bonding bumps can be scattered on top of the whole active layout area. At a high current from a system-level ESD event, the injection from an analog circuit clamp area can disturb the operation of many more connected active devices. A "sneak" latch-up current path can form deep inside the layout of the internal circuit components. Therefore, the clamp latch-up isolation is an important part of the design. These latch-up phenomena are discussed in the next chapter as a logical continuation of the above material.

As it has been demonstrated in the last section, the problem of confidence of the ESD protection window can be quite effectively addressed using the new simulation capability of the unique simulation tool DECIMM^{TM} [19]. A significant and quite important understanding of the impact on the process capability indexes, related yield and reliability of both ESD devices and clamp designs, can be now understood. It can be further verified by an experimental effort directed by simulation results. Thus a methodology for a parametric mixed mode numerical analysis with Monte-Carlo generated input parameters is bridging the gap between the ESD device design and the best practices of semiconductor process technology integration. This new approach is expected to be widely used to estimate and compare *Cpk* parameters of the different process development tools.

Chapter 4
Latch-up at System-Level Stress

Integration of validated stand-alone ESD clamps on chip for system level requirements is not a simple problem. Application specifics and chip functionality need to be thoroughly taken into account to avoid clamp interaction with internal circuit blocks during both system-level ESD stress and normal operation. In high injection conditions induced by system-level ESD current, parasitic devices capable of supporting the conductivity modulation regime may also turn on. From this perspective, a system-level ESD event in power-on conditions can conceptually be treated as similar to latch-up phenomena.

Indeed, in spite of a shorter than standard latch-up [93] time domain, the current through the on-chip ESD clamp in case of system level stress is two orders of magnitude higher. Apparently, this level of injection current may significantly interfere with internal circuit blocks in power-on conditions. Subsequent turn-on of the parasitic structures formed in the layout and upset state of digital, analog, or memory blocks results in an undedicated ESD current path. Parasitic vertical or lateral SCR and NPN structures can be formed in the IC layout and triggered by the injected carriers, especially in case of high voltage circuits. Numerous system-level case studies found a peripheral current conduction physically similar to latch-up events, in spite of their irreversible nature and different time domain.

The following scenario may lead to either reversible or irreversible events. Often, CMOS Core or I/O latch-up is reversible, due to limited current passing through the parasitic SCR and a rather low holding voltage. On the contrary, HV latch-up of a parasitic n-p-n during a BCD process typically results in irreversible burnout. It is rather difficult to identify the burnout as a result of a high detection current during the injection pulse at latch-up test or as a result of the formed permanent latch-up state after the injection pulse. Even a conventional CMOS latch-up structure may not withstand the latch-up operation mode simply because of the weak interconnect design. In this chapter, we will treat latch-up as a physical event of permanent change to the current state (at constant power supply conditions) as a result of short-term current injection.

It is critical that on-chip design takes latch-up into account, since changes to pass the latch-up or system-level ESD tests will inevitably involve a significant

V. Vashchenko and M. Scholz, *System Level ESD Protection*,
DOI: 10.1007/978-3-319-03221-4_4, © Springer International Publishing Switzerland 2014

redesign of the chip layout, in order to substantially increase the space on the chip needed to isolate the parasitic current path.

Thus, understanding major latch-up phenomena in ICs is directly related to understanding the phenomena at system-level stress. It is an important step in dealing with on-chip system-level design. Though these phenomena were described in literature rather broadly [94], physical understanding of the subject is still challenging for many engineers. Due to the direct importance of latch-up to the subject of this book, this chapter not only compiles material directly applicable to system-level design, but also attempts to systematically explain the phenomenon. To achieve this goal, a logical presentation of the physical latch-up phenomenon is done below, using mixed-mode numerical simulation with DECIMM [19, 20].

In the following sections of this chapter, we will differentiate between three major latch-up scenarios that represent physically different phenomena.

Under *Conventional CMOS Latch-up* (Sect. 4.1), a high-current turn-on condition of a parasitic SCR is formed by a PMOS-NMOS inverter pair (or any high-side and low-side connected n-p region pair), as a result of current injection. This phenomenon is represented by two subcases: I/O buffer with internal injection and a core circuit with injection from a remote injector.

Under *HV Nepi-to-Nepi Latch-up* (Sect. 4.2), the turn-on of the parasitic n-p-n structure in a high current state is a result of current injection in one of the pockets. In this case, a nEpi-pSub-nEpi structure is formed in the layout by two n-Epi pockets separated by a P-substrate lateral isolation ring. Both electron injection from a low-side pocket and hole injection from high-side pockets are taken into account.

Finally, the *Transient Latch-up* (Sects. 4.3–4.5), combines a class of physical phenomena where the clamp turns on as a result of short-term voltage overstress. Here, the snapback state essentially represents the latch-up state.

4.1 Conventional I/O and Core Latch-up

4.1.1 Latch-up Simulation Structures

Conventional latch-up physically represents an event of permanent turn-on of a parasitic p-n-p-n structure formed in the layout as a result of a temporary injection event. The permanent turn-on is supported by the power supply with some load resistance providing a current in double-injection conductivity modulation conditions, until power-down or reset is done. While the high-current state of the SCR is the same snapback state as discussed in Chap. 3, the turn-on scenario is different. The turn-on into latch-up mode is the result of carrier injection in the SCR region, rather than the 3-stage phenomena described for the ESD pulse case. The carrier injection for turning on the SCR can be generated both by the internal diode junction in the SCR region, and the external diode spaced apart from the SCR region.

In general, a parasitic SCR p-n-p-n structure is always present in the layout. It can be formed by a close proximity of any n- and p-contact diffusion regions isolated by an nWell at a high-side potential and a low-side-potential-connected pair of n- and p-contact diffusion regions in the pWell. The latch-up capability of this SCR structure depends on the emitter isolation, which determines the minimum holding voltage and minimum holding current in on-state.

In CMOS processes, the most conventional way to form a latch-up-capable structure is with an improperly isolated pair of complementary NMOS and PMOS structures placed in close proximity in the layout. Depending on the layout of the emitter and base regions, the resultant parasitic SCR elementary structure can have a rather low holding voltage, below the power supply VDD level. The SCR pEmitter, nBase, nEmitter and pBase are formed by the PMOS source and nBody and NMOS source and pBody regions, respectively. For the I/O buffer, analog output, or open drain, the remaining PMOS and NMOS drain diffusion regions are connected to the pads. In this case, the injecting diode structures are already formed inside the SCR structure by the corresponding drain-body diodes. When the SCR is formed in an internal digital or analog circuit block, its turn-on can be initiated by injection from the I/O or from the analog circuit block connected to the pad.

For a CMOS process or a CMOS module in a BCD process, despite variations, the *conventional* CMOS latch-up boils down to two main cases—the *I/O* latch-up and the *Core* latch-up.

In the case of the I/O latch-up, one or both of the drains of the PMOS-NMOS pair are connected to the pad. If the pad connected to the PMOS drain is pulled above the power supply VDD level, the forward-biased body diode injects holes directly into the SCR nBase region (Fig. 4.1a). If the pad connected to the NMOS drain region is pulled down below the VSS level, electrons are injected into the SCR pBase due to a forward-biased NMOS body diode (Fig. 4.1b). Since the injection occurs directly inside the parasitic structure, the only effective measure against I/O latch-up is to disable the SCR structure itself. This is usually done by implementing a proper sequence of NMOS and PMOS regions, forming the SCR with a pEmitter-nBase-pBase-nEmitter. In terms of layout design, this means that all I/O-buffer NMOS devices are placed in pWells fully enclosed by pBody diffusion guard rings. At the same time, all of the PMOS devices are placed in nWell(s) that are fully enclosed by nBody diffusion rings. The rings represent the SCR base regions and increasing their length reduces the gains of the parasitic n-p-n and p-n-p responsible for positive feedback in SCR during double injection conductivity modulation. Increasing the ring length both separates the source of injection and increases the holding voltage of the SCR.

During *Core* latch-up, there is no direct connection between the core digital block and the IC pin. For higher logic density, the NMOS pWell and PMOS nWell regions have random body diffusion placement that does not form guard rings. Moreover, the pBody and nBody contact regions are placed within the process design rule margins, which are typically less than 10–20 μm. Thus, core parasitic SCR is formed by the sequence of NMOS and PMOS regions as nBase-pEmitter-nEmitter-pBase, with the biggest emitter-base spacing in the worst case scenario.

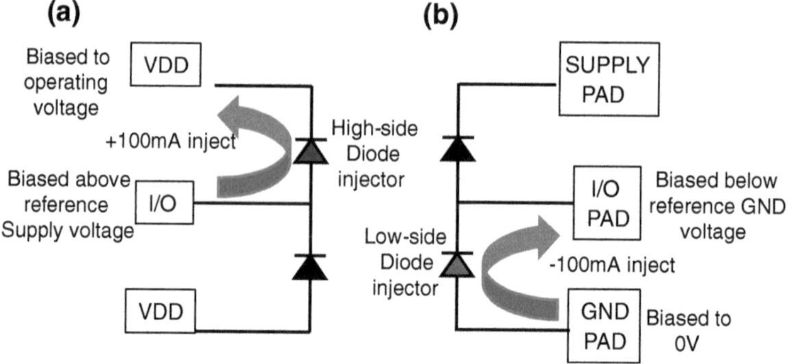

Fig. 4.1 Illustration of the test conditions for the high (**a**) and low (**b**) side injection induced latch-up in I/O buffer

This SCR structure has the lowest holding voltage and requires minimal carrier injection to turn on. In core latch-up, injection carriers are generated at a certain distance from the digital core block in I/O or other analog power device directly connected to the pin and can conduct injection current.

With respect to physical design, there are two different problem statements for both experimental and simulated latch-up rules development. For I/O buffer, the problem lies in determining a length of the p- and n-guard rings that is sufficient to prevent I/O buffer latch-up state formation. This is accomplished with proper placement of the buffer devices in the well regions fully enclosed by diffusion rings (with some exception for the chip periphery).

In the case of core latch-up, the problem is related to finding the minimum space between the I/O injecting diode (or an analog injector) and the core digital (or analog) block that will guarantee absence of latch-up in the core block for a given injection current and initial temperature. It is assumed that the I/O buffer (or power device injector in an analog circuit) is already fully surrounded by the proper guard rings that follow the latch-up design rules for the I/O buffer.

The latch-up test is typically done according to the EIA/JEDEC (EIA/JESD78) Standard [93] for a current of 100 or 200 mA at room temperature or up to 125 °C.

The physical phenomena during latch-up are very easy to understand using a mixed-mode simulation analysis that can represent both external states of the circuit and the internal state inside the semiconductor structure.

The FEM device cross-section (Fig. 4.2a) used to simulate I/O buffer latch-up is represented by PMOS and NMOS structures combined in the same cross-section. In the simulation, varied parameters are the length of the nBody region *Lnbase* and the length of the pBody region *Lpbase* (Fig. 4.2a). A mixed-mode analysis for latch-up can be done in circuits with high-side (Fig. 4.2b) and low-side (Fig. 4.2c) injection. In the circuits (Fig. 4.2b, c), a constant 3.3 V VDD pulse is applied first, followed by a short-term injection current from the current source attached to the common drain connection node in the inverter (Fig. 4.3). During this pulse, some

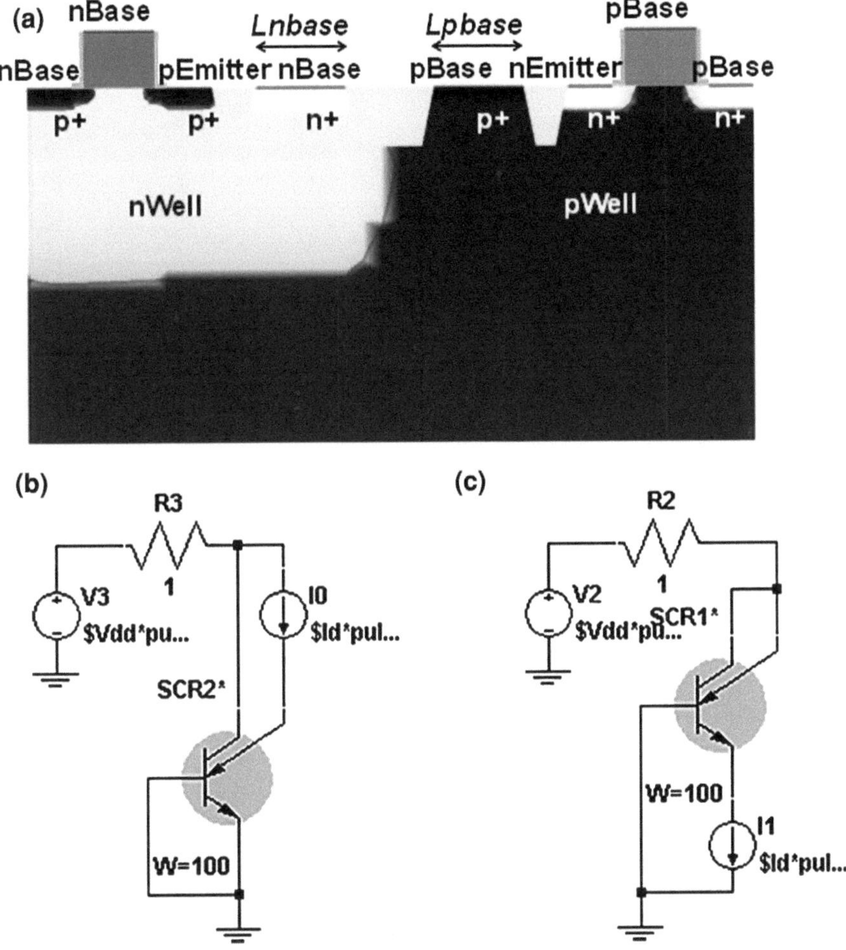

Fig. 4.2 I/O buffer latch-up simulation that includes a full complementary MOS pair (**a**) and a mixed-mode simulation circuit for high-side (**b**) and low-side injection (**c**)

additional detection current from power supply is observed. However, the presence of this current does not necessarily mean that latch-up is imminent.

The informative part of this mixed mode analysis is the change in the current *ISCR* through the SCR when the injection current pulse amplitude is reduced to zero. For no latch-up, the detection current returns to the initial level before injection (Fig. 4.3a). With latch-up, the VDD power supply current supports a new permanently high-current state for the SCR (Fig. 4.3b).

The simulation can be used to generate tabulated data that shows which ring spacing's pass or fail the latch-up test for given injection current level, structure parameters, and temperature (Table 4.1). However, a more advanced method of

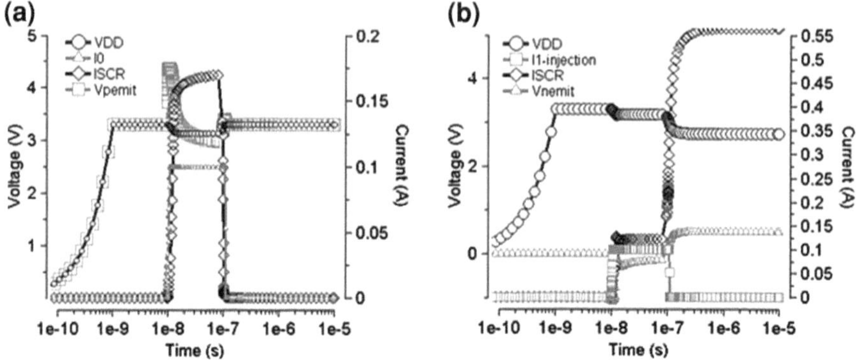

Fig. 4.3 Output of the mixed-mode simulation for the no latch-up high-side injection case (**a**) and for latch-up in the low-side injection case (**b**), where *VDD* is the voltage at the Nbase, *I0* is the high-side injection current, *I1* is the low-side injection current; *Vpemit* is the voltage at the p-emitter, *Vnemit* is the voltage of the n-emitter, and *ISCR* is the power supply current through the entire SCR structure

Table 4.1 Latch-up characteristics for high-side and low-side injection for the I/O structure with 100 μm contact width under 100 mA injection conditions with full CMOS devices and double guard rings structure with n+ ring in nWell and p+ ring in pWell

T(K)	LAAN (um)	LAAP (um)	LANAP (um)	LNbase (um) (N-ring)	LNpase (um) (P-ring)	High side injection	Low side injection
450	0.25	0.25	0.5	0.25	0.25	**Latch-up**	**Latch-up**
450	0.25	0.25	0.25	**2**	**2**	**Latch-up**	Pass
450	0.25	0.25	0.25	**3**	**3**	Pass	Pass
450	**5**	**5**	0.5	0.25	0.25	**Latch-up**	**Latch-up**
450	0.25	0.25	**10**	0.25	0.25	Pass	Pass
300	0.25	0.25	0.5	0.25	0.25	Pass	Pass
300	0.25	0.25	0.25	**2**	**2**	Pass	Pass
300	**5**	**5**	0.5	0.25	0.25	**Latch-up**	**Latch-up**
300	0.25	0.25	**10**	0.25	0.25	Pass	Pass

studying latch-up by means of mixed-mode simulation with DECIMM tool [19] aims at determining the critical latch-up current for each set device parameter. Especially for system-level on-chip design this type of output is expected to be more informative and is more close to the typical JEDEC standard test results.

The automated latch-up simulation mixed-mode circuit can be created with the Latch-up Tester (LUT) circuit element specific to the parameterized DECIMM tool [19, 20]. The LUT combines the power supply voltage and latch-up injection current source in the same circuit block. Then, a series of automated transient simulations with self-adjusting injection current can be performed to rapidly find the critical injection current for latch-up (Fig. 4.4). A sufficient number of current pulses with varying amplitudes of the LUT current source are simulated, tracing the DUT LU curve that satisfies the given conditions.

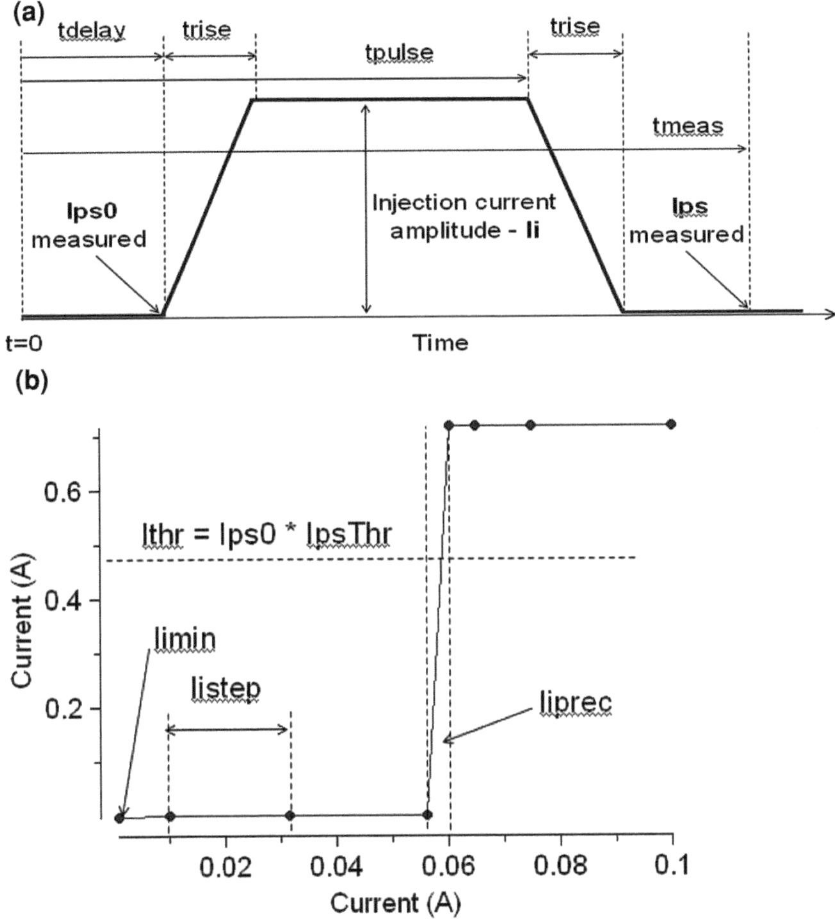

Fig. 4.4 User-accessible LUT time domain parameters (**a**) and automatic curve tracing algorithm on a plot for the power supply current dependence on the injection current (**b**)

The results of numerical simulations for mixed-mode circuits containing FEM devices (Fig. 4.2a) and with an included LUT source (Fig. 4.5a, b) allow us to plot the dependence of the critical injection current upon guard ring length parameters, which, for simplicity, were kept equal (*Lnbase* = *Lpbase*) (Fig. 4.5c). From the system-level design perspective, this analysis is significant in estimating the critical conditions during high current system-level stress. For an injecting current of 100 mA, this dependence shows that the safe guard ring length necessary to avoid latch-up for the profiles in this numerical simulation example is approximately 5 um. This matches well with to the experimental data for a 0.13 μm CMOS process.

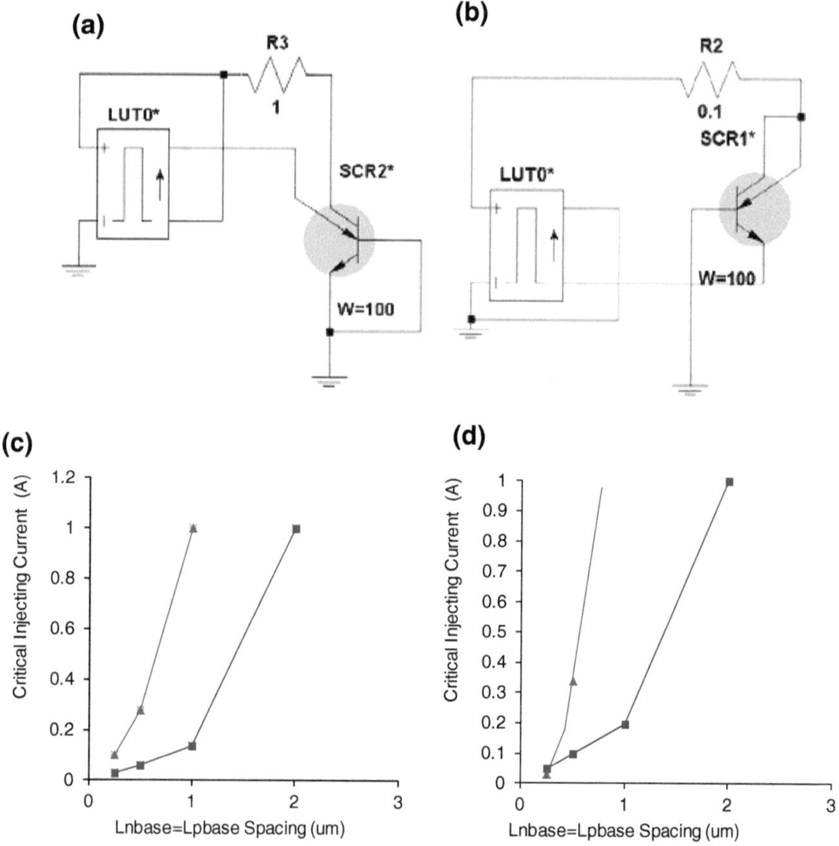

Fig. 4.5 I/O latch-up mixed-mode simulation circuit with an FEM structure (Fig. 4.2a) for cases of high-side (**a**) and low-side injection (**b**) and plotted critical current for latch-up as a function of *Lnbase* = *Lpbase* parameter (**c**, **d**), respectively, for room and high temperature

According to the explanation above, the numerical experiment physically representing core latch-up is different from the one for I/O latch-up. In this case, the inverter with the I/O injecting diode can be represented by the cross-section where the inverter acts as a victim (Fig. 4.6a) and included in the mixed circuit (Fig. 4.6b) as multi-terminal circuit element "CORE". The terminals of the CMOS inverter include power supply "VDD," ground "VSS," pBody contact "PB," and nBody contact "NB". The terminals for the injecting diode include the anode "AN" and the cathode "CA".

To overcome the technical difficulty of physically exemplifying the worst-case-scenario inverter (with the largest possible body-to-source spacing), the separate body terminals NB and PB are connected with additional resistors to VDD and

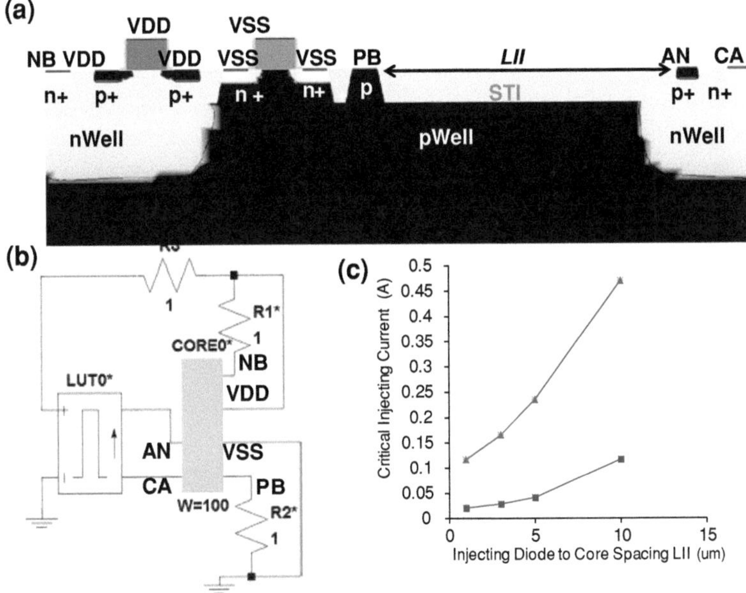

Fig. 4.6 FEM structure for core latch-up (**a**), mixed-mode simulation circuit with LUT (**b**), typical transient characteristics for latch-up, and calculated critical current for latch-up as a function of injector-inverter spacing

VSS rails, respectively. The resistors physically represent the accumulated body resistance formed in the real layout.

The methodology of a numerical simulation of the core latch-up is similar to the I/O simulation methods discussed above. A dependence of the critical current on injector-to-inverter spacing LII (Fig. 4.6c) demonstrates the expected results, predicting a safe spacing of ~ 20 μm between the injecting diode and sensitive core logic block for avoiding latch-up with the chosen simulation parameters. This is in line with experimental data for a 0.13 μm CMOS process.

This new mixed-mode simulation methodology with automated latch-up circuit element demonstrated for the simple CMOS process can be applied to a more complex case. If the CMOS inverter is isolated by a deep nWell or nEpi pocket, latch-up caused by low-side electron injection can occur while this core circuit is at a high potential. Here, the injecting diode has an anode terminal at the ground potential while the cathode terminal is pulled below it, with the substrate injecting holes. Similarly, if the injecting diode cathode is at a high potential with the anode pulled above it, high-side hole injection will be realized.

To link these results to the SCR, the simulated curve trace results for full I-V characteristics are presented in Fig. 4.7. In the high current state the carrier distribution and the current flow lines are the same to the corresponding SCR structure is the same final regime, but obtained by the over-voltage triggering (Fig. 4.7b, d).

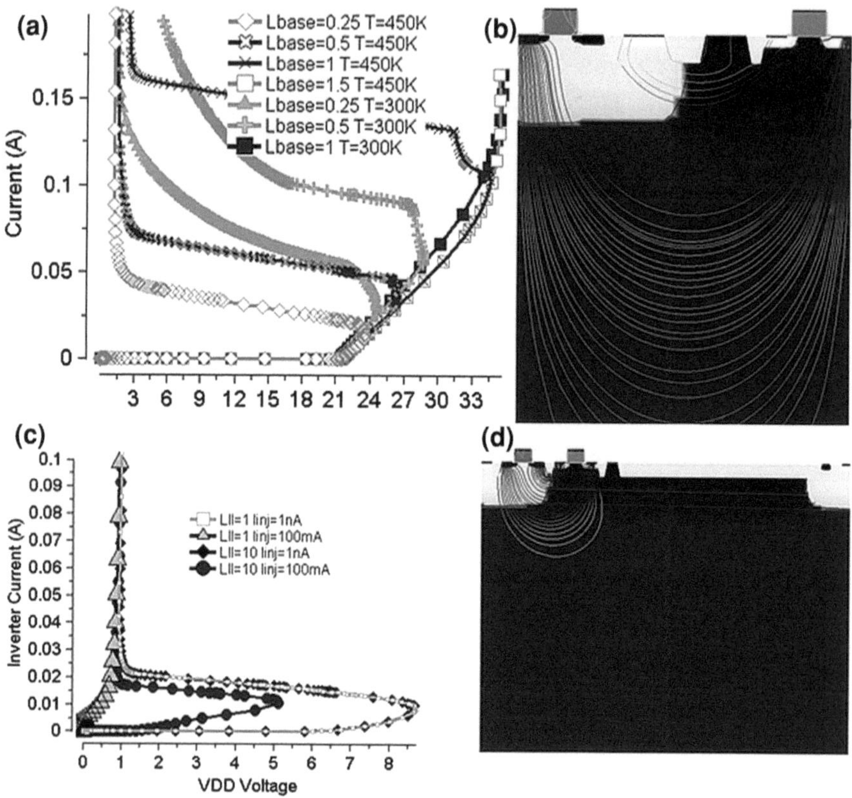

Fig. 4.7 Simulated slow curve trace I-V characteristics and current flow lines in high current state with low holding voltage for the I/O buffer latch-up structures (**a, b**) and core structure (**c, d**), respectively

4.2 High Voltage Latch-up

In the previous section, conventional LV latch-up was presented as the result of a parasitic p-n-p-n SCR structure turn on into the high-current state. The SCR structure is formed unintentionally in the chip layout due to integration of n-type and p-type devices. The turn-on is initiated by an injection current from either the internal forward-biased junction of the SCR itself, or from an external component subjected either to the latch-up test current or a system-level IEC ESD test. In principle the overvoltage results in a similar SCR turn-on effect. This will be discussed in the transient latch-up section below separately.

The physical phenomenon that provides positive feedback for bi-stability under conventional latch-up is the double injection conductivity modulation. An alternative structure that supports a strong positive feedback leading to bi-stable characteristics is a parasitic n-p-n. Unlike in the SCR, positive feedback in the n-p-n is realized at a sufficiently high voltage due to the presence of the high electric field required to support impact ionization at avalanche injection conductivity modulation.

Experimental results show that, similarly to conventional latch-up, the parasitic n-p-n structure can be turned on by either electron or hole injection. The injection is generated by the test current in standard latch-up conditions, or by stress current in the case of a system-level ESD event at power-on. Both stresses can lead to physical latch-up effects despite different time domains. Since the n-p-n is originally at a high voltage, latch-up is realized at relatively high collector-emitter voltages. Therefore, the phenomenon is typically classified as HV latch-up to differentiate it from the conventional LV CMOS latch-up.

In a BCD process, the parasitic n-p-n can be formed by two nEpi regions in the chip layout, with active devices that have n-contact regions connected to separate pads that are subjected to the stress. At a sufficiently close proximity of these nEpi regions, where at least one of the pads is at high voltage, latch-up can occur as a result of temporary injection current. In particular, latch-up can be realized when a grounded nEpi region (pocket) is placed near a nEpi region connected to a HV pad. In the latch-up stress condition, the higher voltage pad represents an nCollector of the parasitic lateral HV n-p-n structure, while the low-side nEpi forms the nEmitter. The base is represented by pSubstrate and p-ring isolation regions between the epi regions. The latch-up stress can be initiated in conditions with no injection current by avalanche breakdown, caused by applying a rather high collector-emitter voltage.

To avoid latch-up, certain prevention layout rules must be developed, experimentally verified, and then followed in the IC design in semiconductor HV processes. These rules are usually determined semi-empirically from limited experimental results. An example of such rules is outlined at the end of this section, after physical phenomena associated with HV latch-up are explained. The latch-up rules primarily focus on preventing HV parasitic n-p-n turn-on by either disabling the n-p-n structure itself or by reducing injection from the junctions in nEpi pockets. For nEpi pockets with high-side p-regions, when the n-p-n structure turns on, the final latch-up scenario may involve a parasitic SCR current path formation that is very similar to conventional LV latch-up.

Thus, the purpose of this section is to provide a condensed overview of the major aspects of such a complex subject as HV latch-up. To achieve this goal, we present mixed-mode parameterized numerical analysis in examples using DECI-MM^{TM} tool [19]. Overall, this material is equally applicable to the BCD and extended CMOS HV processes with deep Nwell isolation.

4.2.1 Nepi-Nepi Latch-up

HV BCD process simulation methodology for latch-up analysis is somewhat similar to the conventional latch-up methods and was described in the previous section. Both high-side (HS) hole injection and low-side (LS) electron injection can be analyzed from the collector-emitter characteristics of the physical FEM latch-up device in conditions of constant injection current. Alternatively, a more accurate transient analysis can be performed with a constant collector-emitter voltage and an injection pulse.

The first method provides information about the dependence of the critical voltage on the injection current level. However, these data may not be easy to extrapolate toward expected latch-up test results. The collector-emitter curve trace at a constant injection current provides only a rough understanding of the critical HV latch-up conditions. It can be projected further by comparing the critical voltage at the beginning of the S-shape I-V characteristics for different injection current and device region parameters. In this case, we essentially analyze the pulsed SOA for the n-p-n at a constant injection current into the base.

The second approach is more accurate. It shows the injection pulse end current state as a function of the injected current and structure parameters. Thus, a deep insight into the associated physical phenomena can be achieved by running mixed-mode numerical experiments with parameterized FEM devices. The same two approaches can be applied to experimental analysis of latch-up in the test structures.

However, both the simulation and experimental analysis of the HV latch-up has a peculiarity related to the self-heating effect. Due to rather high voltage drop and current levels, self-heating cannot simply be neglected as in the case of conventional LV latch-up. The electro-thermal conductivity modulation and current instability effects must be taken into account in addition to electrical current instabilities.

From a design perspective, isolation of the circuits into adjacent nEpi pockets is another angle to consider. Since separation of the pockets involves space on the chip, the problem can be formulated as finding an optimal use of the isolation space between two pockets. In principle, this space can be used for secondary guard rings in the pockets (increasing their size), a pocket Epi extension for blocking Nwell and NBL diffusions, an active guard ring to improve resilience to low-side injection [95], or a combination of the above.

For the complexity of the simulation latch-up analysis needed to resolve the above issues, process simulation with the traditional TCAD approach is impractical. The complexity and volume of simulations become prohibitively large and require lengthy simulation time to make any progress in most practical cases. Complexity occurs mostly due to the large total length of 40–150 μm and up to 100 μm depth of the FEM devices, as well as the need for significant variation in the device structure parameters.

To overcome this challenge, new simulation methodology (Chap. 3) with the parameterized tool DECIMM [19] was successfully applied, using full parameterization of the device geometry and doping profiles in the context of full-featured device-circuit mixed-mode analysis under a broad range of automation capabilities. The outcome of this analysis is used below directly to explain the major regularities of the HV latch-up events. It includes both the thermo-electrical conditions of the standard latch-up test and the adiabatic conditions of the system-level ESD stress.

Real cases of nEpi-nEpi structures with all regions internal to the pockets are rather complex and therefore difficult to use for explanation of basic HV latch-up regularities. Instead, a more generic latch-up case based on a simplified structure is presented in this section first to reveal the main physical effects underlying the complex HV latch-up problem. HV latch-up in a more realistic and complex device structure with all the guard ring regions is compared in the next section to the experimental data for limited structure parameter options variation.

The simplified physical structure combines two isolated HV diodes separated by a grounded pRing (Fig. 4.8a). The diode cathode simultaneously isolates the anode region from the pSubstrate. Vertical isolation is achieved by the nEpi, nWell and n-Buried layer (NBL). Lateral isolation for the 100 V voltage tolerance is provided by a graded blocking junction in the form of an nWell and Epi extension over the NBL to the pWell and P-buried layer (PBL). Thus, this symmetrical structure has only 5 electrodes, defined as the HS diode anode "AH" and cathode "CH", LS diode anode "AL" and cathode "CL," as well as the pSubstrate contact "GND" (Fig. 4.8a).

If the anode of the HS diode is connected to the cathode of the LS diode to form an I/O diode pair, then the two scenarios for carrier injection during a standard JEDEC latch-up test can be seen in the diagrams in Fig. 4.8b, similar to the LV latch-up I/O case. More generic electrode conditions for HS and LS injection in the structure (Fig. 4.8a) are compiled in Table 4.2.

With LS electron injection, the cathode CL of the low-side diode is biased below the substrate potential, causing flow of holes and electrons in the nEpi-pSubstrate junction of the parasitic n-p-n. In the electric field formed by the applied high voltage *Vpower*, the injected electrons drift through the lightly doped substrate toward the HS nEpi.

Intuitively, it can be assumed that in the case of LS injection, the operation of induced latch-up in the parasitic n-p-n is similar to the common emitter bipolar circuit and the characteristics of latch-up are directly related to the characteristics of the BJT structure. Change in the long base region structure, collector-base junction avalanche multiplication properties, and emitter-base junction injection capabilities can impact the critical latch-up conditions. Although some of these assumptions are realistic, operation of the structure in LS injection latch-up conditions is not so simple, as it will be demonstrated below.

The HS hole injection scenario is realized when the HS diode anode "AH" is biased above the HV power supply level *Vpower*. Unlike under LS injection, in this case the carriers are injected inside the HS nEpi region. In these conditions, only a fraction of holes penetrates outside the pocket region and drifts in the electric field toward the LS pocket, further impacting the critical conditions for avalanche injection conductivity modulation in the parasitic n-p-n structure. Apparently, the injected holes are best collected when the entire collector-emitter space is occupied by the pSubstrate ring. Limiting the level of injected carriers from the pocket by rings inside the pocket is also found to be effective.

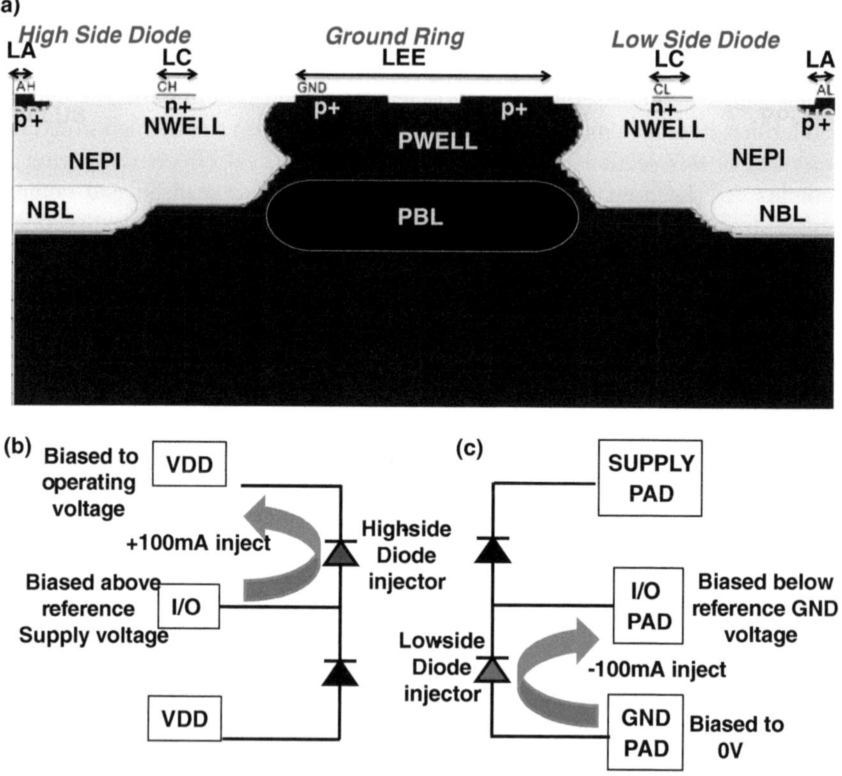

Fig. 4.8 Simplified FEM structure for HV latch-up analysis (**a**) and schematic representation of the conditions for the high- (**b**) and low-side (**c**) latch-up scenario tests

Table 4.2 Electrode conditions of the simplified FEM latch-up structure (Fig. 4.8a) for of HS hole injection and LS electron injection

Injection scenario	HS cathode	HS anode	LS anode	LS cathode	Substrate
High side	*Vpower*	Above *Vpower* in forward current mode	Grounded or floating	Grounded	Grounded
Low side	*Vpower*	*Vpower* or floating	Floating or shorted to LS Cathode	Below ground in forward current mode	Grounded

To analyze both LS and HS induced latch-up cases, the parameterized FEM device (Fig. 4.8a) with different nEpi-to-nEpi region parameters *LEE* was studied using mixed-mode simulation. A circuit for HS injection induced latch-up that recreates the corresponding conditions in Table 4.2 is presented in Fig. 4.9a. In addition to variation of the *LEE* parameter, the structure was simulated for

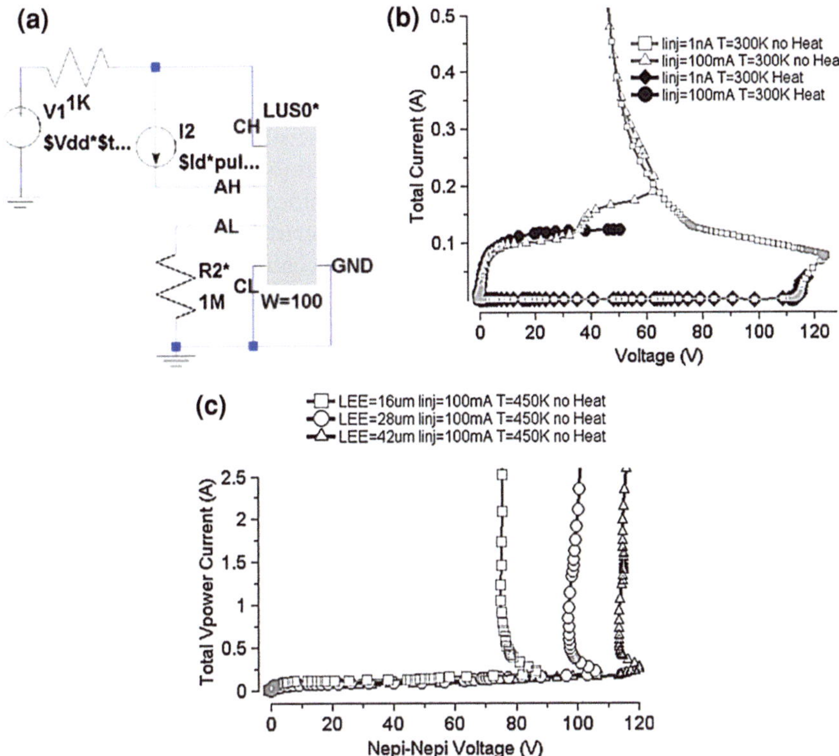

Fig. 4.9 Mixed-mode simulation circuit for curve trace with a constant injection current with HS injection induced latch-up (**a**); simulated I-V characteristics for various anode injecting current levels at LEE = 16 um with and without self-heating effect taken into account (**b**) as well as for different pocket separation LEE parameters (**c**) for isothermal conditions with constant T = 450 K

isothermal and thermal-coupled conditions at constant and initial temperatures, T = 300 and 450 K, respectively. Here, the elevated temperature was chosen 50 K above the typical latch-up requirements ($\sim$400 K) to have a better resolution of the simulated physical effects. The mixed-mode circuit reproduced the slow curve trace (CT) results at a constant injection current level.

Thus, the main conclusions from this HS curve trace simulation analysis for isothermal conditions are in line with basic understanding [5] of avalanche injection in the n-p-n structure. The isothermal critical voltage of the device is a falling function of the injection current (Fig. 4.9b), and increasing the base length results in both a triggering and holding voltage increase (Fig. 4.9c).

However, when thermal-coupled analysis is enabled in the millisecond time domain, the critical temperature of the structure is reached at a much lower voltage than the critical voltages for isothermal current instability (Fig. 4.9b). Therefore, it is logical to expect that the latch-up regime will be physically limited by the thermal runaway, achieving the critical temperature and corresponding local burnout conditions in the device. Moreover, these conditions will be realized not after the stimulating injection current is removed, but during the injection pulse, as a result of the rather high detection current in the nEpi-to-nEpi circuit.

Indeed, the JEDEC latch-up test is defined in the standard [93] in an arbitrary broad range of pulse rise times and widths, from 5 μs to 5 ms. In this time domain, self-heating of the entire die volume becomes significant to be neglected. Therefore, the device is mainly limited by the thermo-electrical effects in the structure, rather than an electrical current instability. On the contrary, the critical conditions during system-level ESD stress are reached in the 100 ns time domain and are therefore close to the electrical physical limits implied by the isothermal current instability. Latch-up phenomena in both electrical and thermo-electrical conditions that include self-heating are analyzed below using parameterized mixed-mode simulation.

In the isothermal critical regime, the dependence of the holding voltage on the pBase length *LEE* (Fig. 4.9c) is in accordance with avalanche injection of the n-p-n device. However, with HS injection the anode electrode AH is above the *Vpower*, and the structure physically represents an SCR with the pEmitter current limited by the latch-up test injection current. This explains the relatively low holding voltage, despite high current conditions and low *LEE* spacings (Fig. 4.9b).

Similarly to HS injection induced latch-up, LS injection induced latch-up can be analyzed using a corresponding mixed-mode simulation circuit (Fig. 4.10a) with the same 5-terminal simplified FEM structure (Fig. 4.8a) for physical latch-up representation. The LS injection induced latch-up phenomena are already rather complex, even in the isothermal case, due to the non-inform regions of the parasitic n-p-n structure.

At no injection current, the I-V characteristics of the FEM structure are similar to those for LS injection (Fig. 4.10b). The critical voltage dependence on the injection current (Fig. 4.10d) is also in line with the general understanding of avalanche injection conductivity modulation. With applied injection current, we observe rather non-linear I-V characteristics (Fig. 4.10b). The non-linearity of the I-V characteristics is due to two different effects.

When the HS anode "AH" (Fig. 4.8a) is not connected, avalanche injection conductivity modulation on the n-p-n structure is realized in two stages. Since the n-p-n structure has a nEpi extension, current-limited S-shape characteristics are observed at first. Here, avalanche multiplication region conditions are formed at the rather diluted nEpi-Psub junction (Fig. 4.10e). Then, after current saturation, the nEpi region is modulated at a voltage of ∼90 V until the electric field maximum is formed at the diffusion junction at some critical current. This junction is capable of providing a much higher current density with a critical regime for negative differential resistance at ∼130 V and 100 mA (Fig. 4.10b).

If the HS anode electrode "AH" (Fig. 4.8a) is connected, it acts as pEmitter of the SCR. The injection current is extracted for the pEmitter at a much lower voltage and thus significantly reduces the threshold voltage for conductivity modulation. This effect is somewhat similar to HS injection and can be understood

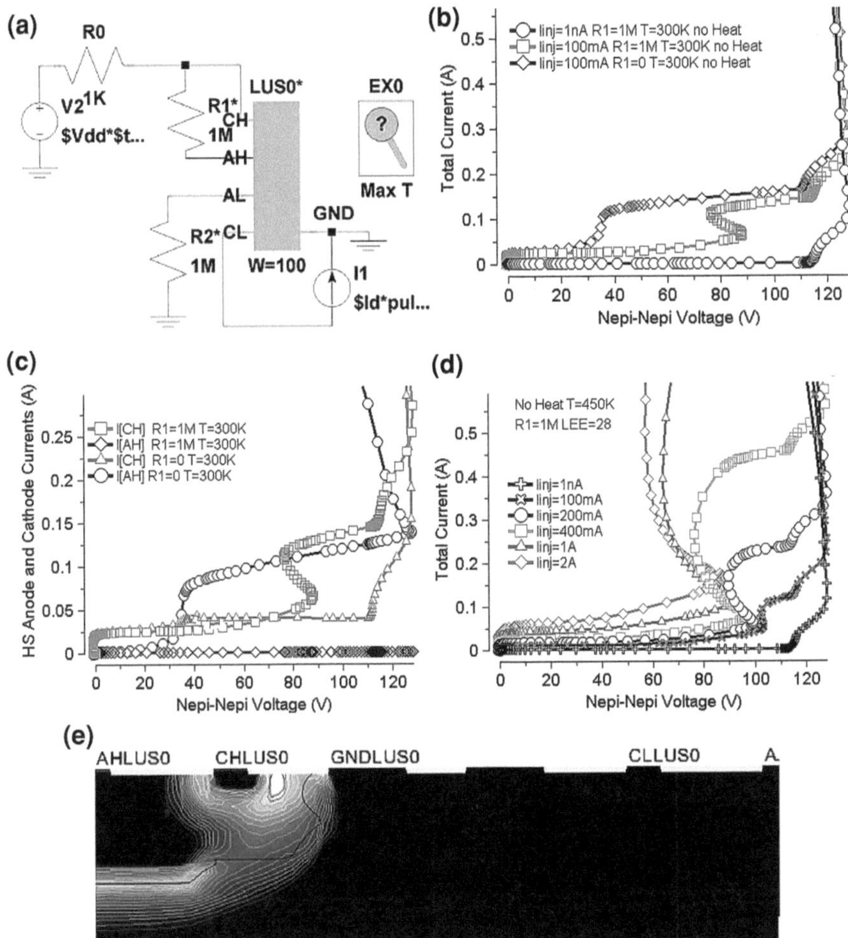

Fig. 4.10 Isothermal mixed mode simulation for LS induced latch-up: circuit for slow curve trace (**a**); simulated I-V characteristics for $LEE = 16$ um and different injection current sand anode electrode connection AH (**b**); separate outputs for the anode AH and cathode CH currents (**c**); slow curve trace characteristics for different injection current levels (**d**) and depth profile for impact ionization in the latch-up state (**e**)

by separating the simulation output of the HS anode "AH" and cathode "CH" current dependencies on the nEpi-to-nEpi applied voltage (Fig. 4.10c). It is apparent that the injection capability of the HS diode anode region depends on the diode's design.

As in HS induced latch-up, the minimum holding voltage is a function of the nEpi-nEpi spacing representing the parasitic n-p-n structure base length. In particular, the results of the standard JEDEC latch-up test (Fig. 4.10d) in nearly isothermal conditions at a current level of ~ 100–200 mA do not cause latch-up of

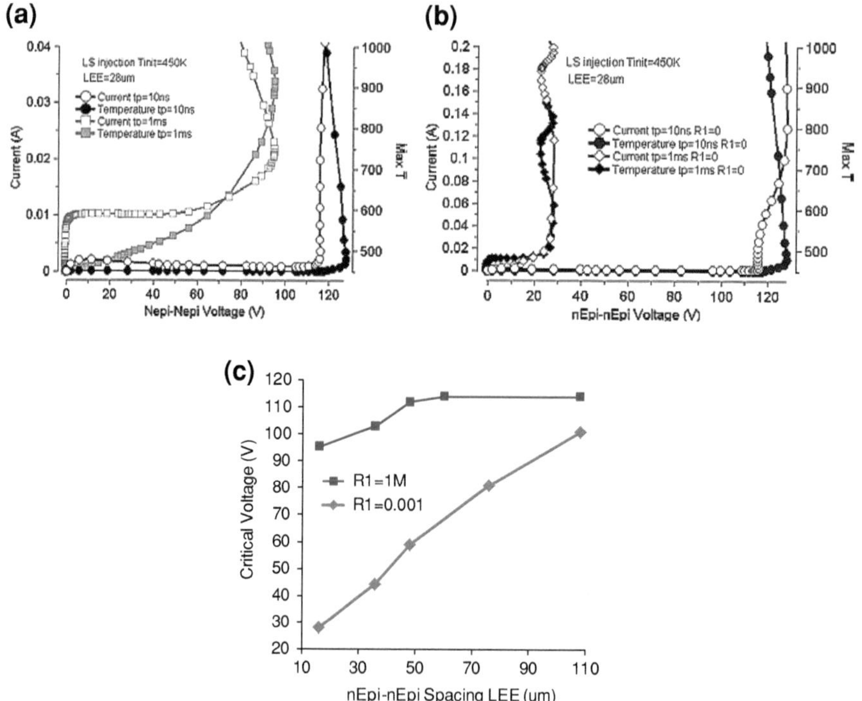

Fig. 4.11 Thermal-coupled mixed-mode simulation for LS injection for different pulse time domains with the self-heating effect; Illustration of the thermal runaway with floating (**a**) and with connected (**b**) "AH" electrode acting as a pEmitter for the parasitic SCR structure and the dependence of the critical voltage on nEpi-nEpi spacing *LEE* (**c**)

the structure in the entire voltage domain up to 100 V. At the same time, the structure will be susceptible to the system-level injection current of several Amps.

A thermal coupled simulation increases the complexity of the latch-up phenomena. Depending on the structure's design, the effect of the connected "AH" electrode acting as a pEmitter is amplified (Fig. 4.11).

The "AH" pEmitter can be part of the active circuit in the nEpi pocket or can be introduced as a ring connected to the nEpi node to partially collect the holes during high-side hole injection. This case will be addressed in an example in the following section.

Though the I-V characteristics obtained by slow curve trace analysis provide a deep insight into the phenomena involved in latch-up, information about latch-up occurrence in transient conditions with a single injection pulse is rather limited. Thus, a comparison of the simulation results with test outcomes is unclear.

A mixed-mode simulation for a circuit with an automated LUT source is more physically accurate. Mixed-mode circuits for transient latch-up simulation at a constant power supply voltage and a single injection pulse are presented in

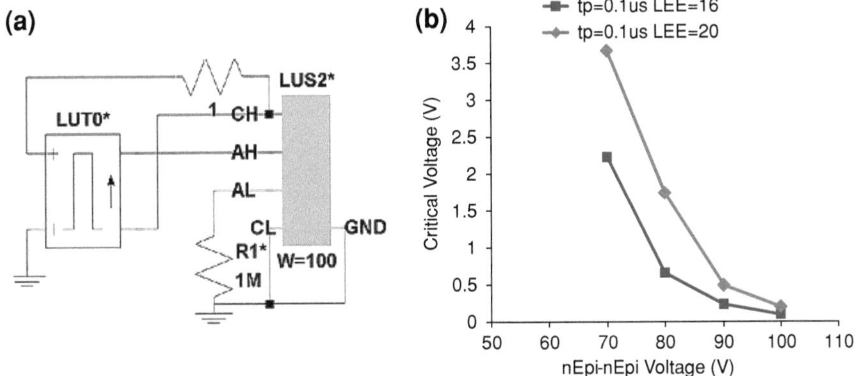

Fig. 4.12 Isothermal HS mixed-mode simulation circuit for automated critical injection current analysis of HS injection induced latch-up (**a**), extracted dependencies of the injection current on the LEE nEpi-to-nEpi spacing parameter in isothermal conditions with T = 450 K (**b**), and thermal-coupled simulation with different injection pulse durations (**c**)

Fig. 4.12a. The outcome of the numerical experiments shows that the critical injection current for latch-up is a function of the power supply current and structure parameters. An example of the calculated critical injection current dependence on *Vpower* for structures with *LEE* = 16 and 20 μm is presented in Fig. 4.12b.

4.2.2 Active Guard Ring Isolation and Experimental Comparison

Experimental data can be acquired using both the curve trace method for the critical voltage and by conducting the standard JEDEC latch-up test and increasing the injection current at a given voltage. The curve trace for LS injection was performed with a constant current source and a slow curve trace of the voltage level. For the standard latch-up JEDEC test, a pulse duration of 3 ms and an injection current of 100 mA were applied, with increase of the step test voltage until the latch-up. The signature of the HV latch-up was irreversible failure mode.

The experimental data for the curve trace characteristics of the LS injection induced latch-up is presented in Fig. 4.13a. Two types of isolation rings between the nEpi pocket regions, a pSubstrate ring and a so-called active ring, were measured. The active ring approach will be explained in details below, in combination with numerical simulation results. The latch-up data for the standard test with the industrial MK2 Thermo Fisher tester (Fig. 4.13b) for LS injection induced latch-up were in correlation with the curve trace experimental data, demonstrating

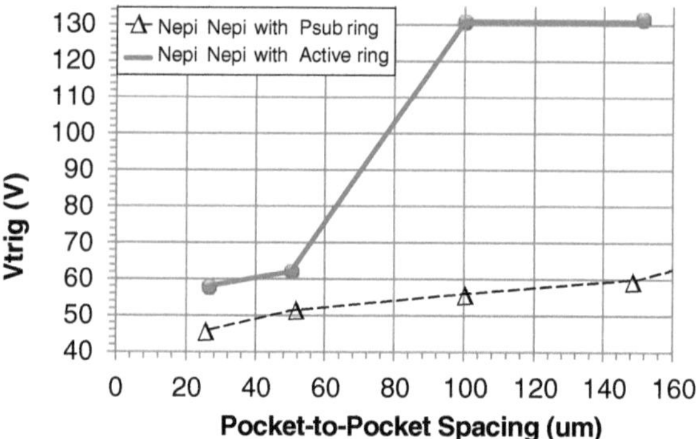

Fig. 4.13 Experimental data for the curve trace and standard latch-up test methodologies for the LS injection induced latch-up [96]

a rather low critical voltage of $\sim$ 50–60 V in comparison with desired 80–100 V target, even for the rather big *LEE* spacing above 150 μm.

There are technical difficulties in performing a curve trace analysis for HS injection induced latch-up, due to the high-side current source connection. Therefore, only the standard test HS injection induced latch-up data are presented in Fig. 4.14. Unlike LS latch-up, the HS injection induced latch-up has demonstrated a significantly higher critical voltage with increase of the *LEE* spacing above 50 μm. Thus, though the same n-p-n device cross-section represents the physical characteristics of the latch-up, a significant difference in the critical voltage is observed between the HS and LS injection induced latch-up cases.

The test structure for the experimental analysis corresponds to the simulation cross-section (Fig. 4.15a) with a respectively high nEpi-nEpi region. A major difference between this structure and the simplified device in the previous section is the design of the HV diodes with a poly RESURF and additional n- and p-rings inside the nEpi pocket, directly connected by the cathode terminal. The original purpose of the p-ring is to collect a part of the injection holes, while the outer n-ring completes the pocket surface isolation after the p-ring. One of the reasons for having the inner pocket p-rings is the resulting low resistive metal connection. The nEpi and inner p-ring share the same metal layer, unlike the pocket isolation pSubstrate ring, which is often connected to the ground potential through a rather highly resistive path.

To understand the miscorrelation between the critical voltages for the HS and LS latch-up, a numerical simulation analysis was performed for the device cross-section in Fig. 4.15a. Since this analysis is applicable to standard latch-up thermo-electrical and ESD test electrical conditions, Fig. 4.15 presents thermal coupled simulations with the Heat equation enabled in the simulation and for the iso-thermal case.

For the given structure parameters with relatively low *LEE* spacing, the critical voltages for HS injection induced latch-up are limited by the self-heating of the device and thermal runaway (Fig. 4.15b). The high detection current in the nEpi circuit as a result of the HS injection becomes a critical factor. The p-ring internal to the HS pocket collects the holes, which results in detection current reduction

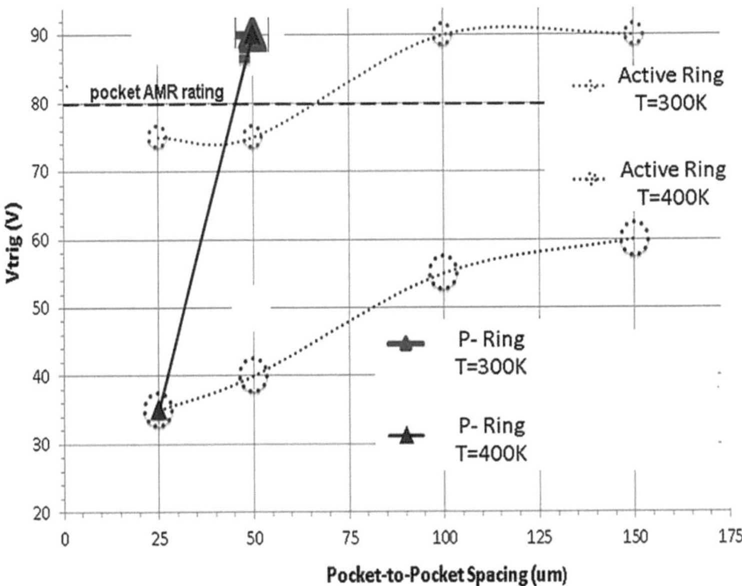

Fig. 4.14 Experimental data for the curve trace of high-side injection (courtesy of Joseph Sheu)

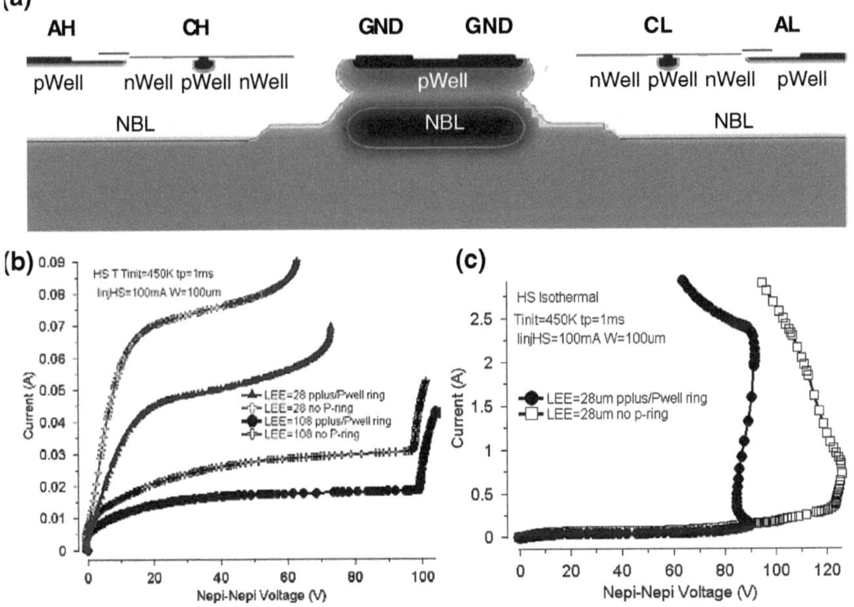

Fig. 4.15 High-side injection induced latch-up simulation. Parameterized cross-section of the FEM structure (**a**) and thermal-coupled (**b**) and isothermal (**c**) curve trace I-V characteristics for a 1 ms voltage pulse (the thermal coupled characteristics are calculated up to the peak temperature $T = 1000$ K in the structure)

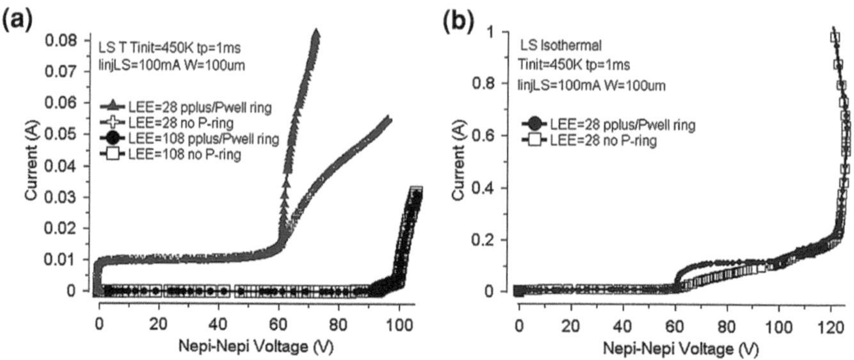

Fig. 4.16 Low-side injection induced latch-up simulation. Thermal coupled (**a**) and isothermal (**b**) curve trace I-V characteristics for a 1 ms voltage pulse (the thermal coupled characteristics are calculated up to the peak temperature $T = 1000$ K in the structure)

(Fig. 4.15b). As a result, the critical voltage usually limited by the peak temperature $T = 100$ K can be achieved at a higher level.

If the Heat equation is disabled, then electrical limitations are observed as a function of the nEpi-nEpi spacing at much higher nEpi voltages (Fig. 4.15d). However, with a higher current density comparable to ESD stress, the HS internal pRing acts as a pEmitter. As a result, the parasitic SCR structure turns on at some critical voltage and current density above the thermal effect limit (Fig. 4.15c). Thus, the parameters for the p-rings in the HS pocket must be carefully chosen. One option is to use a p-ring with only the contact p-diffusion, eliminating the pWell diffusion.

For LS injection induced latch-up, the thermal coupled and isothermal simulation curve trace results (Fig. 4.16a) are also in correlation with the experimental data. With increase of the *LEE* spacing, the detection current is reduced and therefore much higher critical voltages are obtained for the same assumed critical temperature $T = 1000$ K (Fig. 4.16a). With proper design, no latch-up characteristics are observed in isothermal conditions (Fig. 4.16b).

A method of reducing the low-side injection effect was proposed in [95]. An active guard ring was introduced to collect the injected electrons during the drift between the nEpi-nEpi spacing. This solution improves the electron collection by the ring by incorporating an additional deep nEpi region softly connected to the pSubstate through the floating p-region. Thus, a new collection path is created for the injected electrons. The experimental data for the active guard ring were already presented in Fig. 4.13 for LS injection and Fig. 4.14 for HS injection. A major effect of improving the LS injection conditions is the increase of the critical voltage. Conversely, the drawback is the reduction of the critical voltage for the HS injection case.

A cross-section of a structure with the simplified active ring is presented in Fig. 4.17a and used to demonstrate the active guard ring effect. The active guard ring is formed in the middle of the structure by the N-region with the length

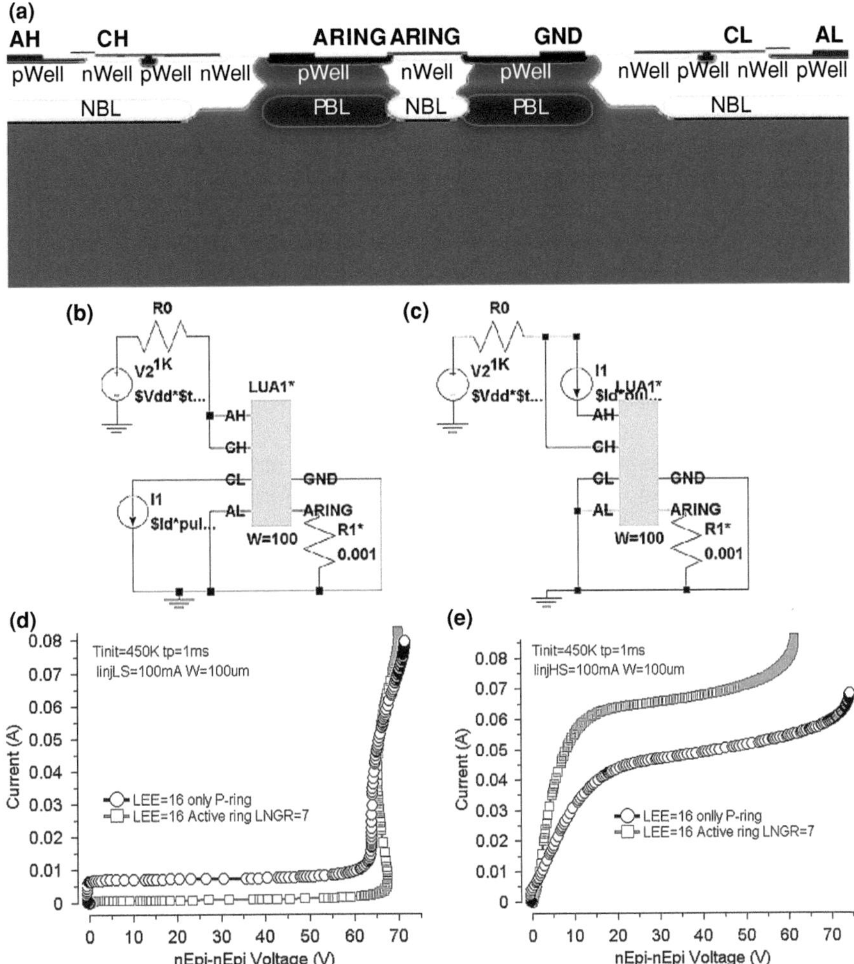

Fig. 4.17 Cross-section for a FEM latch-up structure with the parameterized active ring (**a**) and demonstration of the active ring effect through comparison of the thermal coupled simulation results for curve trace I-V characteristics with a 1 ms voltage pulse and nEpi-nEpi spacing LEE = 16 and 48 μm for HS (**b, d**) and LS (**c, e**) injection. The curve trace is stopped at structure peak internal temperature T ∼ 1000 K

LNring connected by metal to the two adjacent p+-regions. In real applications, all three regions are connected but remain floating (Fig. 4.17c). With an applied nEpi-nEpi voltage, the potential of the n-region of the active guard ring is elevated above ground. This creates a current path for the injected electrons.

As expected, the effect of the active guard ring is substantial for LS electron injection due to the improved collection of the injected electrons. In contrast, in the case of HS hole injection, the active ring provides only a minor reduction of latch-up susceptibility due to the loss of the effective p-region for hole collection.

4.2.3 HV Latch-up Prevention Rules

The major principles for the verification deck can be derived further and formalized for the script coding based upon the combination of the experimental and simulation analysis. The HV latch-up rules can be further implemented in the automated layout verification tools as a part of CAD.

The nEpi pockets at different potentials, containing active devices that are connected to different pads subject to injection or overvoltage latch-up tests, must be separated in the layout. The pocket-to-pocket separation spacing is a function of both the required to pass latch-up test current (usually either 100 or 200 mA) and latch-up test temperature (usually either room or 125 °C). The introduced isolation space between nEpi pockets reduces the gain of the parasitic n-p-n formed by the epi regions and substrate, and therefore results in a voltage increase critical for latch-up.

The circuit blocks inside the nEpi pockets must be completely surrounded by an n-diffusion guard ring at the pocket periphery. This measure prevents the formation of an effective SCR structure by better isolating the high-side p-regions connected to the pad.

Each nEpi pocket with a p-diffusion region connected directly to the pad must be enclosed by a separate p-substrate ring. The p-substrate ring should be not shared by multiple nEpi pocket regions. This rule is targeting the reduction of the hole injection effect that occurs because the substrate p-ring accumulates holes.

This same set of rules should be applied to low-side nEpi placement and to a grounded nEpi presence in particular. Thus, the low-side or grounded nEpi should be separated from the high-voltage nEpi pocket connected to the injecting I/O pin.

The nEpi diffusion ring can be tied to a 5 V power supply pad directly or through a blocking diode.

Resistance between any point on the n-diffusion ring and the supply pad should be limited, for example, below 10 Ohms.

In correlation with these measures, any pocket with an analog or digital internal circuit must be spaced apart from the injecting pocket. The internal circuit must have an n-diffusion ring completely surrounding any pWell tied to the 5 V supply pad.

The n-diffusion ring of the injecting pocket and the n-ring of the logic block should not share the same metal bus connection to the supply pad. Instead, two separate metal busses should be drawn, and they should be sufficiently wide to carry the 200 mA current induced during the latch-up test.

The high-side ground ring of the injecting pocket and the p-substrate ring surrounding the internal circuit blocks should not share the same metal bus connection to the ground pad to avoid a voltage drop in current injection conditions. Indeed, at a bus resistance of ~ 30–40 Ohms, the 200 mA current can create a sufficient voltage drop to turn on the diode junction.

An optional measure for reducing low-side injection induced latch-up is the use of an active guard ring. In this case, the automated rule check verifies that the low-side injector must be surrounded by the active guard ring and spaced sufficiently far away from any HV biased pocket.

In general, for optimized design, a rule deck reference table can be created to define the space requirements dependent on the pocket-to-pocket maximum applied voltage, injection current and latch-up test conditions temperatures.

4.3 Transient Induced Latch-up

By re-phrasing and extending the definition for transient latch-up (TLU) given in [97] it can be formulated as follows. TLU is a formation of a low-impedance current path and the corresponding on-state supported by the power supply as a result of transient overstress that remains at least temporarily after removal of transient overstress conditions. In this definition the major differentiating factor in comparison with conventional latch-up is that the overstress causing TLU is occurring in the time domain faster than 5 μs. This time domain is different from the conventional latch-up test according to the JESD78D JEDEC standard [93] accomplished in the time domain 5 μs–5 ms. The low impedance current is formed and supported by the turn-on of a parasitic thyristor (SCR) or an n-p-n bipolar structure. These parasitic structures can be formed both unintentionally in a given layout and represent a part of ESD structure. The transient overstress can be represented both by overvoltage and injection. The injection can be initiated both inside the structure, thus providing low impedance TLU conditions, and from independent distanced junctions.

In particular TLU can represent an effect of an accidental turn-on of the snapback clamp during normal operation conditions or electrical overstress events when the on-state can be further supported by the applied power supply current. The TLU specific is related to the conditions that a powered-up IC is triggered by a transient pulse with a much shorter rise time and duration than the pulse of a standard latch-up test.

Transient latch-up is pertinent to the topics of this book in its importance to the system-level stress conditions in power-on mode. This section addresses the problem of transient latch-up on the physical level from several different angles. Primarily, a TLU event in a local ESD clamp can be induced by an unexpected temporary electrical regime deviation during normal operation mode or by inter-action between on-chip and off-chip protection network components.

The first subsection introduces an experimental setup for the TLU character-ization of ICs. The following section focuses on the aspects of transient induced latch-up during the normal operation regime and explains the major regularities using a simplified example of the power stage of a LV synchronous buck voltage regulator.

The following subsection discusses TLU from different prospective. In this case, the latch-up event is induced by a system-level ESD stress pulse. A case study of the IC with an off-chip TVS based protection network demonstrates how the pin becomes susceptible to TLU under certain conditions of a system-level event while otherwise passing the standard JEDEC latch-up test.

4.3.1 Test Approach for TLU

In the previous years, many field failures related to TLU have been observed and design proposals were made for the prevention of TLU [97–102] The typical rise time of TLU trigger pulses is a few order of magnitude faster than the minimum allowed rise time in the static JEDEC latch-up test. This allows the use of fast rising ESD pulses a triggering pulse for TLU. Currently no standard method exists which defines a setup for the TLU characterization of ICs. MM testers are used as trigger sources because of their positive and negative swings in the stress waveform [101]. TLP tester and solid state pulsers are used because of the available different rise times and pulse widths [102].

Another possible setup for the TLU characterization is shown in Fig. 4.18. A HMM tester is used as latch-up trigger source. The DUT is powered up with a DC supply to its normal operating voltage. An avalanche diode is placed between the DC supply and the stress path to block the flow of current into the DC supply when the ESD stress is applied [101]. An inductive current probe is connected to monitor the current from the trigger source and the HMM tester to the device under test. A voltage probe monitors the voltage across the stressed pins to detect a possible latch-up state of the DUT.

The setup is used to test the TLU robustness of a real product. The IC is powered up to the typical operating voltage of 10 V defined in the IC datasheet. HMM stress is applied to the supply pins. Like in the static latch-up test the compliance of the DC supply is set to 100 mA. Figure 4.19 shows the voltage between the supply pins of the IC during HMM stress. At a stress level of 2.5 kV the voltage collapses below the supplied voltage. After the decay of the HMM stress the voltage does not recover to the operating voltage and the supply current increases permanently from about 9 to 15 mA. With repeated HMM zaps at the same stress level the supply current increases further until device burn-out occurs which can be measured as an open circuit between the supply pins.

The following subsection is presented to discuss how TLU can occur in different scenarios and design solutions against TLU.

4.3.2 TLU in Case of Switch Pins in Power Trains

Transient induced latch-up occurs primarily due to an accidental triggering of the local snapback ESD clamps during electrical overstress, temporary load change or mismatch, and other short-term deviations from the expected normal electrical regime verified by simulation at the design stage.

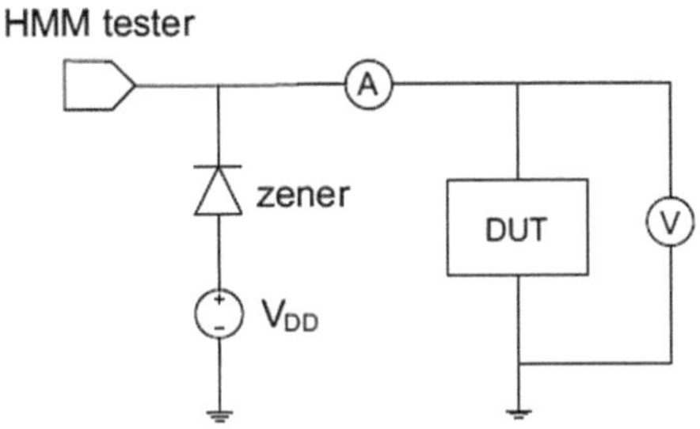

Fig. 4.18 Measurement setup for testing TLU; VDD: DC source; zener: discrete avalanche diode, optional: decoupling capacitor(s) in parallel to the DUT

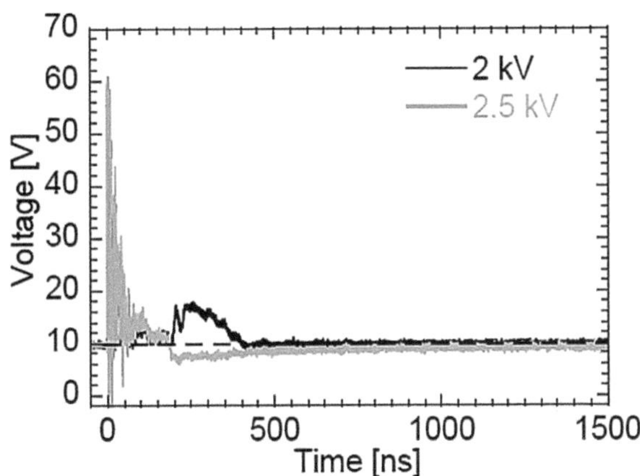

Fig. 4.19 TLU test on a product IC: Voltage between supply pins during HMM stress when the IC is powered up to 10 V

This type of transient latch-up scenario can be explained with an example of the local snapback protection of a switch pin [5]. The detection of the transient current due to gate coupling above the threshold voltage results in the clamp triggering into snapback, followed by the transient latch-up mode. Included in the simplified power train output circuit (Fig. 4.20a) are the external filter components $L_F C_F$, the decoupling power supply capacitor C_P, the bond wire inductance L_W, and the driver clock signals at the gate of the output NMOS and PMOS power arrays. The experimental 5 V circuit demonstrates the irreversible failure during switching after the input power supply voltage VDD is elevated to 6–6.5 V.

A major cause for transient latch-up is the parasitic inductance in the chip package and connections between external components. As a result of the inductive load, the voltage waveforms at the protected node are not ideal, containing large amounts of inductive running voltage noise.

The ringing amplitude increases with increase of the input voltage and a voltage critical for the power NMOS array or local clamp snapback, if applied to the pin, may result in a TLU event. This physical effect is demonstrated using a mixed-mode numerical simulation analysis. The simulation has a load resistor $R_L \sim$ 3.6 Ω, a maximum operation voltage VDD = 5.5 V, and a duty cycle of 50 % to limit the output load current at $\sim (5.5/2)/3.6 = 0.763$ A. The waveforms provided

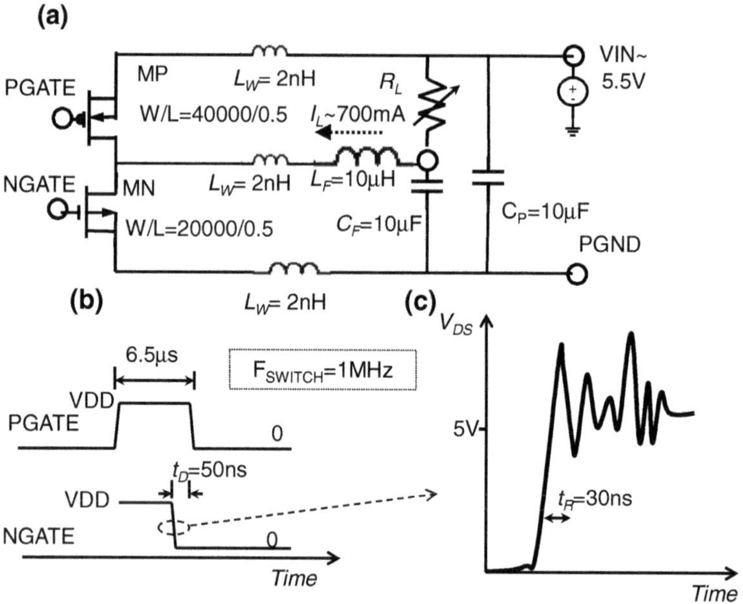

Fig. 4.20 Simplified output circuit block with external components (**a**), and the theoretical waveforms for the driver signals at the gate (**b**) of the output NMOS and PMOS arrays (**c**)

by the driver (not shown) apply a controlling "pgate" signal to the high-side PMOS MP, causing it to be in the low current state when the "ngate" signal is also low. Conversely, the "ngate" signal is provided at a high level only when the "pgate" signal is also high. The non-overlap time is about 50 ns (Fig. 4.20b). For the initial conditions, the output NMOS and PMOS devices are off. Then, the VDD node is set to an arbitrary voltage, followed by a preset of the output capacitor to some voltage using the same initial conditions. This way, both terminals of the inductor are set by the initial conditions realized at the output capacitor.

In the mixed-mode numerical simulation, the switching NMOS array is physically represented by the FEM structure (Fig. 4.21a) with a spacing of up to 10 μm between the pBody contact region and the source. This maximum space is defined in the particular process design rules. The solution of the semiconductor equations for the 2-D cross-section of the FEM NMOS device as a part of the mixed-mode simulation matrix supports the physical effect of avalanche-injection conductivity modulation (Chap. 3). The critical voltage of the NMOS array is a function of both the pBody-Source spacing (*SWS*) (Fig. 4.21b) and the gate bias (Fig. 4.21c).

The TLU effect is directly resultant of the inductive parasitic components in the power supply and ground circuit. Due to this inductive load, the induced voltage overstress of the NMOS array significantly exceeds the VDD power supply level (Fig. 4.22). The amplitude of the transient voltage noise is a direct function of the input voltage. Thus, though the applied input voltage remains within the absolute

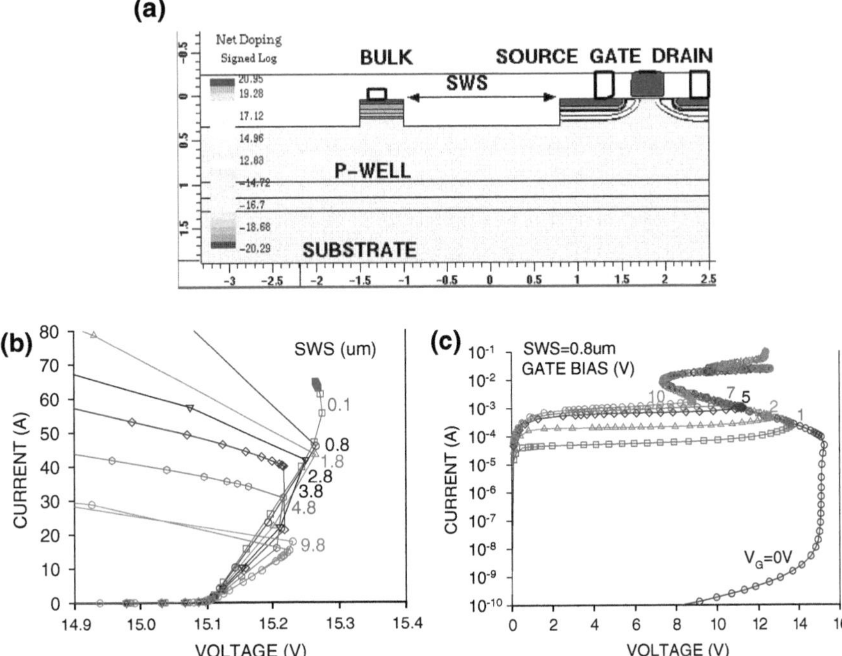

Fig. 4.21 Cross-section of the FEM NMOS device used in the mixed-mode simulation to represent the switching power array (**a**) and the calculated quasistatic isothermal I_D–V_{DS} characteristics for the different *SWS* spacing (**b**) and gate bias V_{GS} (**c**)

maximum rating limits, the reversible regulated input voltage during operation with an inductive load results in a power array TLU event.

Reducing the pBody-to-source diffusion spacing (SWS) to 0.1–0.8 μm causes a ∼ 10-fold increase of the current density critical for snapback mode during switching. This increase can be used to reduce the TLU effects.

A similar TLU scenario due to operation with an inductive load is realized for a power train where a snapback NMOS clamp is used to protect the VIN pin. Because the clamp has a lower triggering voltage, in comparison with the power array, the overvoltage results in a TLU directly due to clamp turn-on. The event precedes a possible TLU in the array.

As an even more simple case, the TLU of a clamp can be demonstrated using the simplified power stage of the asynchronous 5 V buck DC–DC voltage regulator as an example (Fig. 4.23a). In this circuit, the switch pin is protected by a snapback NMOS ESD clamp. Similarly to the previous case, the clamp experiences a voltage overstress above the critical voltage for snapback due to an inductive load at the switch pin. The observed transient latch-up event is due to the snapback of the NMOS ESD clamp and is a function of VIN voltage level (Fig. 4.23b).

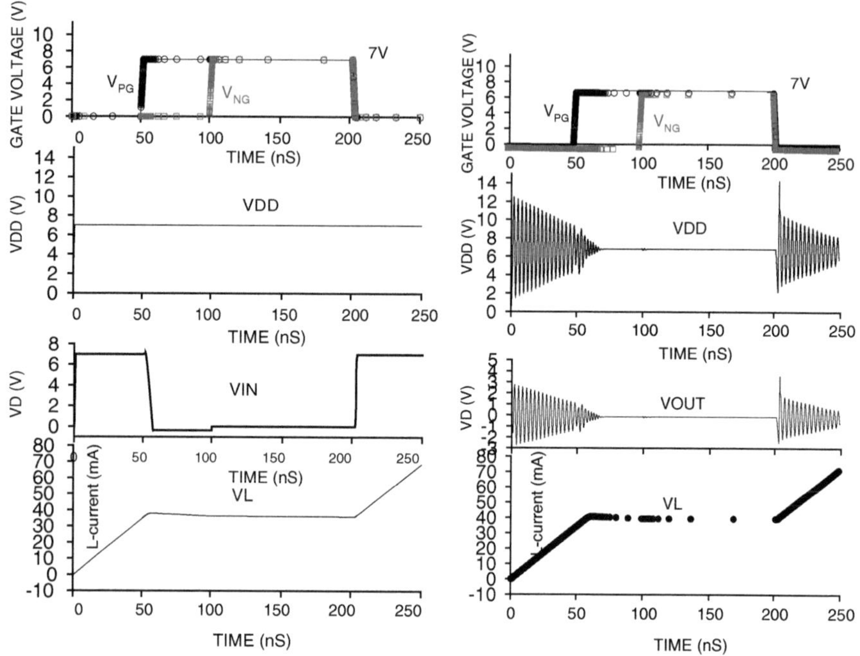

Fig. 4.22 Waveforms of the input gate voltages, VDD node, output node VL, and the current through the inductance without (**a**) and with (**b**) bond wire inductance components included in the simulation circuit for VDD = 7 V

The TLU example cases presented above are directly applicable to system-level events. When a system ESD stress is applied in power-on conditions, the capacitive and inductive coupling and EMI effects may create an unexpected current path. The produced voltage ringing can result in triggering of the ESD clamps or active devices in a way similar to operation with an inductive load. From this perspective, comprehension of the on-chip and off-chip system stability against short-term overstress is critically important.

4.3.3 TLU, Simple Network with Standalone ESD Devices

The fact that an ESD device under system-level stress is in the powered on condition does not necessarily mean that transient latch-up will occur when the holding voltage of the device is below the power supply level. In the powered on-state, an ESD device can support only a minimal current. This physically means that the holding voltage of the ESD device is a function of the device's current. Though current filamentation can significantly change the minimum critical current in 3-D cases, the final latch-up scenario is still a current-dependent event.

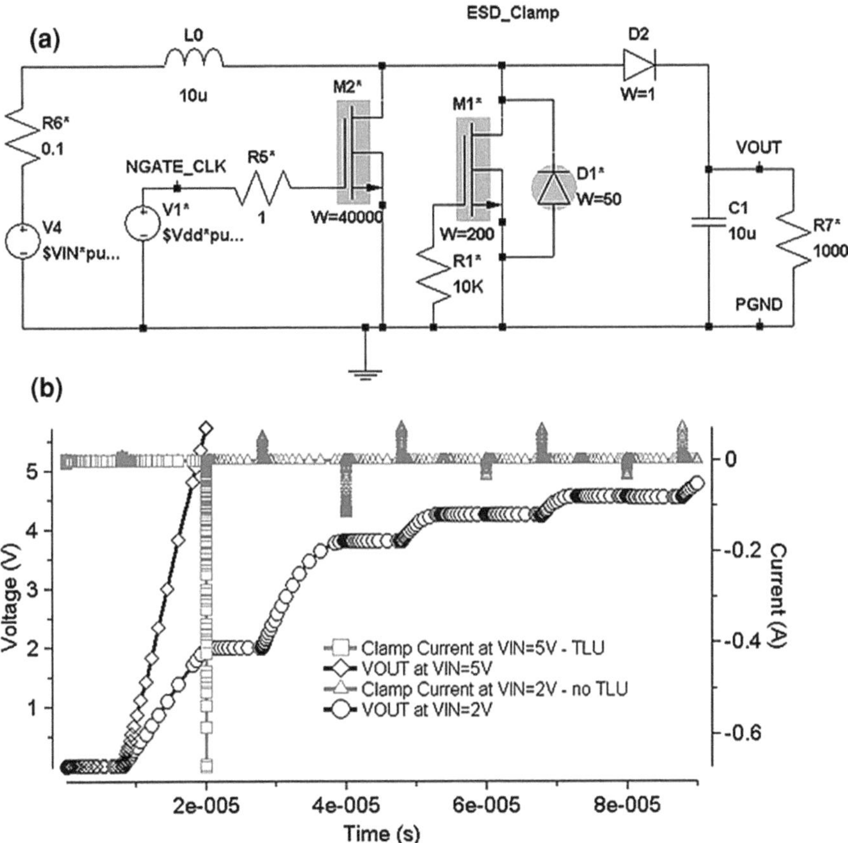

Fig. 4.23 Mixed-mode simulation circuit for the simplified output stage of an asynchronous 5 V DC-DC boost voltage regulator (**a**) and simulation results for the normal operation regime with an input voltage VIN = 2 V and a TLU event at VIN = 5 V (**b**)

A complex example of transient latch-up dependence in an off-chip system-level protection network is described in the next section. The more simple case of a dual-direction DIAC device under a power supply load, represented by the mixed-mode circuit in Fig. 4.24a, demonstrates the load effect on transient latch-up. The circuit in Fig. 4.24a combines the equivalent circuit components for a HMM ESD pulse source applied to the DIAC FEM device in combination with power supply conditions from the voltage source V2.

The two-step simulation analysis includes a slow transient simulation to first set the power supply electrical regime, followed by a fast transient simulation for the HMM pulse. The results (Fig. 4.24c) depend on the load and the ESD design of the DIAC device, so the ESD stress may or may not result in the TLU effect. For example, a TLU is realized when the power supply resistor R7 is equal to 50 Ohms. However, if the resistor value is increased to 100 Ohms, the device returns to the off-state after the

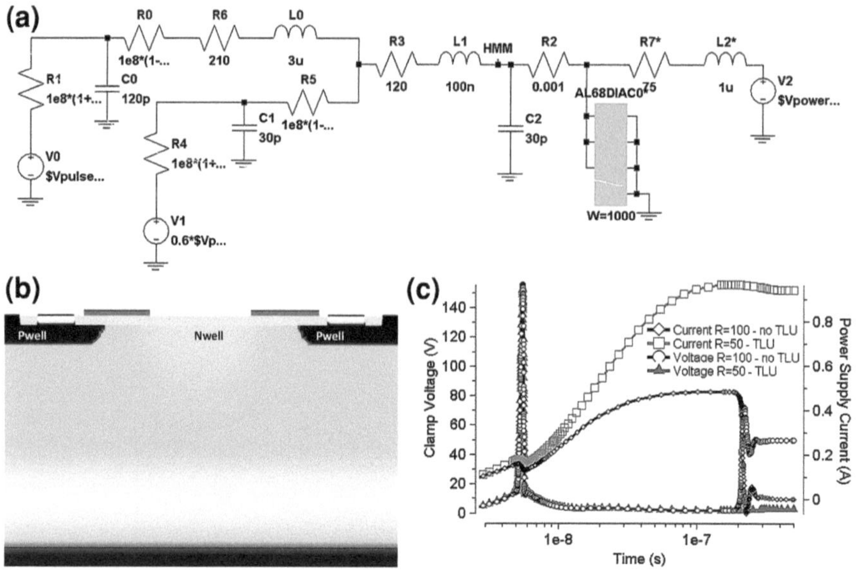

Fig. 4.24 Mixed-mode HMM pulse equivalent circuit for TLU study (**a**) with a DIAC FEM structure (**b**) and the waveforms for the power supply voltage and current for 50 and 100 Ohm loads (**c**)

HMM stress independently of the pulse level and a TLU does not occur. The minimum holding current in the on-state depends on the design parameters of the device that determine the level of positive feedback in the structure. In HV devices, it is rather difficult to increase the absolute minimum holding voltage while preserving the high current on-state capability. Increasing the minimum holding current, on the other hand, appears to be a much more realistic goal.

A cell-level measure for preventing transient latch-up, based on similar principles, involves modification of the ESD device itself. In the following example, the DIAC cell is modified by distributing the poly ballasting regions along the I/O and GND n+ and p+ injectors (Fig. 4.25a). The experimental results for this solution, based on a 50 V tolerant DIAC cell, demonstrate that increasing the holding voltage produces an effect equivalent to additional resistance in the power supply combined with the minimum holding current of the ESD device (Fig. 4.25b).

4.3.4 *TLU. Impact of the On- and Off-chip Protection Networks*

Since TLU in a powered-up IC is triggered by a transient pulse with a much shorter rise time and duration than the pulse of a standard latch-up test [93, 97] it may not be detected during the standard JEDEC latch-up test [97]. At the same

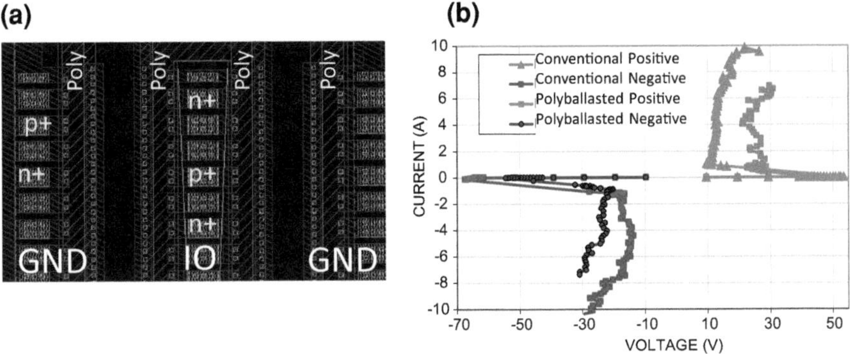

Fig. 4.25 Layout view of a poly-ballasted DIAC cell with interdigitated emitters (**a**) and comparison of the experimental TLP results for conventional and poly-ballasted DIAC cells (**b**)

time a system-level ESD stress with a rise time in the sub-nanosecond time domain can namely create the conditions for TLU. The final event can be a combination of over-voltage and current injection that exceeds the specifications of the standard JEDEC latch-up test and can trigger parasitic structures with a current path different from this test [97, 102].

The exact TLU scenario and internal block sensitivity depends on the design of the off-chip and/or on-chip ESD protection. In the following particular TLU case study, measurements and simulations demonstrate the transient interaction of the parasitic SCR in the IC with both off-chip and on-chip protection network components. The latch-up-sensitive IC is protected with a TVS-based off-chip protection network (Fig. 4.26) that consists of power supply decoupling capacitors and a 3.3 V TVS diode. The entire network combines the off-chip components and an on-chip parasitic p-n-p-n structure formed by a CMOS circuit and an active clamp (PC).

For adequate analysis of the circuit (Fig. 4.26b), the DC supply voltage is applied through a bias-T setup that includes a large inductance of 700 μH and a parallel capacitance of 0.2 μF to prevent the flow of ESD current into the supply. The bias-T with a capacitor in series to the RF signal replaces a blocking diode typically used to enable the application of negative ESD stress and limits the impact of the output circuit of the DC supply on the latch-up behavior of the tested circuit.

The electrical characteristics of the active network components are compared in Fig. 4.27. During a standard latch-up test, the voltage at the tested pin is held at 1.5 times the 3.3 V maximum operating voltage for this 0.13 μm CMOS logic block. Therefore, the SCR triggering voltage of ∼ 15.5 V (Fig. 4.27) cannot be reached. The TVS diode's clamping voltage is above both the SCR and active clamp levels under system-level ESD stress conditions. By design, the TVS diode is expected to provide the main current path during a system-level ESD stress event, since the SCR voltage remains below the triggering threshold and the active clamp is capable of conducting only a limited current.

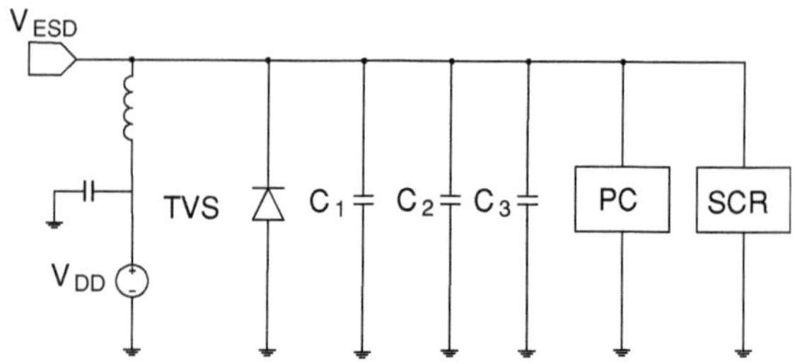

Fig. 4.26 Simplified circuit for an IC with an off-chip protection network for the VESD pin used for the TLU case study analysis

Fig. 4.27 TLP I-V characteristics for the SCR (width = 50 um), active clamp (NMOS width = 800 um) and TVS diode

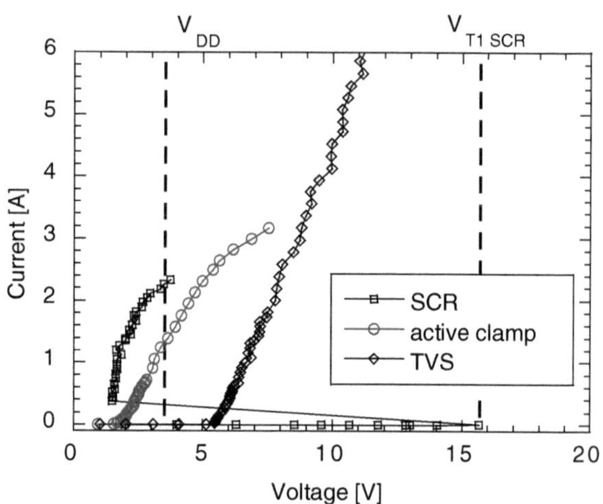

For the initial voltage peak, the mixed-mode simulations with the DECIMM tool [19] for the voltage waveform of the standalone SCR and the on-wafer HMM measurements at IEC61000-4-2 stress were in agreement (Fig. 4.28). However, for accurate analysis, it is critical to take into account the capacitors and the parasitic components of the TVS diode (Fig. 4.29).

No TLU is observed during positive and negative stresses if the SCR is tested with the TVS diode in parallel and no off-chip or on-chip decoupling networks are present. At some HMM stress levels both the SCR and the TVS diode turn on. During positive stress, SCR snapback results in turn-off of the TVS diode [103]. Due to the bias-T network between the DC supply and the circuit, the output capacitor of the voltage supply is not discharged before the HMM current fully decays. Hence, the SCR turns off before the supply current can significantly rise. During negative stress, both the TVS diode and the SCR act similar to forward-biased diodes and share the HMM stress current after the triggering.

On the contrary, the circuit is highly sensitive to latch-up with the off-chip protection network. Standalone SCR latch-up is realized during positive stress at 4 and at 2 kV negative HMM stress. The latch-up is peculiar to the oscillatory voltage waveforms created by the network at the SCR node.

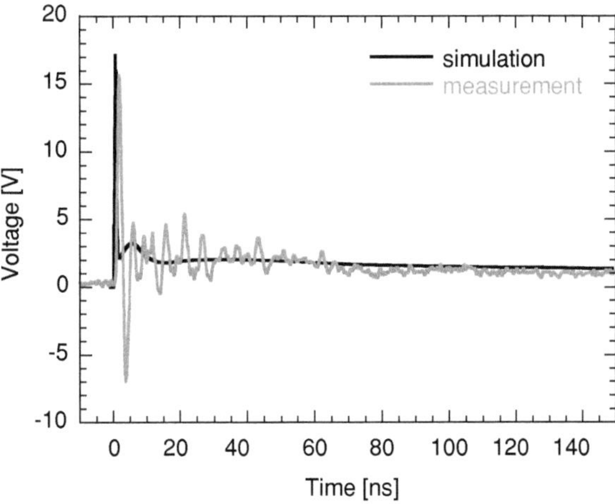

Fig. 4.28 Comparison of the simulated and measured voltage waveforms across the SCR for a 1 kV HMM

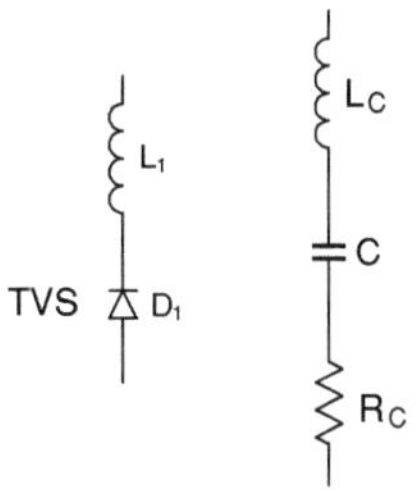

Fig. 4.29 Equivalent circuit for the off-chip TVS diode and capacitors including L1 to represent overshoot and LC and RC parasitic components

Despite both positive and negative stresses, the off-chip network mainly conducts the current during the first few nanoseconds of the HMM pulse, though a small residual current flows into the on-chip devices due to the parasitic inductance component of the decoupling capacitors and the TVS diode. Therefore, at some positive or negative HMM stress level, the SCR turns on and remains in the latched state.

The SCR turns on at a significantly lower HMM stress level during negative HMM stress. This is because of a two-stage domino turn-on effect that occurs due to the oscillating voltage at the SCR node. The negative ESD pulse forces the supply voltage below the zero ground potential. At the negative voltage, a substantial amount of carriers is injected in the SCR body regions due to the forward bias of the pWell-nWell blocking junction (Fig. 4.30a). Then, during the immediately following fast positive voltage swing, the effective turn-on voltage of the SCR with injected carriers becomes significantly lower than the standalone triggering voltage (Fig. 4.27). As a result, the SCR turns on into double injection conductivity modulation mode (Fig. 4.30b). The low-impedance path created by the on-state SCR causes a fast discharge of parasitic capacitors in the system, followed by a fast change of the voltage polarity at the supply pins. The pWell-nWell junction of the SCR becomes reverse-biased and sweeps out the excess minority carriers (Fig. 4.31).

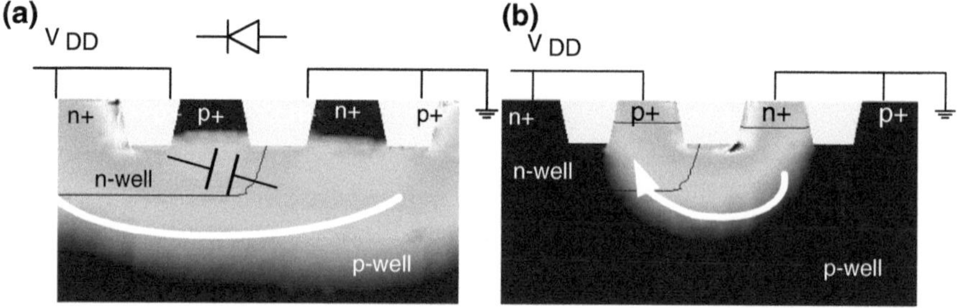

Fig. 4.30 Simulated current density in the SCR during TLU (**a**) after 1.5 ns and (**b**) after 3 ns; the white arrow is the current flow, the HMM stress level is at −2 kV HMM, the VDD is at 3.3 V

Fig. 4.31 Simulated current and voltage of the SCR during negative stress, causing TLU at an HMM pulse of −2 kV and a supply voltage of 3.3 V (**a**), and the current waveforms for different circuit (Fig. 4.26) nodes (**b**)

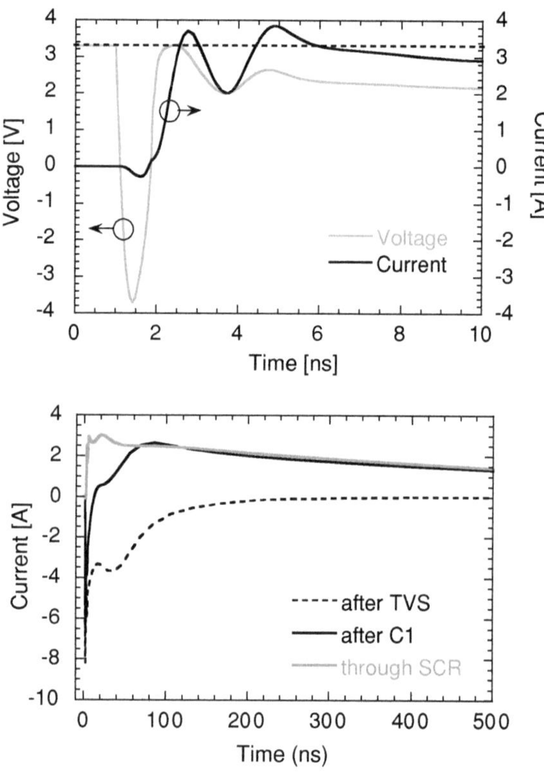

A different TLU scenario is observed during positive HMM stress when an active clamp is not included in the circuit (Fig. 4.26). In this case, a TLU is realized when the voltage across the off-chip protection network exceeds the SCR triggering voltage of 15.5 V, due to a voltage drop on the parasitic inductance of the network components. This voltage level is reached at a HMM stress level of +4 kV if no active clamp is present.

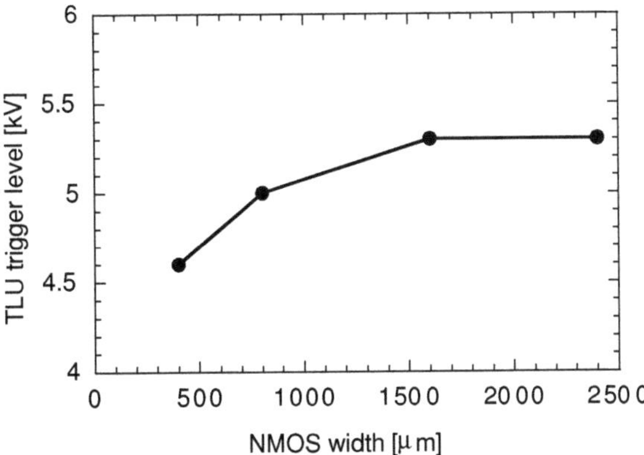

Fig. 4.32 Trigger levels during negative HMM stress vs. the width of the NMOS in the active clamp

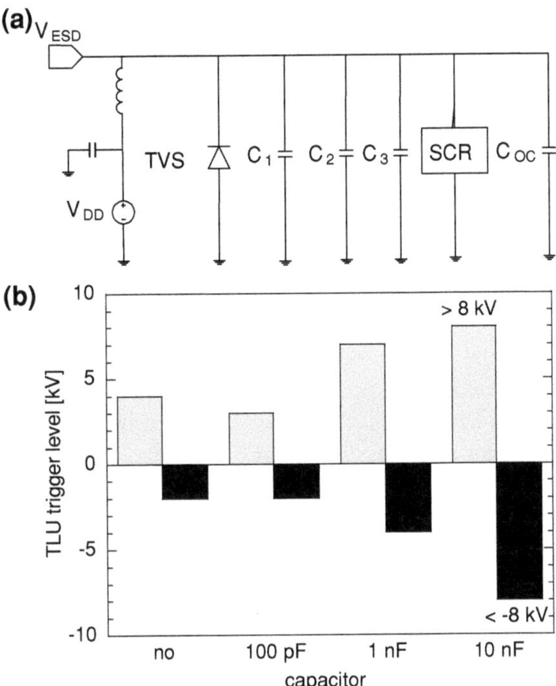

Fig. 4.33 Schematic of simulation setup with an on-chip decoupling capacitor (C_{OC}) (**a**), and TLU trigger levels for different sizes of the on-chip decoupling capacitor (**b**)

No SCR triggering in TLU mode is observed at a positive stress of up to 8 kV HMM as long as the on-chip active clamp is in place. The active clamp limits the peak transient voltages to a level below the SCR turn-on voltage.

During negative stress, both the SCR pWell-nWell diode and the body diode of the active clamp NMOS array conduct the stress current. Due to lower injection levels, the critical TLU triggering voltage is improved from -2 up to -5 kV, depending on the width of the active clamp array (Fig. 4.32).

On-chip capacitors can significantly impact the triggering behavior and failure of ESD clamps in an unpowered IC [104]. The TLU scenario can be expected to change if a parallel on-chip capacitor (C_{OC}, Fig. 4.33a) is added to the SCR. However, significant passing level improvements were only observed for capacitor values above ~ 10 nF, which are unrealistic for on-chip integration.

Thus, an off-chip protection network alone may not be enough to prevent the risk of TLU. Sufficient secondary on-chip protection or decoupling may be more effective measures, as well as optional digital and analog circuit blocks implemented more robustly for latch-up (Sect. 4.1). The mixed-mode simulation approach with accurate parasitic components of the off-chip and on-chip networks provides a way to verify these design solutions.

4.4 Application Example

This section demonstrates how the theoretical background presented in Chaps. 3 and 4 can be applied to real design examples. The design of transceivers for automotive systems is used as an exemplary background for discussion of these essential aspects.

4.4.1 LIN and CAN Transceivers

Over the last two decades, the automotive industry has adopted the use of highly integrated smart-power ICs. These ICs not only contain a wide range of analog, power, and digital functions, but often include pins with system-level protection capabilities. The digital core and standard analog building blocks (operation amplifiers, comparators, data converters, and voltage/current references) are now integrated on the same silicon die and possess the capabilities to control motors, switches, and solenoids generate both switching and linear power supplies for internal external and circuitry. Examples of such system-level pins in this integrated circuit, which occupy only a small fraction of the Si die area, are CAN and LIN transceivers.

The on-board automotive network includes many systems (Fig. 4.34), combining CAN and LIN. The CAN (Controller Area Network) covers mostly front-end modules, which include dash board instruments, antilock brakes, and cruise control. CAN is one of the most commonly used communication pins for automotive applications [105]. The CAN has speeds from 125 kbps to 1 Mbps so systems are able to communicate simultaneously at any time, with arbitration in place to ensure that the messages are understood. The LIN (Local Interconnect Network) contains a low-cost serial bus with <20 kbps and a LIN master initiating a response from "slave" systems. Thus, the LIN is mostly for back-end modules like safety lights, door locks, power windows, and climate control.

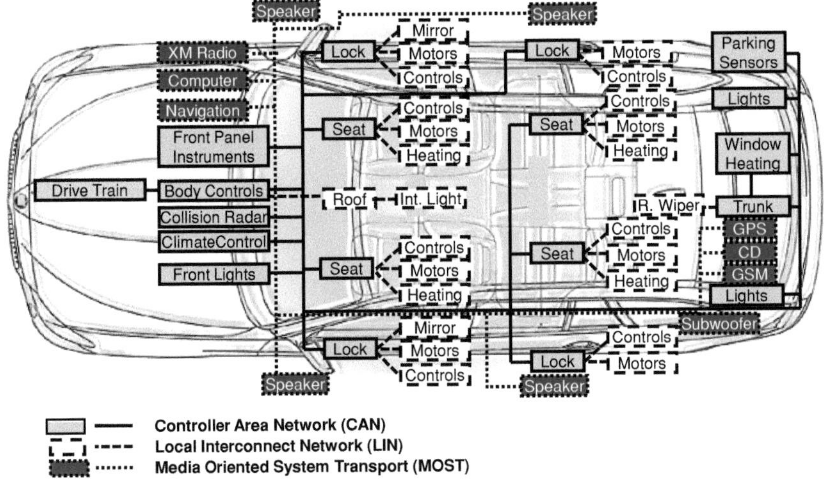

Fig. 4.34 Example of the composition of automotive systems

The typical CAN requirements CANL/CANH at the system level are ±6 kV contact and no air gap spec ISO 7637 1/2/3a/3b. For 12 V automotive systems, the pins' DC voltage tolerance is expected to be within the dual-direction range of ±40 V, involving additional tests. These specific automotive system tests include a bus short to battery; a DPI (Direct Power Injection); and injecting a common-mode AC onto the CAN bus during transmission to check the output for glitches or jitter.

These automotive systems exist in an extremely harsh environment that combines factors of system interoperability, system-level ESD, and EMI. Additionally, the high quality of IC products must be guaranteed by the zero DPPM (defective parts per million) failure rate target. These requirements impact the entire integrated component design, starting with selection of the silicon process features such that cost goals are maintained.

Though transceivers serve a critical communications function for automotive systems, their functional circuitry alone is not sufficient to define specific process technology for system-level ESD [106]. Higher levels of integration of analog and digital functions require high-density logic blocks, high-precision analog devices and HV power-optimized devices, while low cost requirements are still in place.

Thus, these system-level ESD protection challenges must be addressed at the level of free on-chip ESD device solutions tolerant of short-circuit faults from the battery and ground, shorted load conditions, and higher-than-expected transient voltages along the CAN bus (caused by chokes that are used to improve the radiated emissions for EMI compatibility).

Module pins can be exposed to an unprotected environment as a result of mishandling during assembly or maintenance. Failures due to ESD exposure must be eliminated by increasing protection toward zero DPPM on the chip in particular.

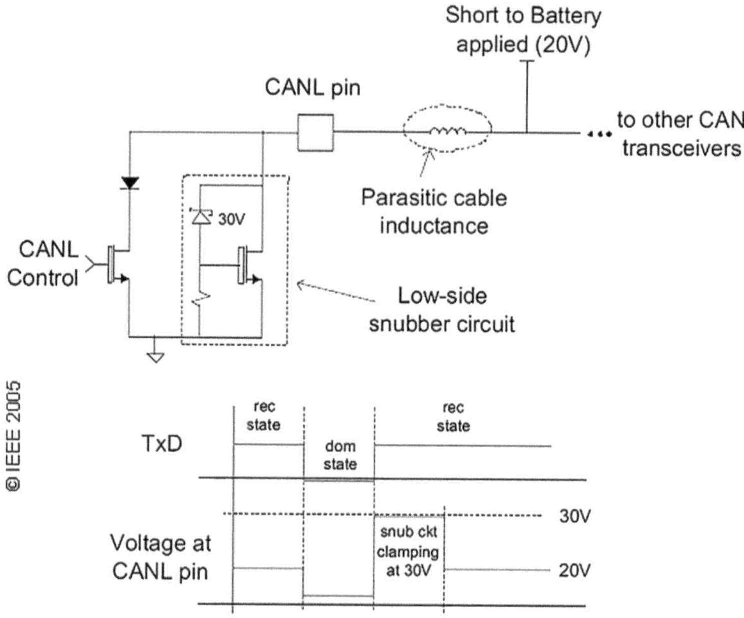

Fig. 4.35 Circuit showing the short circuit condition and ESD clamp used to prevent a voltage transient [106]

During short-to-battery test conditions, an excessive transient voltage is generated by the inductance of the CAN bus (Fig. 4.35). The driver on the CANL pin transitions into the current limit, establishing a current in the inductor. When the driver turns off, the CANL node has high impedance, while the bus inductance maintains the current with the corresponding voltage rising until the electromagnetic energy is fully dissipated.

In this case, if an SCR ESD clamp is used for local protection, damage can occur due to transient latch-up (Sect. 4.3) induced by the transient voltage spike. When short circuit conditions are applied from a low-impedance source, the SCR will continue to conduct until it is destroyed. Therefore, a more appropriate solution is either an active clamp referenced by large footprint avalanche diode or lateral PNP, DIAC or avalanche diode clamps with holding voltages above 20 V (Chap. 3).

Chokes are used to minimize radiated emissions and improve the immunity of the receiver. For CAN applications, the choke is a transformer that uses in-phase windings. The choke is specifically designed for blocking higher-frequency alternating current (AC) in an electrical circuit, while allowing lower-frequency or DC current to pass. Unlike the wiring of the bus, CAN chokes can have very high inductance of up to 100 uH. This inductive component can store substantial electro-magnetic energy that needs to be handled by an ESD clamp in the corresponding non-ESD-test conditions.

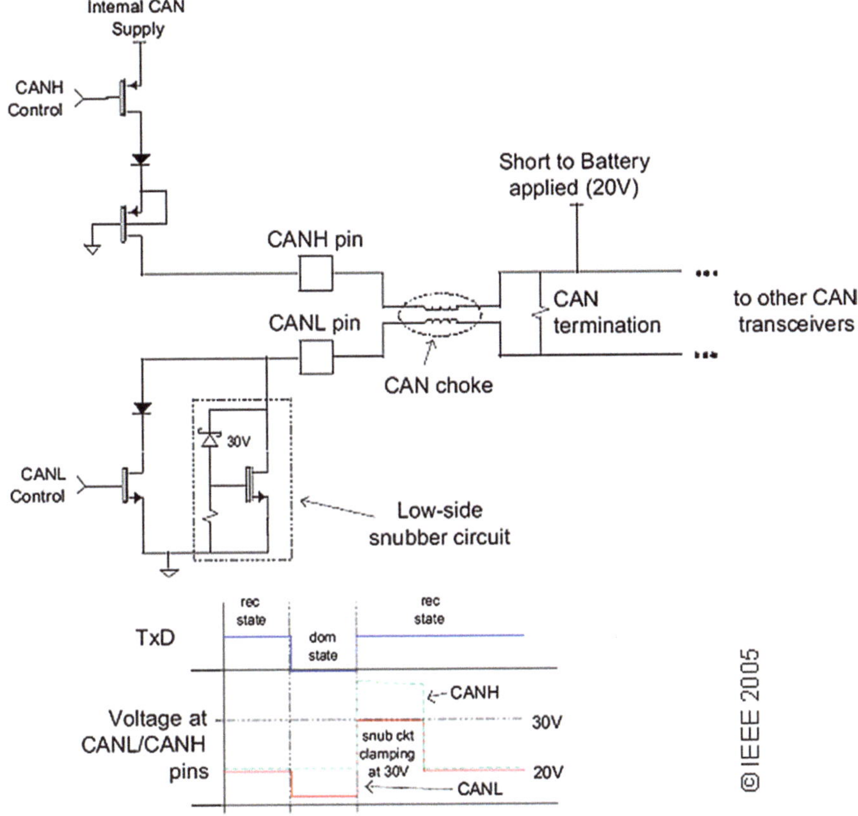

Fig. 4.36 CAN short circuit with a CAN choke [106]

Figure 4.36 illustrates a 20 V short occurring along the bus side of the choke. When the driver goes into the dominant state, the current builds up in the choke on the CANL side. Since most transceivers are designed with a current limit, this part of the waveform does not impose any problem for the device.

However, upon entering the recessive state, the CANL pin starts to transition to a positive voltage, since the current in the transformer cannot immediately change. The voltage increases until the polarity across the transformer changes, at which point the current begins to decay.

However, this change in current is mirrored in the other winding, which raises the voltage on the CANH. Under these conditions, it can be shown that the change in voltage is approximately equal for the CANH and CANL. The problem is that the CANH begins at the short circuit voltage (20 V in this example). If CANL is allowed to rise to 30 V as dictated by the "snubber" network, then the CANH will also rise to 30 V, resulting in a final voltage of 50 V. If this voltage level is above the triggering voltage of the ESD clamp, it must be lowered.

One way to solve this issue is to provide an additional snubber network on the CANH pin. Although the amount of energy needed for the CANH winding is much smaller, it is important to maintain electrical matching between the CANH and CANL for good signal quality.

Another common issue is the loss of ground. Under these conditions, the ground of the module is disconnected from the chassis while connection to the battery is still maintained. However, the CAN bus pins will continue to be referenced to the chassis ground, which appears as a very negative voltage with respect to the SBC ground. Since the battery voltage can be as high as 40 V under load dump conditions, the CAN bus pins must be designed to traverse as far as −40 V in high impedance conditions. Thus, dual-directional solutions are required to meet this specification (Chap. 3).

4.4.2 CAN Transceiver Case Study

The CAN, LIN and other communication pins in addition to IEC, ISO, and EMC ESD tests need to pass IBEE-Zwickau test [107, 108]. EMC IBEE-Zwickau test for automotive applications includes both a direct and an indirect stress. In the direct test the IEC gun is directly applied to the integrated circuit pins while in the indirect test, the IEC stress is applied through a 100 nH common-mode (CM) choke transformer [107]. Different waveform shapes of these tests complicate qualification of the ESD protection on IC level [107, 108].

An example for an entire spectrum of application problems related to the system level on chip design for automotive CAN pins has been recently presented in [109]. The study has demonstrated the specific failure mode under inductive IEC stress related to the inductor core saturation driving the increase of the rise-time from 1 to ∼20 ns. The inductive system level stress with EMC IBEE-Zwickau for CAN pin pair through the CM choke transformer (Fig. 4.37a) has been allied to IC pins protected with +45/−30 V tolerant dual-direction ESD clamp based upon back-to back stacked NLDMOS-SCR clamp components (Fig. 4.37b).

In spite previously verified components performance for the low rise-time ESD pulse, the clamp performance for the slower rise time did not meet the expectations. For the automotive qualification pulses both an underpass for Choke test (Table 4.3) and the windowing effect at IEC test (Table 4.4) have been observed.

A similar underpass was detected for another automotive test when CAN pins are stressed by IEC/ISO pulses through a loan coaxial cable (5 m). This Through Cable Discharge Event (TCDE) is different from the Cable Discharge Event (CDE). In CDE, the cable is charged to a certain potential first and then discharged through the part leading to longer time stress [110]. A TCDE mimics different events on pins that are connected to the source of stress through a cable. The test is defined to mimic automotive applications in which pins of different electronic parts are connected through a cable harness that runs around the car.

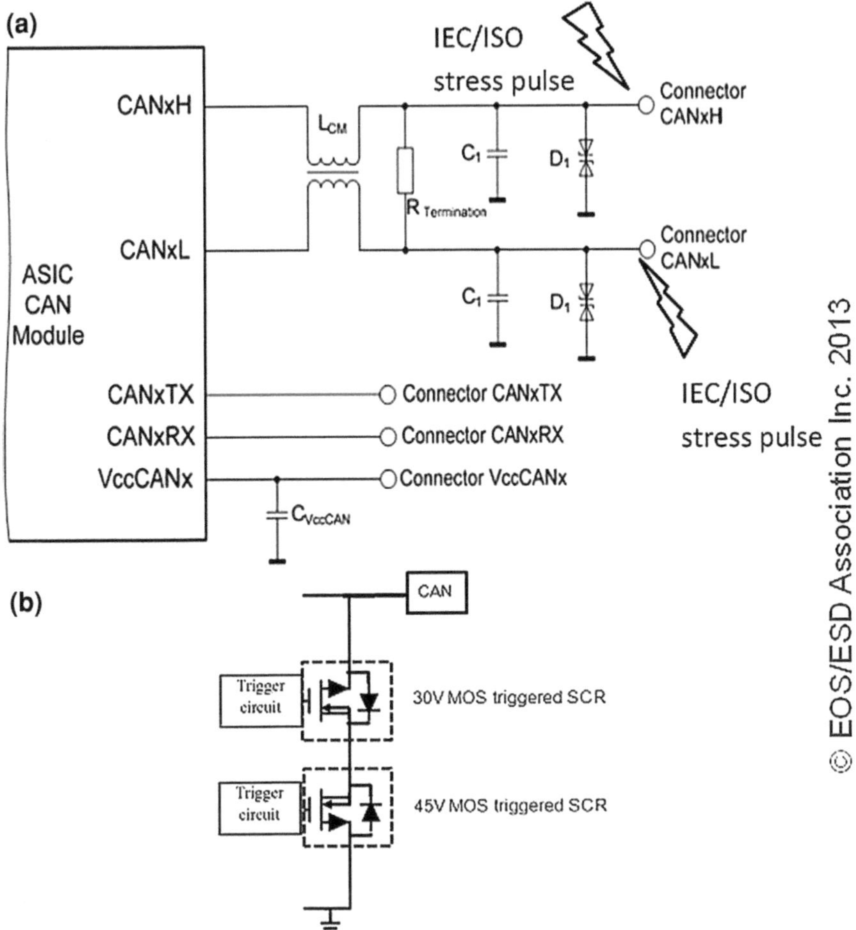

Fig. 4.37 CAN pin schematic with CM Choke LCM = 100 μH as part for the automotive ICs qual (EMC IBEE-Zwickau) (**a**) and the clamp schematic (**b**) [109]

Table 4.3 The IEC test results of the CAN pin w/and without CM Choke [109]

	Without choke (kV)	With choke (kV)
Positive IEC	>8	>8
Negative IEC	>8	−3

The system setup to evaluate the ESD performance of TCDE using in [109] used a 5 m RG58 cable grounded at both ends and has demonstrated correlation of the negative underpass with the regular IEC and Zwickau test (Fig. 4.38).

Due to higher rise time a significant non-uniform turn-on effect has found to be responsible for only partial NLDMOS-SCR clamp components turn-on. With the

Table 4.4 The single IEC strike test results show a failure window between −3 and −5 kV [109]

IEC strike level (kV)	Results
−1	Pass
−2	Pass
−3	Fail
−4	Fail
−5	Fail
−6	Pass
−7	Pass
−8	Pass
−9	Pass

Fig. 4.38 Block diagram for the Through Cable Discharge stress for CAN pins verification and the stress results for a 5 m RG58 cable grounded at both ends [109]

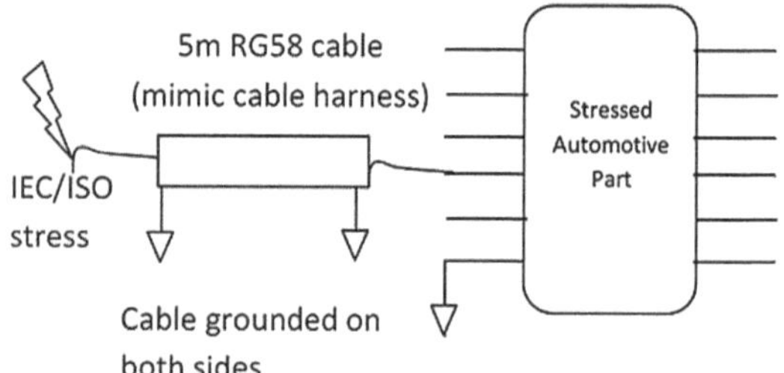

	Direct	TCDE
Positive ISO (2kΩ/330pF)	>15kV	>15kV
Negative IEC (2kΩ/330pF)	>15kV	-2kV

© EOS/ESD Association Inc. 2013

specific CM choke inductance value of 100 μH and the measured IEC pulse rise time was ∼25 ns. In spite of that this value was found 4 times less the simulated value for the ideal inductor, it was still represented a significant increase of the rise time in comparison with the non-connected choke (Fig. 4.39a). Meantime, the miscorrelation between ideal and real choke inductance upon current level has been explained by the cobalt choke core effect saturation (Fig. 4.39b).

Upon additional slower rise time TLP data for the back-to-back bi-direction NLDMOS-SCR clamp negative pulse the non-uniform turn-on has been confirmed. The increase of the rise rime in [109] was leading in the change of the ESD protection device turn-on uniformity. The turn-on with different on-state resistance was an indicator of only partial turn-on of the NLDMOS-SCR device fingers (Fig. 4.40a). The effect is similar to the non-simultaneous multi-finger turn-on described in the above sections of this book for DIAC structure. The effect was observed under rise time increase of the both TLP (Fig. 4.40a) and HMM (Fig. 4.40b) phase I-V characteristics.

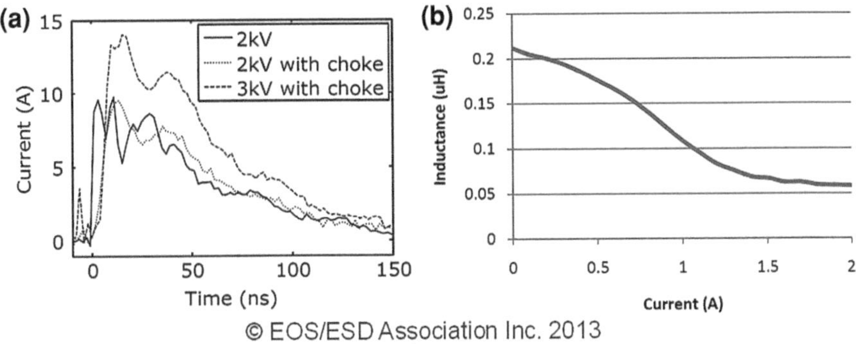

Fig. 4.39 Measured current waveforms during IEC strikes (**a**) and inductance as a function of current demonstrating the core saturation effect responsible for faster *dI/dt* [109]

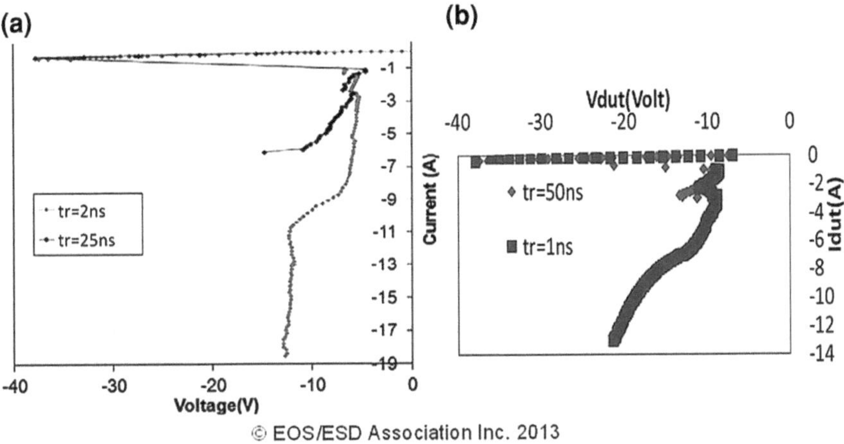

Fig. 4.40 Bidirectional ESD clamp and I-V characteristics for the negative TLP pulse with of 2 and 25 ns. **a** and negative HMM stresses with 1 and 50 ns rise-times [109]

The fix design measure to improve their ESD design was suggested in [109] in form of ballasting of the dual direction device clamp components on the finger level. Instead of the original common node connection of the high side and low side NLDMOS-SCRs (Fig. 4.41a) components the connection has been done on the individual finger level (Fig. 4.41b). ESD performance of the clamp been resigned to improve the uniformity of the turn has been demonstrated both by the long rise time TLP (Fig. 4.41c) and test system level test result.

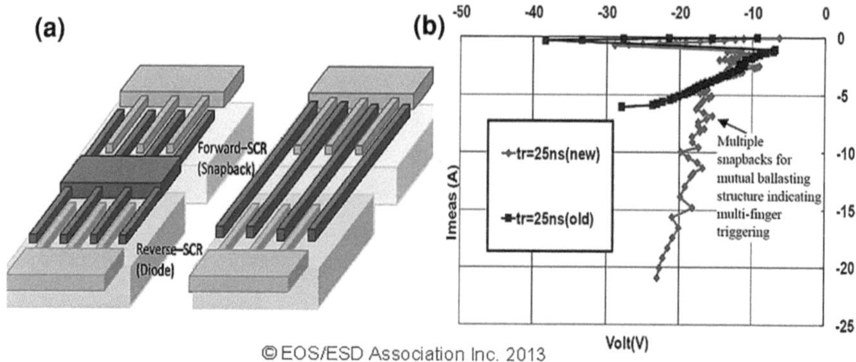

Fig. 4.41 Schematics showing the metal connection of the middle node between the forward-SCR snapback-mode and the reverse SCR (diode-mode) (**a**) is the conventional connection with common metal in the middle node (**b**) the mutual ballasting technique with each finger of the forward-SCR connected to the corresponding finger of the reverse SCR. Comparison of the new and the old SCR layout with long rise-time TLP of 25 ns [109]

4.5 Summary

This chapter has demonstrated that there are many on-chip design aspects that make chip integration of a validated stand-alone ESD clamp for the system level pulse. Application specifics and chip functionality need to be thoroughly taken into account to avoid clamp interaction with internal circuit blocks during both system-level ESD stress, normal operation and specific system level tests that can induce both transient latch-up and apply pulsed conditions that are hard to validate on the standalone clamp level.

Subsequent turn-on of the parasitic structures formed in the layout requires understanding of major latch-up phenomena in ICs. To avoid latch-up, certain latch-up layout rules must be developed, experimentally verified, and followed in semiconductor HV processes. These rules are usually determined from limited experimental results and primarily are focusing on preventing HV parasitic n-p-n turn-on by either disabling the n-p-n structure itself or by reducing injection from the junctions in nEpi pockets.

Transient latch-up (TLU) is an effect of accidental turn-on of the snapback clamp during normal operation conditions or electrical overstress events when the on-state can be further supported by the applied power supply current. Transient latch-up is pertinent to the topics of this book in its importance to the system-level stress conditions in power-on mode.

Over the last two decades, the automotive industry has adopted the use of highly integrated smart-power ICs that not only contain a wide range of analog, power, and digital functions, but often include pins with system-level protection capabilities, for example, CAN and LIN transceivers. The on-board automotive network for communication pins in addition to IEC, ISO, and EMC ESD tests need to pass IBEE-Zwickau test, as well as Through Cable Discharge Event (TCDE). These tests produce the stress pulse waveforms that are in significant deviation of the component level ESD stress waveforms. The slower rise time may result in non-uniform turn of the ESD protection solution and thus require special design measures to overcome the issues.

Chapter 5
IC and System ESD Co-design

The trend towards high level integration of the system functional blocks on-chip is eliminating the barrier between the components and systems. System-on-chip (SoC) and system-in-package (SiP) designs now often combine a variety of analog and digital circuit blocks that can directly interface with system ports and therefore may require system level ESD protection capability. This creates a design paradigm shift toward the specification of the system-level ESD requirements for selected IC pins. In general, compliance with the component-level low energy ESD (CDM, MM, HBM) standard tests does not guarantee the withstanding of even a fraction of the high energy ESD transients introduced by the system-level pulses with 30 kV pre-charge level. To avoid the impact on real products reliability, especially at smaller form factors of the portable systems, the IC product specifications now often include system level requirements. With the scaling of the semiconductor processes down to the advanced technology nodes, both active devices and interconnects become less and less high current capable. As a result of this the on-chip protection design for system-level ESD stress becomes increasingly challenging and often must be combined with the off-chip protection as a more cost effective approach.

The ESD test methodologies enabling a protection design for component- and system-level ESD stresses were discussed in Chap. 2, demonstrating how the transient device response during ESD stress can be extracted from captured voltage and current waveforms. These test methodologies combined with the understanding of the on-chip ESD protection design (Chaps. 3 and 4) are now used in this Chapter to develop an effective and robust ESD co-design approach for the system-level IC pins. This new approach combines transient device characterization on test boards and on-wafer setups with device and circuit simulations. The goal of such combination of simulations and on-wafer characterization is to enable the design and verification of the system-level ESD protection solutions at the early stage of the IC design and prior to the final system design or definition.

The first section of this chapter introduces the available off-chip ESD protection devices followed by an overview of the available simulation tools for ESD protection design. Methodologies and examples for the device and circuit modeling are provided. The simulation models are used for two key system-level ESD design methodologies: datasheet based design and the co-design. The required

V. Vashchenko and M. Scholz, *System Level ESD Protection*,
DOI: 10.1007/978-3-319-03221-4_5, © Springer International Publishing Switzerland 2014

input for each methodology is discussed together with their advantages and disadvantages. Through several case studies the recommendations for the design of ESD protection structures for system-level IC pins are outlined. Finally the chapter is concluded with a comparison, benchmarking and discussion of the introduced design methodologies.

5.1 Off-Chip ESD Protection with Si TVS Components

The overall off-chip ESD protection strategies have been already introduced on a higher level in the introductory Chap. 1, highlighting the basic principle for the off-chip protection. In this section the off-chip protection is presented at a different depth mainly focusing on off-chip protection methodologies with Si-based transient voltage suppressor (TVS) diodes.

Many different types of TVS are used for the off-chip/board-level ESD protection. Depending on their application active or passive devices are in use. Active devices like TVS diodes protect an IC by shunting the ESD current after the breakdown of a semiconductor junction. The discussed devices are able to withstand more than one ESD discharge and thus are able to protect the IC to several ESD discharges throughout its life time.

This section introduces the most common off-chip ESD protection devices. Where available the typical quasi-static or transient device characteristics are discussed together with some application specific properties like the device capacitance.

5.1.1 Silicon TVS Device Structure

This section presents some brief material on device-level design of transient voltage suppressors (TVS) [111–114] with the purpose to differentiate between the on-chip ESD device solutions and TVS devices on the conceptual level, rather than to pursue a comprehensive TVS device design review.

Single TVS devices available on market are usually represented by either two-pin single-direction TVS or dual-direction TVS (Table 5.1). The dual-direction version of the device is simply obtained by combing two uni-directional diodes in the same chip, with a common floating p-substrate and some reverse path diodes. In multi-port TVS devices, the uni-directional devices are co-packaged (Fig. 5.1).

TVS process technology targets an extremely low cost and low mask/process steps count. This vertical device process technology is somewhat similar to discrete diode or transistor components. Typically, it uses a highly doped n- or p-substrate that forms the corresponding cathode or anode of the uni-polar TVS diodes. The top epi- and buried layer regions form the blocking junction with the substrate, targeting the breakdown and clamping voltage characteristics.

Table 5.1 Comparison of single-node TVS diodes for high speed applications from various manufacturers

TVS model	Clamping voltage @ 1 A (V)	Capacitance to ground (pF)	Leakage current (µA)	Package
PESD5V0F1BLD	11	0.4	0.1	SOD882D
ESD7481MUT5G	10	0.25	0.05	X3DFN2
TPD1E05U06	10, unidirectional	0.42	0.01	2X2SON
Rclamp0531T	12	0.5	0.1	SLP1006P2T
ESD101-B1-02	8	0.1	0.05	TSSLP-2-4

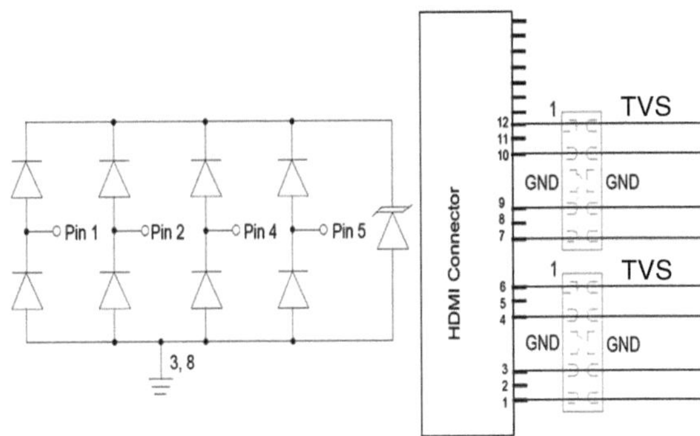

Fig. 5.1 Example of TVS array application for HDMI port protection

The TVS device itself is either based on vertical avalanche diode architectures with an optional depletion region for parasitic capacitance reduction using a lightly doped thick epi-region (Fig. 5.2a, b), an SCR-type vertical device (Fig. 5.2c, d) or a punch-through BJT (Fig. 5.3).

Thus, the TVS device architecture in general is incompatible with the BCD process and cannot be integrated as a module, primarily due to a significant increase of costs. However, TVS can be co-packaged with analog IC.

In general, the desired physical location for TVS placement is directly at the ports, antennas or some other PCB periphery, rather than at IC pins. Use of TVS is intended to limit electromagnetic interference induced by a system-level ESD pulse over the entire system, not just at an IC pin. Thus, the generic functionality here is different for the on-chip system-level solutions in highly integrated mixed-signal IC and discrete TVS used in the system.

Often, TVS use is driven by last-minute ESD compatibility issues of the system and is a quick "patch" fix squeezed onto the PCB. In subsequent sections it will be shown that TVS can be also used in a more systematic way in the frame of an ESD co-design methodology.

Overall, TVS combine a low level of integration, an extremely small footprint and a small pin count. The high level overview of the TVS devices for different types has been already presented in Chap. 1 from their application perspective. In this section the focus on the TVS components is done from the co-design and

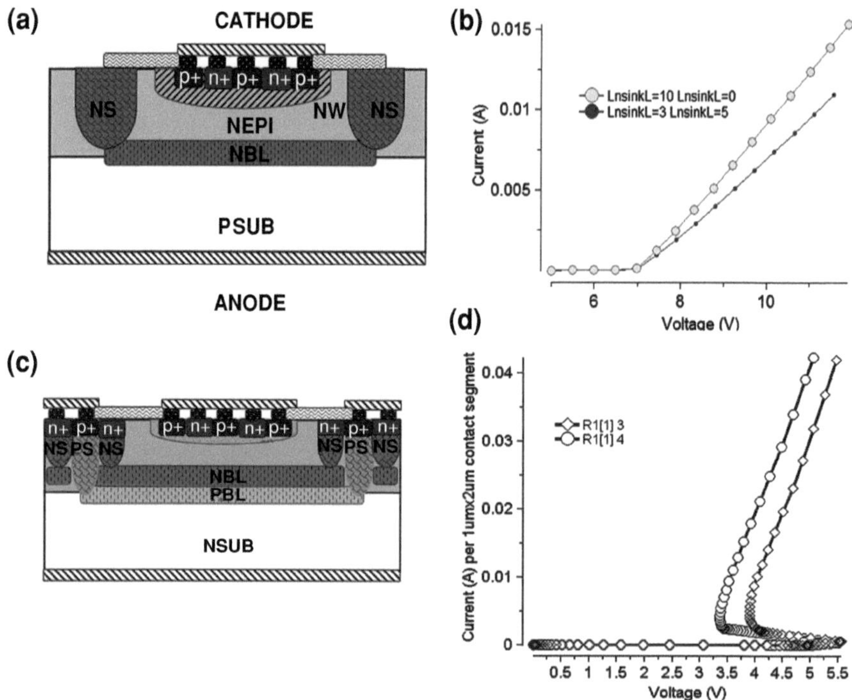

Fig. 5.2 Examples of avalanche diode based (**a**, **b**) and SCR-based (**c**, **d**) TVS structure cross-sections with simulated I-V characteristics

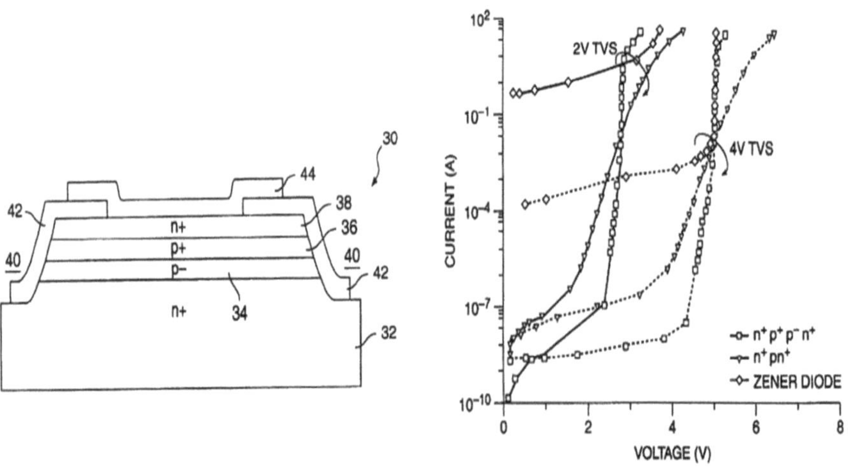

Fig. 5.3 Low-voltage punch-through transient suppressor, employing a dual-base structure [92]

transient characteristics perspective. The material below presents from this point of view only Si based TVS devices.

Traditionally Si TVS for the ESD/EMI protection have been treated in the bill of material as discrete diode components. During the last decade due to emerging systems, Si TVS applications bridged the gap between the discrete and integrated (IC) products. In combination with advanced packaging technologies these semiconductor components became integrated multi-functional devices rather than just single functional discrete components. Depending on its specification the protection devices are either of general purpose or targeting a limited range of applications based on their performance and form factor. There are two major classes of protection devices: protection against ESD and EMI and clamping type of products for the protection of data or signal lines.

The Si TVS devices include thyristor-based snapback or (sometimes called crowbar) devices and non-snapback clamping avalanche diode based devices. If during EOS/ESD stress the voltage across the TVS diode exceeds its breakdown voltage, the TVS devices turn-on and provide a low impedance path for shunting the applied ESD/EMI current. Thereby snapback TVS device provide a high current and energy level protection against both ESD pulses and transient power surges. The relevant protection standards include IEC61000-4, IEC61000-5 and telecommunication standards.

The low energy TVS are primarily targeting ESD IEC61000-4-2 standards, EMI (electromagnetic interference) and EMC (electromagnetic compatibility) IEC61000-4-3/6/8 standards, with the typical applications like hand-held and mobile devices, consumer electronics, computing, high speed data, and signal lines. Both high and low energy TVS can be integrated or co-packaged with discrete active and passive devices or ICs. Non-snapback Si TVS components are mostly used in low voltage circuits providing lower energy level in comparison with snapback TVS such as consumer equipment.

Si TVS diodes exist as unidirectional or bidirectional devices and in huge varieties regarding form factors and packages. Furthermore, TVS diodes are offered either as single devices or arrays. A unidirectional TVS diode operates as a rectifier in the forward direction like any other avalanche diode, but can handle much higher peak currents. In reverse it behaves like a single forward-biased diode. An important electrical parameter of TVS diodes is their capacitance. TVS diodes are wide devices to carry the large currents during system-level ESD stress. The large device width increases the (junction) capacitance of the diodes. As a result some TVS diodes can have capacitances up to 100 pF. However, TVS diodes with capacitances below 1 fF are available for high-speed interfaces like USB or HDMI. These TVS diodes can contain several single diodes in series, including p-i-n diodes, to lower the overall capacitance.

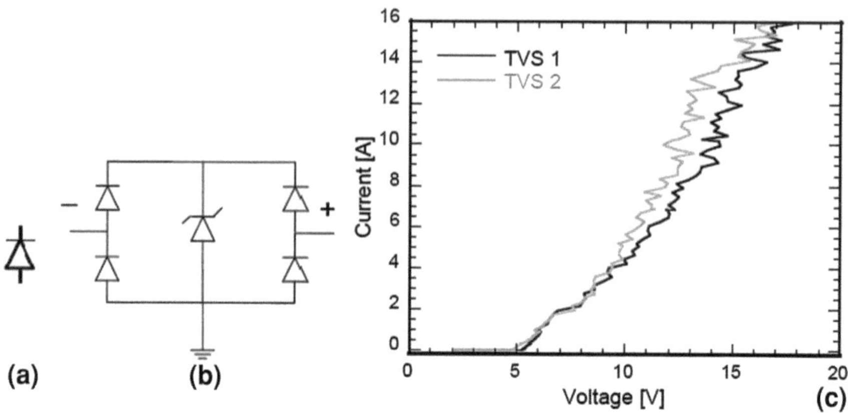

Fig. 5.4 Schematic of the two TVS diodes TVS 1 (**a**) and TVS 2 (**b**) with the parameters (Table 5.2) and comparison of their 100 ns TLP I-V characteristics (**c**)

5.1.2 Silicon TVS Characteristics

The equivalent schematics of two example TVS diodes are shown in (Fig. 5.4a, b). The corresponding 100 ns TLP I-V characteristics are presented in Fig. 5.4c. The most important datasheet parameters are summarized in Table 5.2, including V_{BD}: DC breakdown voltage (at 1 mA), V_{ESD}: robustness during IEC61000-4-2 contact discharge, C_{TVS}: capacitance input to ground.

Although both TVS diodes have the same DC breakdown voltage the measured TLP I-V characteristics are different. The triggering voltage under TLP stress is lower for TVS 1. However, the higher on-resistance of TVS 1 causes a larger voltage drop at higher stress level. From the measured transient characteristic for TVS devices (Fig. 5.5) the voltage waveform across the TVS2 diodes for 4 kV HMM stress has a higher overshoot. The overshoot during turn-on is directly related to the turn-on speed of the TVS diode and is an important parameter for the ESD protection design. It must be taken into account at selection of a TVS diode for the specified design window. Depending on the design both TVS diodes can deliver some advantages. If the higher overshoot can be tolerated by design then TVS 2 is the preferred off chip component since it provides lower clamping voltage than TVS 1 and thus reduce the level of the secondary current at the IC pin.

5.2 System-Level ESD Design Modeling and Simulation

Circuit modeling and mixed-mode simulations are key steps of the system-IC co-design methodology derived in the last sections of this Chapter. Respectively a model representation of the active and passive devices in the system level ESD

Table 5.2 Datasheet parameters of two TVS diode examples

Parameter	TVS 1	TVS 2
Manufacturer	On semiconductor	Texas instruments
Model	ESD5Z2.5T1	TPD2EUSD30A
V_{BD} (V)	4.5	4.5
V_{ESD} (kV)	30	8
C_{TVS} (pF)	145	0.7

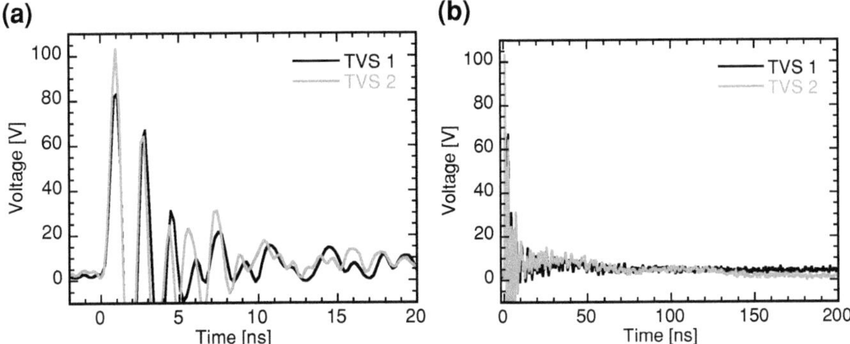

Fig. 5.5 Voltage across TVS 1 and 2 during IEC61000-4-2 stress: **a** first 20 ns and **b** full stress duration; characterization when mounted in a test board, stress level: 4 kV

network is an important milestone for any design. Application of the simulation tools and extraction methodologies is directly used to compose the ESD protection off-chip component models that can be further used to support the system-level protection design. The consistent approach covers all related aspects starting from the equivalent circuit of the test setups, behavioral models for off-chip and on-chip components combined with a proper net list representation of the PCB-level network.

5.2.1 ESD Tester Model

Modeling and simulations during ESD design require an adequate representation of realistic ESD stress pulse sources, realized in different simulation environments. This includes the extraction of suitable models for the existing ESD testers. Typically an ESD tester can be represented either by simulation circuits known from previous studies or by extracting a lumped element model [115] from a real tester.

The example HMM tester model used in this section is based on the schematic and component values of the discharge circuit of a real industrial HMM tester

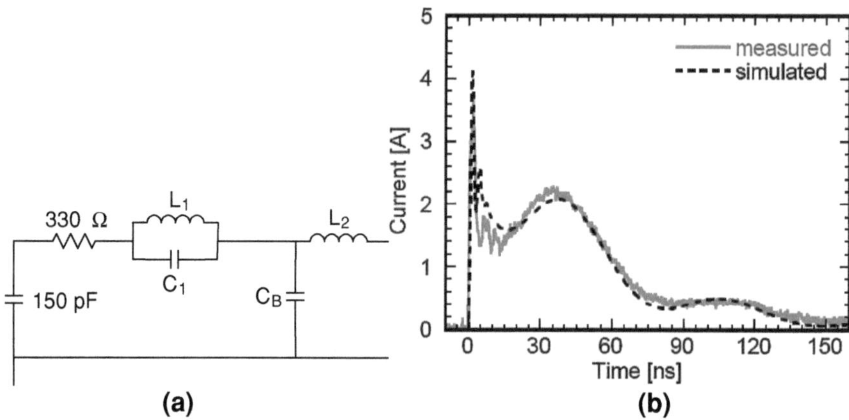

Fig. 5.6 HMM tester model in SPICE; **a** simulation circuit and **b** comparison measured and simulated current into short; HMM stress level: 1 kV

(Fig. 5.6a). It provides a good agreement between the simulated and the measured current waveforms. The SPICE modeling language implementation makes this model flexible for inclusion in different simulation environments including field solvers and mixed-mode simulators.

This simple circuit includes the 150 pF capacitor initially pre-charged to a stress level voltage as an initial condition for the transient simulation. The charge redistribution in time through the $L_1C_1C_BL_2$ network creates the double peak waveform with the parameters that closely corresponds to the system level IEC pulse (Fig. 5.6b).

5.2.2 Representation of ESD Devices with Behavioral Models

The conventional MOS transistor compact models, like for example BSIM3V3, are usually not suitable for compact ESD simulation (see Chap. 1). These models do not include the components for the conductivity modulation conditions of high impact ionization rate and injection. At the same time extraction of the precisely calibrated ESD compact models [116–117] (Chap. 1) might be a too complex task for many users' timeframe. In addition, running simulation of complex circuits may cause convergence issues for the snapback and high current region during ESD stress.

An alternative and a more simplified approach described below do not require implementation of the compact models with complex parasitic bipolar elements as a part of the model. In spite of the impact on the accuracy of the models and corresponding simulation results the analysis is rather adequate for the system-

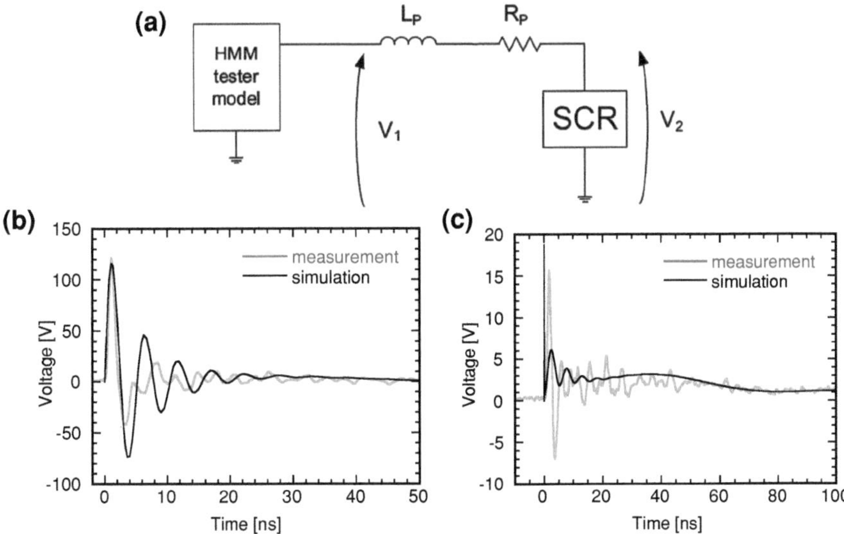

Fig. 5.7 SPICE simulation circuit for on-wafer with SCR behavioral model including the probe parasitic inductance L_P (2×20 nH) and resistance R_P (2×0.4 Ω) (**a**) and comparison of measurements and simulations results for the case with parasitic circuit components included (**b**) and eliminated (**c**) for 1 kV HMM stress

level ESD protection design. It can be directly used to study the transient interaction between on-chip and off-chip ESD protection networks.

Several case studies are used in this book to demonstrate this approach. In the first example an SCR device provides the low holding voltages ~ 1.5–2 V due to double injection conductivity modulation in high current ESD stress regime. The SCR model can be also used to adequately represent the circuit latch-up or transient latch-up susceptibility. To model the SCR structure a behavioral SPICE model of a standalone, discrete SCR in available SPICE libraries is used for example from discrete component vendors. The advantage of these models is the mimicking of a basic operation of the parasitic bipolar junction transistors. By manual fitting of the behavioral model transient parameters, for example the dV/dt ratio for the device turn-on, the SCR component representation by a model can be done. The probe needle parasitic are important to include in the SPICE simulation for a HMM simulation of a SCR on wafer (Fig. 5.7). These parasitic components can be extracted using the calibration methodology for HBM on-wafer testers as described in Chap. 2.

As a result of such approach the comparison of the simulated and measured voltage waveforms for the cases with and without on-wafer parasitic components demonstrate the effect (Fig. 5.7b, c). In this example the SCR was implemented by modifying the SPICE model of a discrete thyristor broadly available in the corresponding design kits for these components.

A relatively good agreement between measurements and simulations (Fig. 5.7) provides a confidence that a similar approach can be applied for the off-chip SCR-based TVS components. The observed faster rise time and slightly higher voltage overshoot in the simulated voltage waveform in Fig. 5.7c can be attributed to bandwidth limitations of the voltage probe used in the measurements.

If a more accurate analysis is required, another practical alternative to the behavioral model fitting method is the use of TCAD device finite-element models introduced into the mixed-mode simulation (Sect. 1.5). The limitation of this approach is related to the challenge to accommodate large circuits and the possible impact on convergence of the simulation.

The next case study is brought below to demonstrate that the ESD behavioral models are not only required to represent the primary ESD protection on-chip or off-chip components. For example the simulation of the a grounded-gate snapback NMOS (SNMOS) becomes important when the operation of internal circuit protected by the clamps based on these devices needs to be taken into account as a part of IC-component co-design for system level specifications. The SNMOS ESD devices are often used for an open drain output in small current drivers with added self-protection capability. Thus during a system level ESD event, applied to the pin with the SNMOS, the simulation of the current sharing scenario between the on-chip and off-chip protection becomes important. A precise calibrated SNMOS compact model was reviewed in Sect. 1.4. As an alternative within the scope of this section SNMOS can be represented by a more simplified behavioral model. This model includes an ideal reverse breakdown voltage diode and serial resistance to match the on-state resistance of the SNMOS with the experimental data. This represents the device holding voltage characteristics.

In spite of such significant simplification the approach provides a sufficient accuracy in case of dV/dt triggered NMOS devices with strong gate coupling that is typical for small current drivers. This represents the worst case scenario at least. The experimental TLP I-V characteristic of the SNMOS is adequately mimicked by the simplified behavioral model. The device shows no snapback due to strong dV/dt effect, while the holding voltage ~ 6 V can be modeled simultaneously with the breakdown voltage under the added series resistance ~ 0.8 Ω. The approach further provides an acceptable matching between the circuit simulation and the measurements obtained with HMM testing (see Fig. 5.8).

5.2.3 TVS Diode Models

As key components in system-level ESD protection design, the TVS diodes are usually represented by either uni- or bi-directional devices. Both single and multi-port compatible arrays components are offered by multiple vendors. Depending on the internal device type the TVS can be represented in a circuit simulation both by a rather simple or a more complex model.

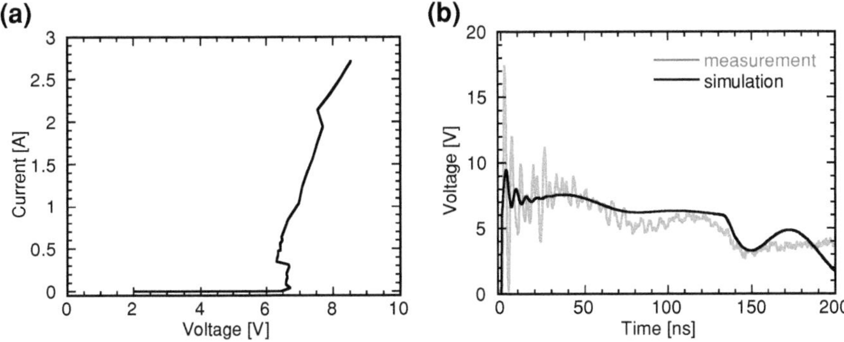

Fig. 5.8 TLP I-V characteristics of the snapback NMOS clamp with high dV/dt effect (**a**) and comparison of the experimental voltage waveform and circuit simulation results with the behavioral model (**b**) for 1 kV HMM stress

In the majority of cases the internal TVS device structure in the clamping voltage mode can be represented by the avalanche diode in reverse-biased breakdown mode or by a device supporting a conductivity modulation regime in snapback mode. In both cases the extraction of the model parameters for the high-current region is based on the results from TLP testing. The TLP characteristics are informative both for forward and reverse current modes.

In case of uni-directional TVS components with an internal avalanche diode structure the conventional diode compact model can be used to represent both the ideal diode in forward bias and avalanche breakdown mode with added serial resistance to match TVS on-state characteristics. The approach is mainly adequate if the diode I-V characteristic in breakdown conditions is linear in the high current region which is typical for most of the Si TVS diodes. From TLP measurements the triggering or breakdown voltage and the on-resistance are extracted to be introduced into the SPICE circuit. In [118] several diode strings were used to model the breakdown voltage.

The bi-directional TVS diodes often simply combine two back-to back diodes similar to the ESD clamps explained in Chap. 3. In each stress polarity one of the TVS components provides the breakdown current path, while the other TVS diode conducts the current in forward-biased mode. The fast rise time of the system-level ESD stress current may cause a forward-recovery effect. It results in a formation of a significant overshoot during turn-on [119]. The forward recovery effect needs to be considered in bi-directional TVS diode models [120]. According to [113] TVS diodes can show a significant forward recovery during turn-on.

The overshoot for unidirectional TVS devices can be represented by matching a series inductance to measurement data. Figure 5.9 shows the simulated and measured voltage across a uni-directional TVS diode. A good agreement between the measured and the simulated results is obtained.

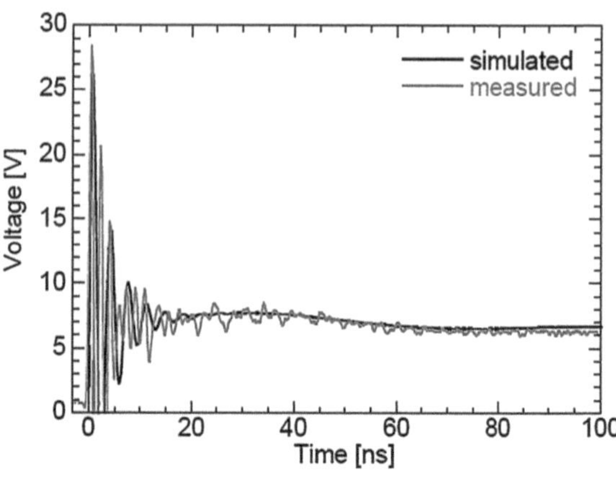

Fig. 5.9 Measured and simulated voltage waveform across a TVS diode for the HMM stress level 1 kV

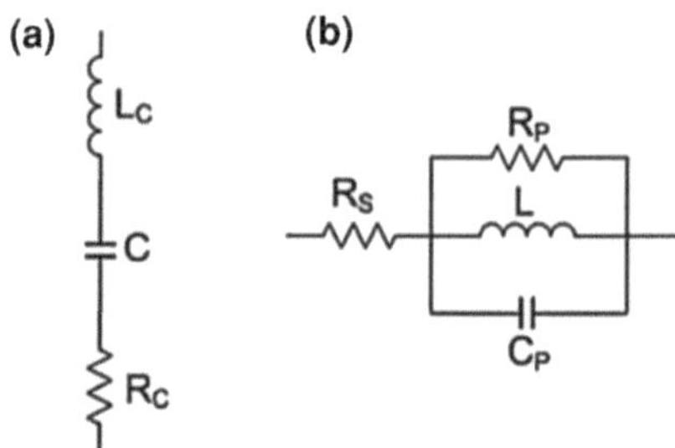

Fig. 5.10 Equivalent schematic of real passive devices: **a** capacitor and **b** ferrite bead

5.2.4 Modeling of Board-Level Passive Components

The circuit models for passive devices include the equivalent circuit for the parasitic components. For example a capacitor model (Fig. 5.10a) includes the series connection of the capacitor with a parasitic inductance and a parasitic resistance. While the resistance is usually relatively small, the inductive component may reach 1–3 nH for surface mount devices and more than 10 nH for axial-leaded components. The presence of inductive components creates the conditions for a significant voltage drop when a high ESD current is conducted by the capacitor. An example of the equivalent circuit for a ferrite bead (Fig. 5.10b) includes the parallel connection of the inductor with the parasitic resistor and capacitor as well as additional serious resistor circuit component.

The extraction of the parasitic capacitor circuit components can be assisted with the step response of the component. Alternatively a discrete capacitor can be analyzed using HMM current and voltage waveforms if the stress is directly applied to the component. The calibration methodology similar to HBM on-wafer setups (Chap. 2) can be applied to obtain the data for extracting the parasitic inductance and resistance of the capacitor for example by fitting the equivalent

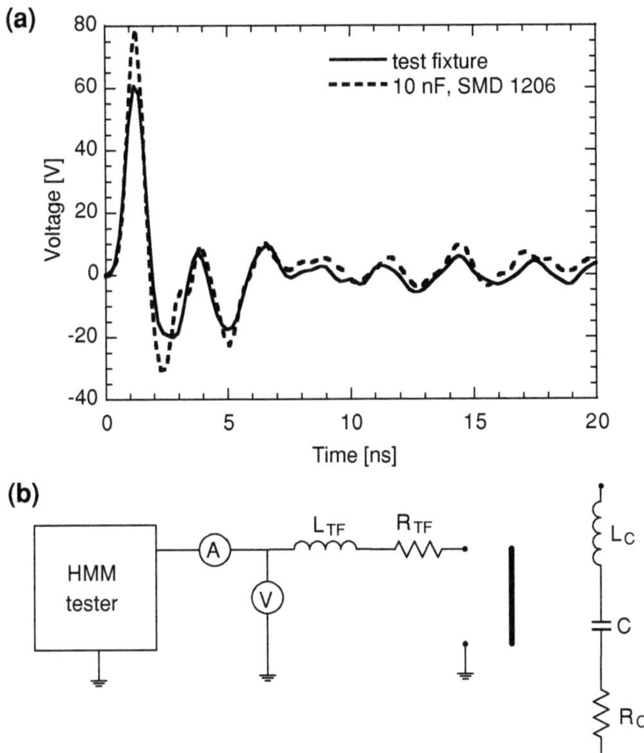

Fig. 5.11 Measured voltage waveform (raw data), captured from of a 10 nF SMD capacitor and the test fixture at HMM stress level 1 kV (**a**) and the block diagram for the measurement setup used for extraction of the parasitic elements of the off-chip capacitors with the parasitic inductance L_{TF} and parasitic resistance R_{TF} of test fixture, parasitic inductance L_C and resistance R_C of capacitor itself (**b**)

circuit. An example of the voltage waveforms captured for a discrete 10 nF SMD 1206 capacitor 1 kV HMM pulse is presented (Fig. 5.11a).

The obtained high peak voltage on the measured voltage waveform includes an additional voltage drop on the required test fixture (Fig. 5.11b). Therefore the extraction of the capacitor parasitic needs to be done in two steps. First the parasitic of the test fixture is extracted by applying HMM stress to a short circuit. Then the calibration methodology similar to HBM tester (Chap. 2) is applied to obtain the parasitic of the test fixture. The HMM data capturing is repeated with the capacitor connected to the test fixture. The parasitic of capacitor and test fixtures are de-embedded. In a third step the parasitic components of the test fixture need to be subtracted to obtain the accurate capacitor circuit components.

Although the extracted equivalent circuit components may look negligibly small, their impact at higher system-level ESD stress level is significant. For example there can be 30 % more voltage overshoot during system-level ESD

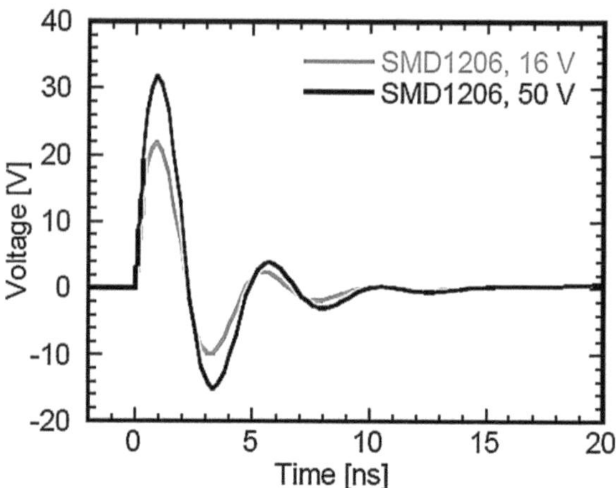

Fig. 5.12 Simulated voltage waveforms across two different voltage-rated 1 µF capacitors during 8 kV IEC61000-4-2 stress

stress when two capacitors in the same package type are compared (Fig. 5.12). The higher voltage rated capacitor has a larger parasitic inductance which results in a 10 V larger voltage peak during ESD stress. This is significant when performing accurate system-level ESD simulations and when designing an off-chip protection network.

5.2.5 Mixed-Mode Simulation

The mixed-mode simulation approach (Sect. 1.5) is an equally useful and powerful tool for the system level design. The degree of freedom to include the representation of the active on-chip and off-chip active and ESD devices by the Finite Element (FEM) models tremendously improves the accuracy of the simulation and removes many limitations imposed by the simplified behavioral, even of accurate compact models.

In case of IC with system level protection requirements, co-designing with the mixed-mode simulations combine the same advantages of traditional circuit simulations with the accuracy of semiconductor transport equations solution in the FEM-based device simulations. Since the semiconductor carrier transport equations are solved together with the circuit equations, complex device-circuit interactions of ESD designs can be studied. The use of FEM device models is similar to the on-chip design and in this case the need for compact models for the off-chip active components can be eliminated too. The extension of the simulation combines the on-chip device models with off-chip/PCB-level devices and parasitic.

An example of the mixed-mode HMM circuit simulation with an on-wafer nLDMOS-SCR includes the HMM tester model with lumped-elements, the nLDMOS-SCR FEM device based on the original process information and the parasitic components of the on-wafer setup (Fig. 5.13a). Figure 5.13b compares

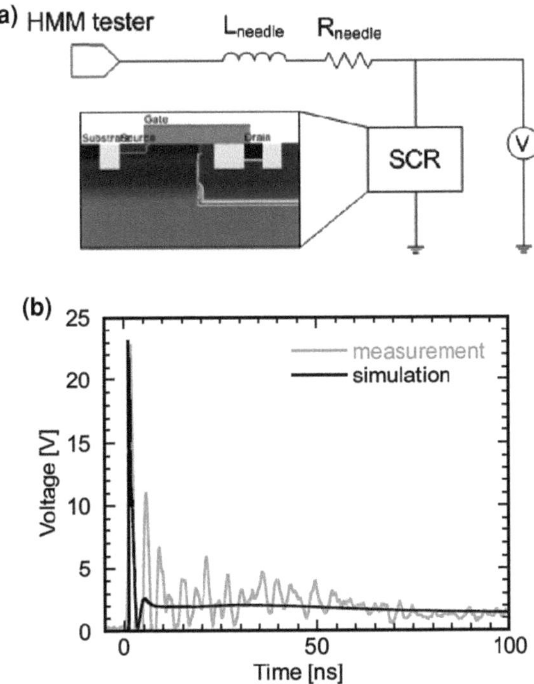

Fig. 5.13 Mixed-mode circuit for HMM simulation of on-wafer nLDMOS-SCR (**a**) and comparison of the measured and simulate voltage waveforms for stress level 1 kV (**b**)

the obtained voltage waveform with measurement data. The HMM voltage waveform measured with a *Kelvin* setup is found in a good agreement with the simulation in the time domain for both the turn-on and the on-state of the nLDMOS-SCR (Fig. 5.13b).

5.3 Datasheet-Based System Level ESD Design

The datasheet based design is mainly brought in this section for comparison. The remaining part of this chapter introduces the *IC-system co-design* methodology for which the material of this book was originally composed for. Overall, the datasheet based design methodology represents the most straightforward and intuitive way to develop systems without any extra steps.

Typically, in the datasheet based approach the appropriate voltage and signal tolerance TVS are simply placed at all the ports of a system. In the example of the mobile phone (Fig. 5.14) the TVS protects not only all user data interface ports, antennas and charging circuits but even the display and the speaker. Some TVS components are also used for the internal protection.

The protection with TVS includes the implementation of a robust mechanical path with connector shielding and strong PCB traces to the ground, a main high

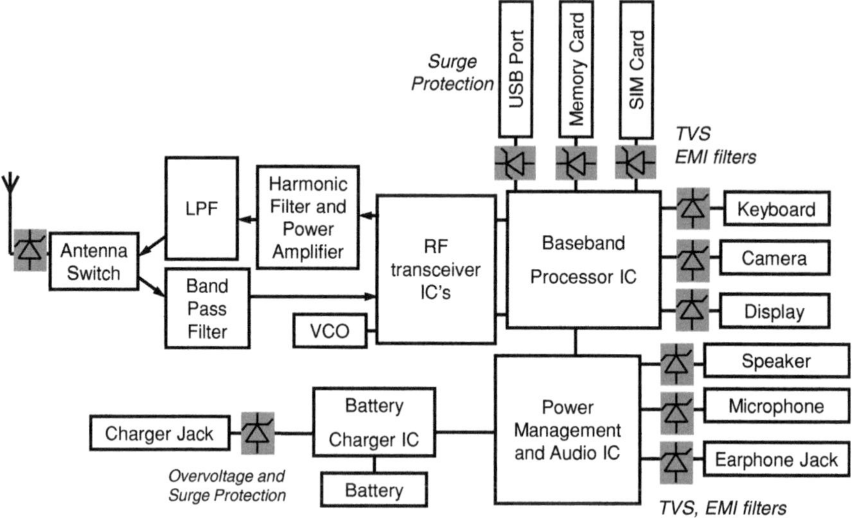

Fig. 5.14 Example of the TVS application for a mobile phone system

Fig. 5.15 Main high current electrical path for the TVS and through the IC pin with the on-chip ESD protection structure

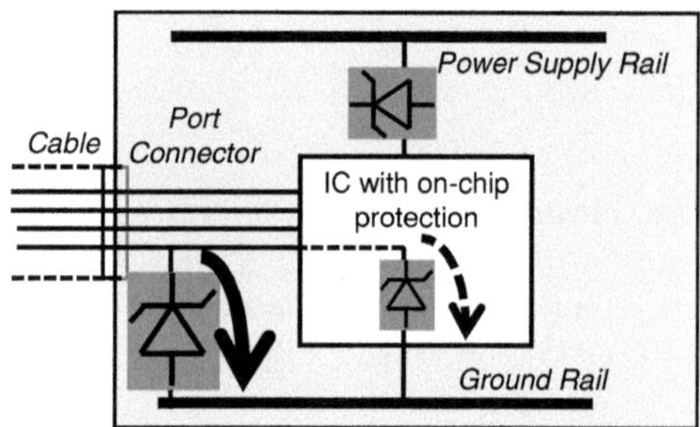

current electrical path for the TVS and a consideration of the current path through the IC pin of the on-chip ESD protection structure (Fig. 5.15).

The design of a system with ICs that include system level ESD protection at the interface pins as a primary ESD protection stage is not simple. It requires a special effort taking into account the transient voltage drop at the system interconnects. For example, in case of an IC pin not immediately interfacing with the port but placed inside the system, the inductive transient voltage drop causes a significant elevation of the port voltage that can be undesirably coupled to other system components and can result to unexpected effects (Fig. 5.16).

The design information for discrete TVS, IC and passive components is extracted from corresponding datasheets published by the vendors. Thus the approach is mostly based on previous system designer experience and does not represent a co-design methodology fulfilled with the meaning explained in the following sections. In particular the IC protection design may be neither

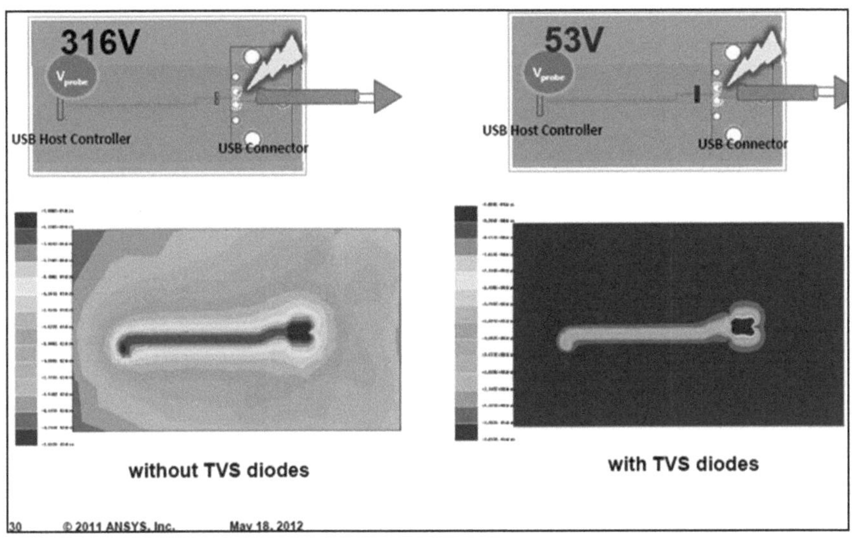

Fig. 5.16 Example of inductive interconnect loading for high-speed signal USB with TVS diode [91]

application specific nor adapted and optimized in the system development design flow. Thus this approach is mainly based on the composition of the off-chip ESD protection to ensure appropriate stress limitations at the IC pins and in an application board itself.

Respectively, within the datasheet-based system design the protection components are selected based on the specification of the IC pins—maximum operating voltage, absolute maximum ratings, noise and signal integrity related parameters. Those parameters are for example in case of IC components, the capacitive load which a data line or antenna port can tolerate. Unfortunately these IC parameters are often specified in the datasheets representing only the functional operating range time domain, rather than the ESD transient characteristics in the corresponding time domain.

IC datasheets usually provide only the passing level during component-level ESD qualification with the stress models HBM, MM and optional CDM.

Similarly the typical parameters of system-level ESD protection TVS devices may be specified without presentation of the real clamping voltage waveform or for the microsecond time domain of the surge pulse, rather than a system level ESD pulse waveform. Thus the datasheet information might be misleading to derive the transient triggering characteristics during ESD stress, the clamping voltage at a given ESD transient current, the linearity of the on-resistance, and the parasitic capacitance. The datasheet information for the passing level of the standalone component itself must not be used as the only factor in the design decision making process, as it was shown in the first section of this chapter.

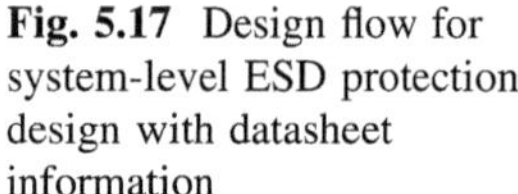

Fig. 5.17 Design flow for system-level ESD protection design with datasheet information

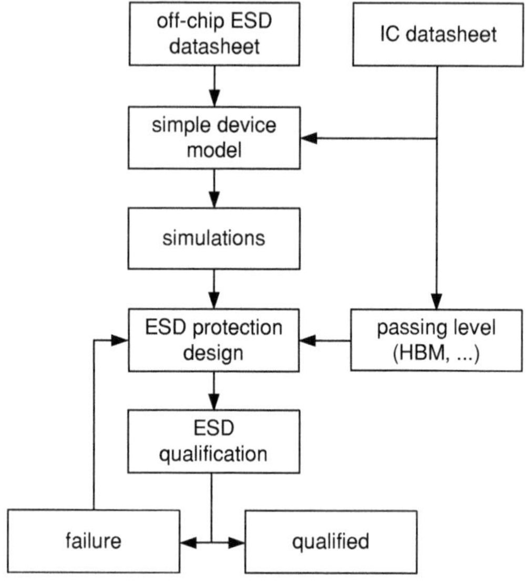

The datasheet system ESD design flow can be illustrated by the block-diagram in Fig. 5.17. After the datasheet research simplistic semi-empirical device models are created using the extracted datasheet parameters and then further used for circuit simulation analysis. In this workflow the on-chip ESD protection of the IC is replaced by a corresponding simplified equivalent ESD clamp characteristics similar to the combination of the diode and resistor, described above. This is done to represent on the behavioral level the contribution of the IC ESD protection network to the ESD current path [113, 118].

The main purpose of the simulation analysis is to compare the calculated residual current at the IC pins with the pin ESD passing level in the IC datasheet. For physical verification of the ESD robustness the system is tested using the corresponding qualification standards, for example IEC61000-4-2. Typically, if the specified passing level is not achieved, more off-chip ESD protection components are used to "patch" the design.

The datasheet-based design methodology has a number of disadvantages. Typically IC suppliers only specify the minimum required passing level like the commonly used 1–2 kV HBM stress in their datasheets. The absolute failure level of the IC pins is not provided and therefore the minimum passing level must be assumed as the worst case scenario. Moreover for the component passing level standards the test is accomplished in the power-off conditions for different pin combinations based on for example the JEDEC specifications. In general, these conditions are not similar to the system level stress, when power on-conditions for different pins may not only bring an alternative current path, but also disable RC-triggered active clamp ESD protection solutions.

A confident system level design requires a very conservative approach if an IC is only treated as a "black box" with attached datasheet parameters and an unknown type of on-chip ESD protection network. Using only the datasheet information easily leads to over-design and hence can compromise the figures of merit of the system's performance. To enable a more advanced approach the following section introduces a major improvement of the conventional system-level protection design. The co-design approach matches the on-chip and off-chip ESD protection design to obtain an efficient and reliable overall ESD protection solution.

5.4 IC-System ESD Co-design Concept

The IC-System co-design concept is a logical outcome from the understanding of the system ESD protection network as at least two stage circuit with the primary current path provided by an off-chip TVS and the secondary current path provided by the on-chip ESD clamps. The capability of an IC pin to safely conduct a fraction of system ESD current is an advantage when certain IC pins are required by specification to withstand a certain passing level of the system level pulse verified in the laboratory environment. It is also quite straight forward that a more optimal system design can be achieved if the system design is done by simulation with precise electrical ESD characteristics of the off-chip and on-chip protection components.

The major differentiation for the IC-System co-design concept from the datasheet based design (even with IC that includes system level specified pins) is based on the inclusion of the analysis of precise models and characteristics of the off-chip and the on-chip ESD protection structures. This approach can be illustrated by an example of a high-frequency amplifier with an off-chip protection network made of TVS and decoupling capacitors. The circuit includes the matching inductor L_{ESD} to maintain the application bandwidth of the circuit connected to the RF input port (Fig. 5.18). The amplifier power supply is set by the voltages V_{CC} and V_{EE} with corresponding decoupling capacitors C_{VCC} and C_{VEE} for power supply decoupling. The on-chip input ESD protection is realized using the rail based network with ESD diodes D_{ESD} and power clamp PC.

When ESD stress pulse is applied at the RF input port of the application board the main discharge current path is provided by the off-chip protection TVS components used before the input capacitor. In this case only a small residual current is expected to be conducted by the on-chip protection diode and power clamp network.

This example will be further used to illustrate two co-design methodologies: System-Efficient ESD Design (SEED [114]) that mainly relies on the additional TLP data as a co-design methodology input and an improved SEED that in addition incorporates a HMM data analysis in the co-design workflow.

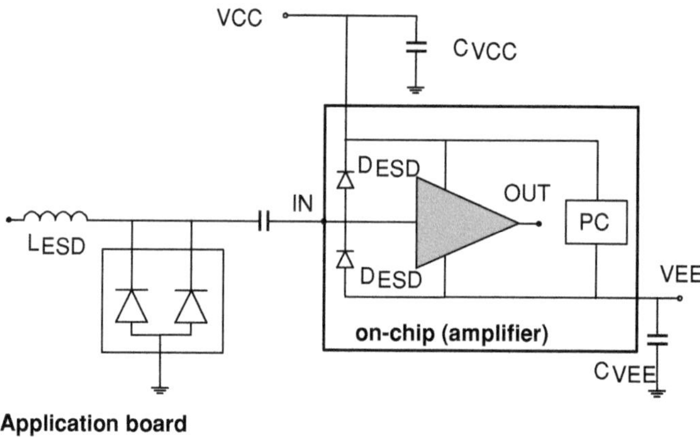

Fig. 5.18 Example of the IC-system ESD co-design

5.4.1 The Co-design Methodology with TLP Data

With the release of the White Paper 3, parts I and II [114, 121] by the Industry Council on ESD target levels a principle co-design methodology has been introduced. It is called System-Efficient ESD Design (SEED). The concept was proposed to be applied for the design of off-chip ESD protection solutions meeting the standard system-level ESD specifications.

The SEED methodology is based on TLP characterization and allows bridging the existing gap between system and IC ESD protection design. The quasi-static TLP I-V characteristics are collected both for IC and for off-chip active and passive components to provide more adequate initial information for the system-level ESD design building blocks. The optional vfTLP testing analysis can be incorporated as a part of the methodology for example in case of sensitive inputs with thin gate oxides or the need to take into account the reliability to fast transient pulses. In [122] the CDM peak current was used as a measure of the USB transceiver input robustness to account for the physical effects produced by the first HMM peak. However, this strategy relies on a not adequate physical basis due to a generally different CDM current path and location. From this perspective a correlation between the two pin HMM and TLP/vfTLP stresses is expected to be more appropriate.

Once the TLP characteristics of the system level network building blocks are collected, the extracted parameters from the pulsed I-V characteristics can be used to compose simple models. These models include the more realistic pulsed I-V characteristics and therefore are expected to provide a more optimal outcome for the design as accomplished by the similar steps as the datasheet based design. The overall design flow steps include now TLP characterization of the system network components, model fitting, transient simulations to study different design parameter variations, application board design and final verification (Fig. 5.19a). The obtained TLP I-V characteristics for the on- and off- chip components are

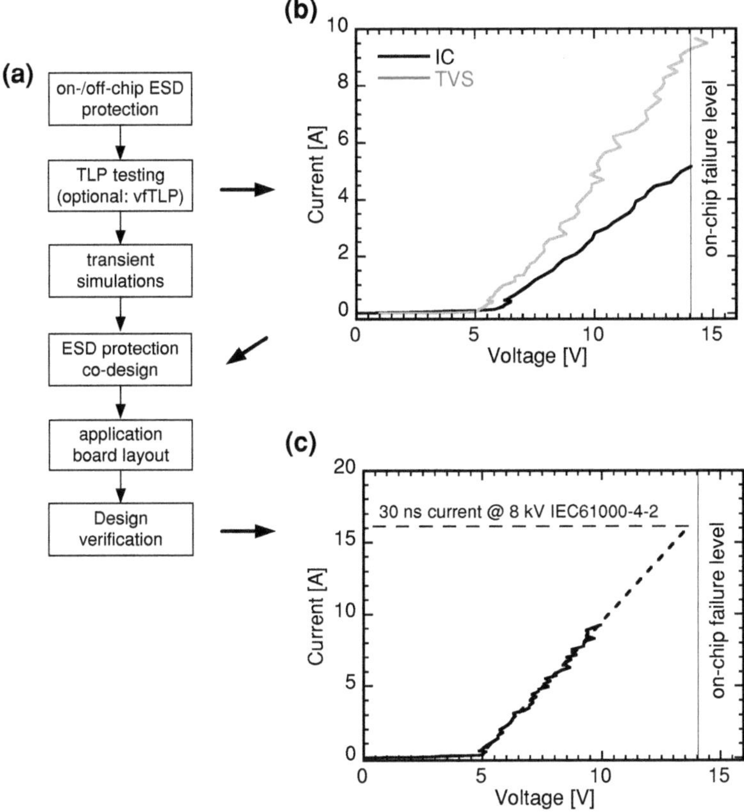

Fig. 5.19 Illustration to SEED design flow (**a**) with TLP I-V characteristics of on-chip protection (IC) and off-chip TVS (**b**) and final TLP I-V of the protected design measured into system input (**c**)

compared to identify the design window (Fig. 5.19b). The final design of for example a target protection level of 8 kV IEC61000-4-2 is preliminary validated by the TLP measurement for the corresponding 30 ns current level of ~ 16 A.

If necessary, in a more advanced way the TLP I-V characteristics are used to extract model parameters to be used in ESD transient simulations for comparative analysis of the system design options and optimization with the additional off-chip components. The off-chip components that can be added to the circuit (Fig. 5.18) can include isolation resistors [124], passive components like ferrite beads [113, 123] and common mode filter [122], as well as alternative or additional TVS components. By collecting TLP data, the target current and voltage at the IC pin for the safe level becomes precisely known in the ESD pulse domain. Thus, a higher level of confidence for the design can be expected. A preliminary verification of the design can be done by measuring the TLP I-V characteristics of the accomplished design (Fig. 5.19c).

In the example, the protection target is expected to conduct safely the current after 30 ns during an ESD stress of 8 kV IEC61000-4-2. For thermal failures, this current can be approximated to be equivalent to a 100 ns TLP current of 16 A. Due to the different pulse waveforms, this approximation will always result in a higher failure level in comparison with IEC61000-4-2 stress.

In spite of all attractiveness and simplicity, the co-design flow based on TLP data has its drawbacks. Since TLP essentially provides only a quasi-static characteristic in the 100 ns time domain, the important information of a voltage overshoot is missing. Because of the fast rising transient system level stress current, this part of the ESD stress cannot be neglected. For example in [122] the TLP-based methodology predicted for an USB2 protection solution design a protection level of ~ 10 kV IEC, while the demonstrated experimental robustness stress is about 14.5 kV. The large difference between simulations and measurements was attributed to the missing or less accurate models of some of the board parasitic.

Thus, in spite of the progress in comparison with the datasheet-based design the co-design IC-system methodology can be further improved by taking into account the transient characteristics over the entire ESD pulse time domain. The case study of the next section will demonstrate that the TLP I-V characterization alone is incomplete to be used as input for the co-design approach. The transient information needs to be added to take into account possible transient interaction between the ESD protection on IC-level and the ESD protection on board-level. The resolution of the problem is provided by the incorporation of a HMM measurements step into the co-design flow.

5.4.2 IC-System Co-design with Additional HMM Testing

The TLP-based design level of confidence can be re-enforced with HMM characterization to enable a more accurate circuit simulation step in the co-design flow. A case study for the system-level ESD protection of a high-voltage tolerant IC pin combines the internal circuit implemented in the LV CMOS process as a part of System-on-Chip (SoC) solution. Typical examples include line drivers, USB interfaces, and display drivers. The HV tolerant IC pins are expected to operate at higher supply voltages than the LV technology used in the IC product fabrication.

In this case the conventional ESD protection solutions based on the standard process with the low voltage gate oxide device must be replaced with junction-based free devices. The presented example includes a test board with board-level components and an on-chip protection HV tolerant device (Fig. 5.20). The indicated variables on the circuit diagram include the system-level ESD current I_{HMM}, the residual current into the on-chip protection device I_{DUT} and the voltage across the on-chip protection device V_{DUT} with the insert of the test board view for the connection of the off-chip components to the used on-wafer setup. The example is focused on the interaction between on-chip and off-chip devices using a measurement and simulation based analysis.

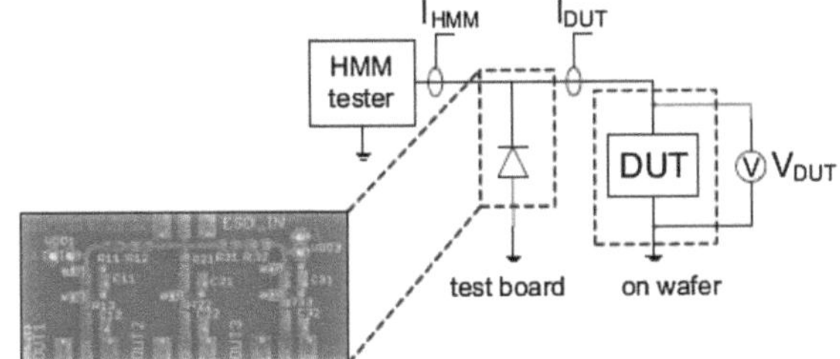

Fig. 5.20 Schematic of setup for system-level ESD experiments with off-chip components

The HV-tolerant SCR ESD protection device is implemented in a 130 nm CMOS process technology with standard CMOS devices in the PDK tolerant to 1.2 and 3.3 V. The junction-based SCR has a rather high triggering voltage of ~ 15.5 V. Due to absence of a thin gate oxide it can tolerate the high voltage without any impact on the long term reliability parameters. The SCR device is measured standalone on-wafer. The probe needles and probe holder parasitic extraction methodology (Chap. 2) is included in the analysis to determine their influence in the setup during ESD stress.

A dedicated double layer test board (Fig. 5.20) connecting the off-chip components to the DUT has been manufactured using FR4 as board material. The top layer contains the PCB traces and the footprints for off-chip components. The bottom layer represents the ground plane and is connected with plated via to the top layer. To emulate a typical application board, the board traces are not designed with specific impedances defined by their width. The board with the off-chip protection devices is connected via SMA connectors to the probe needle holder and probe needles to make the electrical connection to the DUT.

The TLP and vfTLP tester HANWA T-5000 are used as stress sources for the extraction of the I-V curves. The system-level ESD stress source is a HMM tester HANWA HED-W5000 M. Next to the current probes, a high impedance passive voltage probe is connected to the SCR in a KELVIN configuration to capture the transient behavior of the SCR during different stress conditions and with different off-chip configurations.

The TLP I-V characteristics of the on-chip and off-chip ESD protection devices were compared to select a suitable off-chip protection (Fig. 5.21) for the protection device in the 5 V tolerant application. The selected TVS diode had turn-on at ~ 6.2 V under the junction capacitance ~ 80 pF and the HMM/IEC61000-4-2 robustness ~ 30 kV.

The triggering voltage of the TVS V_{T1} is ~ 6.2 V which is much lower than the SCR triggering voltage $V_{T1} \sim 15.5$ V. Namely the TVS is expected to provide the primary current path to dissipate the system-level ESD stress up to a current level of ~ 5.4 A until the triggering of the SCR. At a higher current level, the SCR is expected to trigger into snapback mode thereby taking over the ESD current path from the TVS diode. In the beginning of the snapback regime the current through the SCR quickly exceeds the device high current capability (Fig. 5.21). Therefore this scenario inevitably results in the thermo-electrical burnout of the SCR.

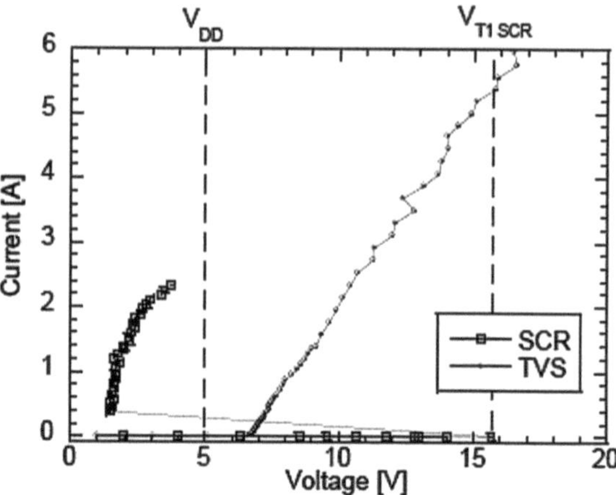

Fig. 5.21 Comparison of the measured 100 ns TLP I-V characteristics of the SCR and TVS components used in the example

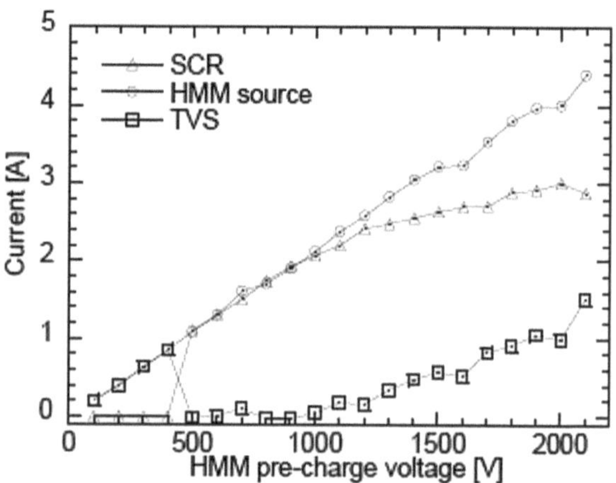

Fig. 5.22 Measured HMM stress current after 30 ns: HMM stress current, current through SCR and TVS diode

However, the HMM measurements of the circuit for the SCR device in parallel with the TVS diode demonstrate a different result. The SCR turn-on is observed at a rather low stress level below the estimated current of ~ 5.4 A. The measurements of the transient HMM current after 30 ns point out that the main current path is provided by the SCR rather than the TVS (Fig. 5.22).

Such triggering of the SCR at low HMM stress level cannot be explained based only on the measured TLP I-V characteristics. Transient characteristics of the TVS diode related to the turn-on speed and voltage overshoot formation become critical too. In part this can be resolved by the vfTLP data acquisition for the standalone TVS components. The 2 ns vfTLP I-V characteristics were measured with 200 ps rise time and extracted in two different averaging windows for the current and voltage readings.

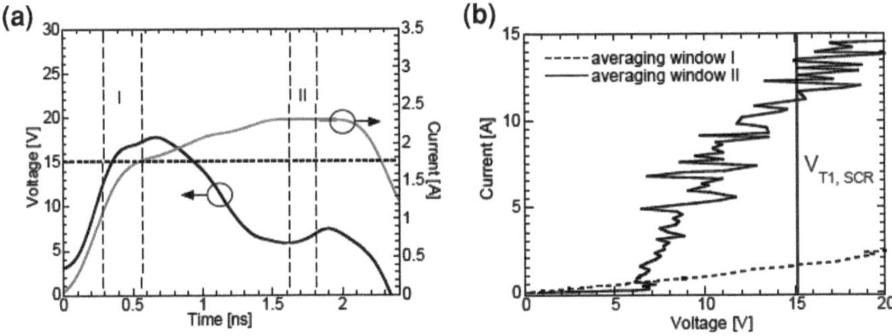

Fig. 5.23 Experimental 2 ns/200 ps transient voltage and current **a** vfTLP I-V characteristics **b** for standalone TVS diode extracted for two averaging windows I: 0.3–0.6 ns and II: 1.7–1.9 ns

The first averaging window (I) is defined at the beginning of the vfTLP pulse, while the second (II) is capturing only the end of the vfTLP pulse with more settled transient voltage and current into a quasi-static regime (Fig. 5.23a). The corresponding vfTLP I-V characteristics demonstrate the essence of the physical effect related to the TVS turn-on speed. The I-V plot for the window (I) demonstrates a rather low current that the TVS conducts at the beginning of the pulse. This supports only $\sim$1.6 A at the triggering voltage level V_{T1} of the SCR (Fig. 5.23b). This low current is in a good agreement with the measured HMM peak current of $\sim$1.65 A when the SCR triggers. Apparently, including these transient data into the TLP-based co-design approach covers an important effect of the interaction between the on-chip and ESD off-chip protection device.

The turn-on of the SCR at low stress level directly impacts the ESD robustness of this system test case with a parallel TVS component. Overall, the expected failure level from the purely TLP-based design would be $\sim$2.9 kV. There is a factor of 1.15 between the TLP failure current and the 30 ns current at HMM failure level [125]. However the real failure level with added TVS diode is only 2.1 kV. Most of the second pulse current goes through the SCR and only a small fraction is conducted by the TVS. At the HMM stress level $\sim$2.1 kV, the residual current through the SCR causes irreversible breakdown (Fig. 5.24).

Similar to the triggering of the SCR at low stress level, the sharing of the 2nd pulse current at higher stress level between SCR and TVS diode cannot be explained only with TLP I-V curve data. Transient simulations are required to gain the understanding of these results.

To study the transient behavior of on-chip and off-chip protection devices the circuit simulation with HMM tester were set according to Sect. 5.2.1. The TVS component was represented by a standard SPICE model matched to the datasheet and TLP characteristics. To approximate and model the overshoot of the TVS diode an inductance is added in series and its value matched to measurement data obtained from the standalone TVS diode mounted in the test board.

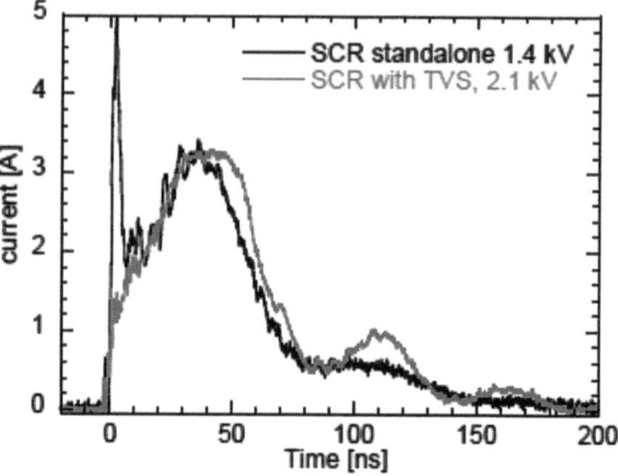

Fig. 5.24 Current through SCR with and without TVS diode in parallel; stress level: failure level

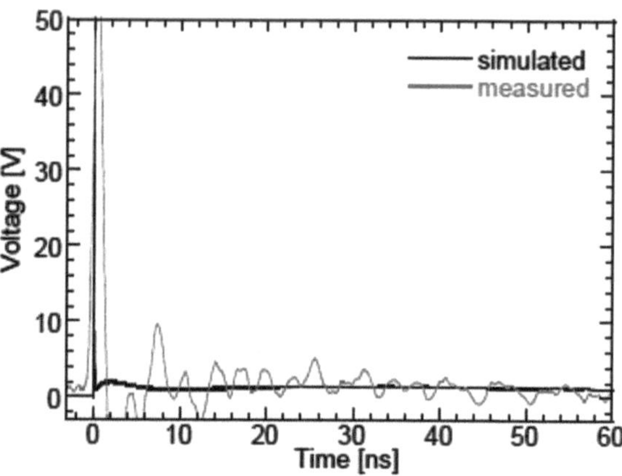

Fig. 5.25 Simulated and measured voltage across the standalone SCR including the parasitic of probe needles and probe needle holder for the 1 kV HMM stress

The SCR model was extracted by modifying a SPICE model of a commercial discrete SCR component. This behavioral model captures the triggering, the on-state resistance and the holding current. It provides an adequate approximation of the real device behavior in the transient ESD pulse domain. The comparison of the simulated and measured voltage across the standalone SCR includes the parasitic of the probe needle and holder. A good agreement for the voltage after snapback was verified (Fig. 5.25), while the difference in the overshoot amplitude was attributed to bandwidth limitations of the passive voltage probe used for the measurement.

To verify the accuracy of the simulation setup, first the SCR triggering has been simulated with the TVS diode in parallel. A good agreement between the simulation and measurement results has been found for the current through the SCR at two different stress level (Fig. 5.26a and b).

For the transient analysis the voltage and current waveform characteristics have been extracted in different circuit locations for different pre-charge levels.

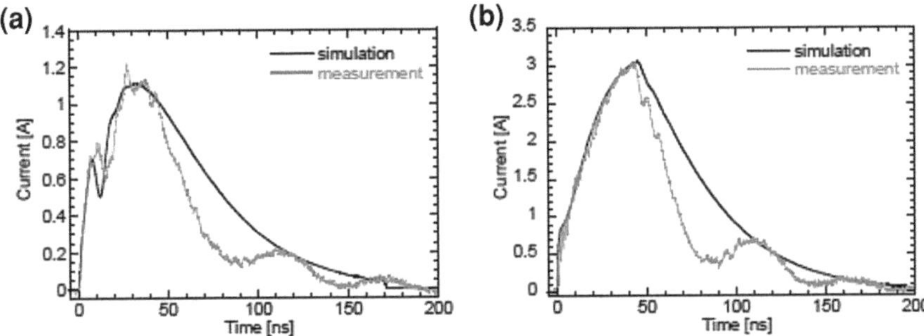

Fig. 5.26 Measured and simulated current through SCR with TVS in parallel test case for the HMM stress level 0.5 (**a**) and 1.5 kV (**b**)

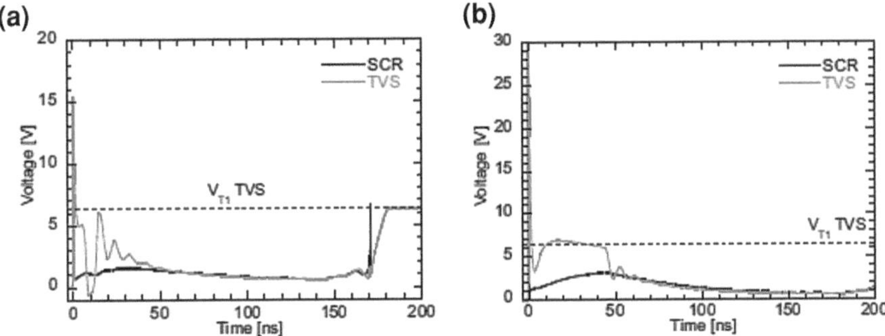

Fig. 5.27 Simulated voltages across TVS diode and SCR for 0.5 (**a**) and 1.5 kV (**b**) HMM stress level

The voltage across the TVS and the SCR for two pre-charge levels demonstrate that the TVS turns-off at low stress level (Fig. 5.27a). The low holding voltage of the SCR, in spite of additional voltage drop across the probe holder and needles, is insufficient to keep the TVS in on-state. At higher stress level the TVS remains in the on-state during the entire duration of the second current peak of HMM pulse (Fig. 5.27b).

The scenario of the current redistribution between the SCR and TVS during HMM stress in the test case is directly influencing the end result. At low HMM stress level the current through TVS diode and the SCR test case components is observed only during the first HMM pulse current peak (Fig. 5.28a), while the current path for second current peak of the HMM pulse is fully provided by the SCR circuit component. For the higher HMM pulse amplitude, the TVS contributes to the current path during the second HMM pulse peak (Fig 5.28b).

A simple fix of the circuit can be achieved by an additional resistor which limits the residual current through the SCR to a safe value. The methodology [124] to calculate the required resistance value for the peak current of 8 kV HMM can be

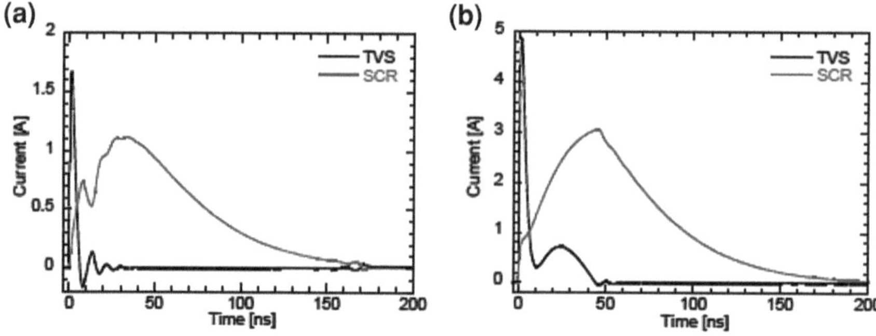

Fig. 5.28 Simulated currents through TVS diode and SCR for the 0.5 (**a**) and 1.5 kV (**b**) stress levels

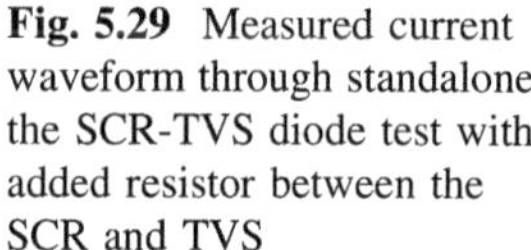

Fig. 5.29 Measured current waveform through standalone the SCR-TVS diode test with added resistor between the SCR and TVS

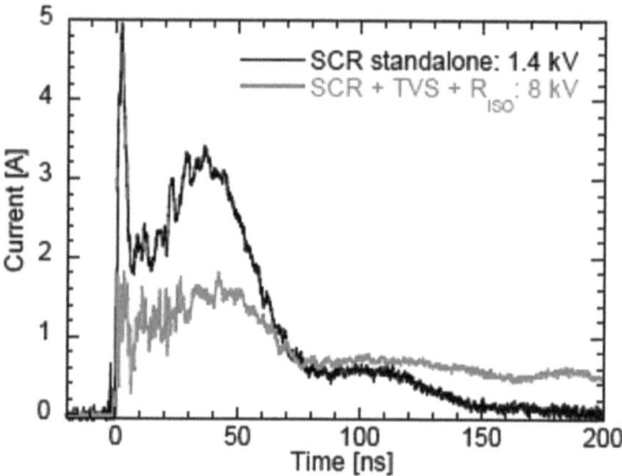

used assuming the safe current level for SCR around 1 A. The estimated resistor of $\sim 7.3~\Omega$ is sufficient to fix the HMM stress compatibility of the test case for a protection level of ~ 8 kV. With the added isolation resistor the residual current through the SCR stays at a safe level (Fig. 5.29).

The detailed analysis of the TVS-SCR case study reproduced in this section clearly demonstrates that the TLP-based (and SEED) IC-system co-methodologies are limited in their present form. Taking into account transient information is required in addition to the quasi-static TLP I-V data acquisition and the corresponding co-design steps. This ensures a protection design which takes into account the transient interaction of the on-chip and the off-chip ESD protection devices related to these devices turn-on speed and the IC non-linearity of their characteristics. Depending on the applied HMM stress, the off-chip dedicated protection current path through the TVS off-chip components can be taken over by the low-holding voltage snapback ESD operation of the on-chip ESD protection devices or even by the transient latch-up susceptible parts of the circuit. Thus a

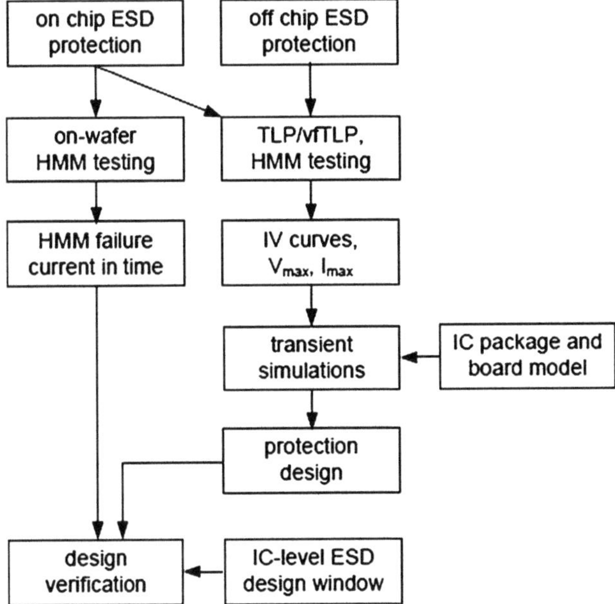

Fig. 5.30 Design flow for system-aware ESD co-design using on-wafer measurements and transient simulations

more detailed *transient* analysis that includes HMM pulse measurements and/or simulations is an essential step of the IC-system co-design.

5.4.3 Co-design Flow with TLP and HMM Testing

Combing the basic understanding demonstrated in the above section's example a more advanced co-design IC-system flow is suggested. In general it includes the TLP and HMM characterization steps according to Sect. 5.4.1, on-wafer measurements and transient simulations done in order to account for the transient behavior of both off-chip components and off-chip ESD protection devices and internal circuits (Fig. 5.30). At first, the quasi-static and transient device parameters of the off-chip and the on-chip protection devices are captured with HMM testing, TLP, and optional vfTLP testing. The HMM failure level is captured to obtain the robustness of the on-chip ESD protection to system-level ESD stress for later verification of the designed ESD protection solution.

The obtained device information is used to extract the parameters for suitable device models used for further transient simulations. With these added behavioral or compact models for the application board and IC components the transient interaction of the off-chip and on-chip protection devices is further analyzed in a "closest to reality" environment. The protection solution is designed and verified by comparing the simulated current through the on-chip ESD protection device in the entire ESD pulse time domain with the measured failure current levels recorded during initial standalone HMM testing. The TLP characteristics are used to extract the quasi-static maximum voltage and current included in the design flow, while HMM is used to extract transient information combined with the pass-fail verification of the protection solutions. This co-design methodology is expected to improve the overall design and to eliminate an unexpected transient device behavior during the early phase of system development.

5.5 System-Aware On-Chip ESD Protection Design

The purpose of this section is to outline major principles for the component-level ESD protection design of the IC pins that are directly connected to the system ports. The major differentiation to the on-chip system level devices in Chap. 3 is that in this section the on-chip design protects only for a fraction of the total system level ESD discharge current. This principle outline is accomplished through a set of case studies to derive each particular aspect of the system-aware on chip ESD protection design as a result of on-chip and off-chip components interaction.

5.5.1 Experimental Setup for the Case Study

In the case study the devices under test (DUT) were measured on-wafer with a corresponding extraction of the probe and holder needles equivalent circuit parameters. The parasitic inductances and resistances are extracted from HMM measurements on a short load connected to the on-wafer setup as a DUT. The algorithms described in Chap. 2 are applied for the de-embedding.

TLP and vfTLP I-V characteristics were obtained with the TLP tester HANWA T-5000. The on-wafer tester HANWA HED-W5000 M was used for HBM and HMM pulse generation based on discharge circuit similar to ESD guns with the stress waveforms in compliance to the IEC6100-4-2 system-level ESD standard [126]. A high impedance passive voltage probe was connected directly to the DUT in a Kelvin setup to capture the transient voltage waveforms (Fig. 5.31).

The measurement setup and the presented design methodologies (Sect. 5.4) are used to study two typical component-level ESD protection scenarios: analog circuits and protection of IC manufactured in advanced CMOS technologies with additional system-level ESD protection components.

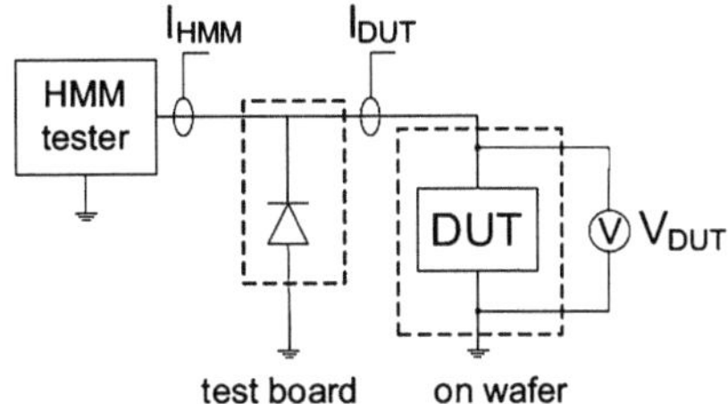

Fig. 5.31 Diagram for the measurement setup with test board and DUT on wafer with indicated the system-level ESD stress current I_{HMM}, the current through DUT I_{DUT} and the voltage across DUT V_{DUT}

Table 5.3 ESD test results for the three ESD clamps for HMM and 100 ns TLP stresses

	Device type	HMM (kV)	TLP (A)	V_{T1} (V)	V_H (V)	R_{ON} (Ω)	I_{T2} (A)
Clamp 1	SNMOS	1.5	2.7	6.6	6.25	1.3	2.7
Clamp 2	LVTSCR	1.7	2.7	5.7	1.6	2.2	2.7
Clamp 3	nLDMOS-SCR	1.8	2.8	15.9	1.5	0.5	2.8
TVS				5.5		0.9	>10

5.5.2 Selection of ESD Clamps for External IC Pins

Assuming some degree of freedom to select the on-chip ESD protection solution, different clamp types can be used at the IC design stage. While different ESD on-chip solutions may provide similar passing level for component level ESD tests and, perhaps, even have a similar footprint, their operation for the system level pulse waveforms in the two state network circuit can be different. Thus, some on-chip ESD clamps can be more preferable for the co-design than the others. To demonstrate this, three typical ESD clamps were compared to identify the best protection option from the perspective of the advantages for the system design using experimental and simulation results.

These three representative ESD clamps for the IC protection pin protection include the SNMOS (Clamp 1), a LVTSCR (Clamp 2) and an nLDMOS-SCR (Clamp 3). All three device options have satisfied the requirements for the component-level ESD robustness in a 90 nm CMOS technology. The major figures of merit for the ESD operation were extracted from TLP I-V characteristics and compared with the TVS off-chip component (Table 5.3). The TVS diode has a DC breakdown voltage of $\sim$5 V, a junction capacitance of 105 pF and an IEC61000-4-2 protection capability up to 30 kV. The quasi-static device parameters extracted from the TLP characteristics (Table 5.3) demonstrate that both Clamp 2 and 3 work in snapback operation, while the on-state resistance of Clamp 2 is the highest of the four devices. It exceeds $\sim$2 times the corresponding TVS parameter. Clamp 3 has two times less on-state resistance in comparison with the TVS component.

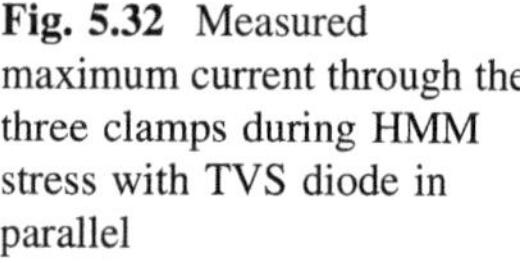

Fig. 5.32 Measured maximum current through the three clamps during HMM stress with TVS diode in parallel

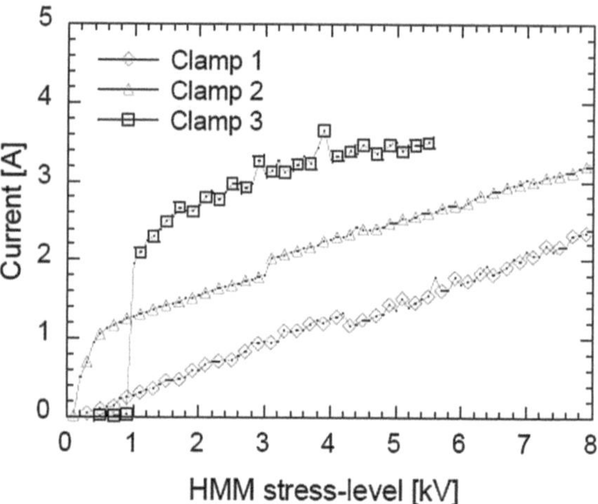

The TLP data summary were complemented with transient HMM test results for each clamp in parallel with the TVS. The comparison demonstrates that the HMM current through each ESD clamp is substantially different when the TVS is placed in parallel (Fig. 5.32). For example Clamp 3 turns on only at a stress level above 1 kV HMM due to its higher triggering voltage V_{T1}. Clamp 1 and 2 are engaged practically at any stress level since they have the triggering voltages only slightly above the TVS breakdown and the clamping voltage according to Table 5.3 results. At the same time due to lower on state resistance, Clamp 3 conducts a higher current and fails already at 5.5 kV HMM pulse. In contrast the HMM stress level above 8 kV can be withstand by the TVS-clamp test cases when either Clamp 1 or Clamp 2 is used. Thus, in spite of the original impression that Clamp 3 is the more suitable on-chip component at least in comparison with the SNMOS clamp, the experiment results are actually opposite.

This difference becomes even more apparent in the HMM pulse waveform comparison at the same stress level. A higher current through the on-chip Clamp 3 is observed (Fig. 5.33). After 50 ns from the beginning of the pulse the entire HMM pulse current is conducted only by Clamp 3 rather than the TVS component. Since Clamp 3 is not designed to withstand the primary system level discharge—the test case Clamp+TVS fails.

If the voltage waveforms are added to the analysis it becomes apparent that Clamp 3 triggers into snapback mode and the clamping voltage falls below the TVS clamping voltage (Fig. 5.34) thus taking over the ESD current. The formed domino-effect is initiated only after the Clamp 3 turn-on voltage is achieved. Therefore at lower HMM pulse level—no Clamp-TVS interaction is visible.

The equivalent circuit schematic for the Clamp-TVS test case for the HMM stress operation with components in on-state is shown in (Fig. 5.35). The elevated transient voltage across the TVS above the breakdown voltage is due to the inductive parasitic of the test board and test setup (Eq. 5.1).

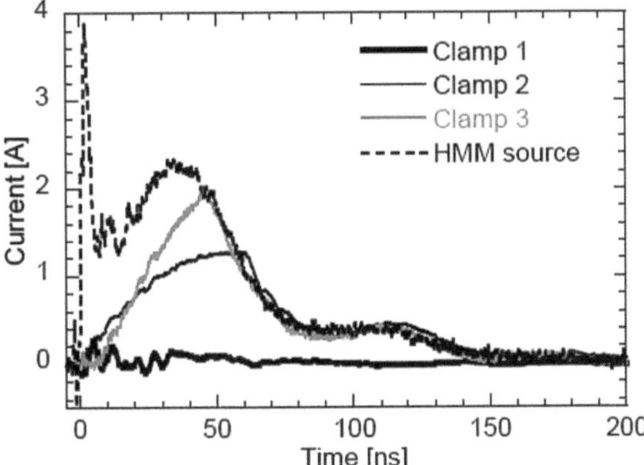

Fig. 5.33 Measured current through the three clamping devices with TVS diode in parallel in comparison to applied HMM stress current, HMM stress level: 1 kV

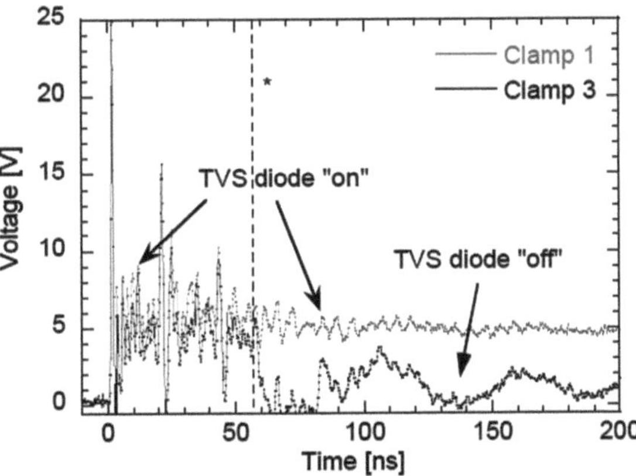

Fig. 5.34 Measured voltage across TVS-Clamp test cases with Clamp 1 and 3 for HMM pulse 2 kV the time moment of TVS turn-off for the Clamp 3 case marked at 57 ns

Based on a simple analysis of this circuit the reverse breakdown voltage V_{BDTVS} of the TVS is lower than:

$$V_{BDTVS} < L_{setup}\frac{dI_{Clamp}}{dt} + I_{clamp}(R_{setup} + R_{ON,Clamp}) + V_{hold,Clamp}, \qquad (5.1)$$

where I_{clamp} the current through the ESD clamp, L_{setup} the parasitic inductance of the probe needle holder and probe needle, R_{setup} the parasitic resistance of the probe needles, $R_{ON,Clamp}$ the on-resistance of the ESD clamp and $V_{hold,Clamp}$ the on-state (holding) voltage of the ESD clamp.

The HMM current (Fig. 5.34) increases a second time after ~ 10 ns thereby causing a voltage drop across the parasitic inductance. For the time range from 20 to 35 ns and the used wafer prober, an additional impedance of 2.7 Ω is formed in series to the on-wafer ESD clamp.

The higher clamping voltage of Clamp 1 and the parasitic voltage drop of the on-wafer setup keep the TVS in on-state during the entire HMM stress duration at

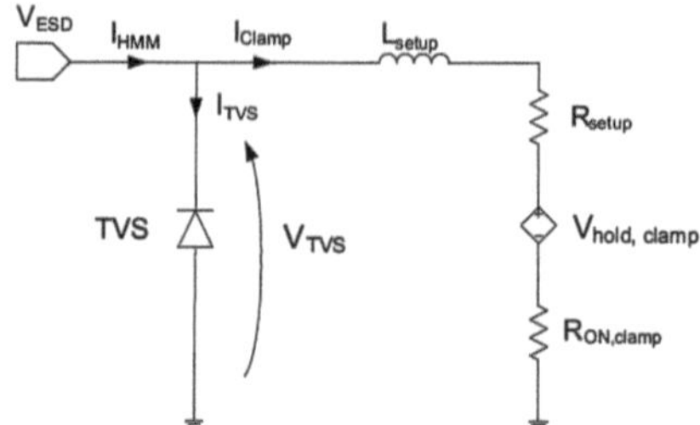

Fig. 5.35 Circuit diagram for TVS-Clamp test case with indicated inductance probe needle holder L_{setup}, probe needles resistance R_{setup}, on-state ESD clamp resistance $R_{ON;Clamp}$, on-state (holding) voltage of the ESD clamp $V_{hold;Clamp}$, the voltage across TVS diode V_{TVS}, the total ESD stress current I_{HMM}, the TVS current component I_{TVS} and clamp current component I_{Clamp}

all stress levels. Consequently the TVS shunts more current. In case of Clamp 2 the holding voltage is low too. However the relatively high on-state resistance causes a sufficient voltage drop to maintain TVS on-state conditions. Clamp 2 conducts the lower transient current.

Thus based on the on-wafer measurement, Clamp 1 and 2 are the most desirable ESD clamp options for the on-chip protection of an IC pin, while Clamp 3 with a low holding voltage and its low on-state resistance fails to provide the appropriate level of robustness of test case TVS-Clamp.

To estimate the individual clamp contribution behavior in an IC package which is mounted in an application board, the simulation analysis was further conducted. The known parasitic of the measurements setup and the modeling approach described in Sect. 5.2 were included the in circuit simulation analysis. The TVS diode was represented by a standard diode compact model with the adjusted breakdown voltage and on-state resistance according to TLP measurements. The voltage overshoot at diode turn-on was accounted by an additional serial inductance in the sub-circuit that represented the TVS. Since the TVS diode is unidirectional and normally reverse-biased during ESD stress the forward recovery effect can be neglected.

The SNMOS-based Clamp 1 was also represented by a reverse-biased diode model with the clamping voltage matching TLP data. The device snapback characteristic was neglected. The SCR-based ESD Clamps 2 and 3 were represented by means of behavioral models derived from the compact models of a discrete thyristor component. The device parameters include the dV/dt factor, the triggering and the holding voltage, as well as the holding current.

The described simulation setup was validated and verified by simulation of the current waveforms through Clamp 3 in the on-wafer setup. The results are comparable with the measurement data (Fig. 5.36).

For the evaluation on package the on-wafer setup parasitic were replaced with the equivalent model of a DIL IC package pin (Table 5.4) extracted from an IBIS model [133]. In addition to the IC package, a parasitic inductance of a 5 mm board

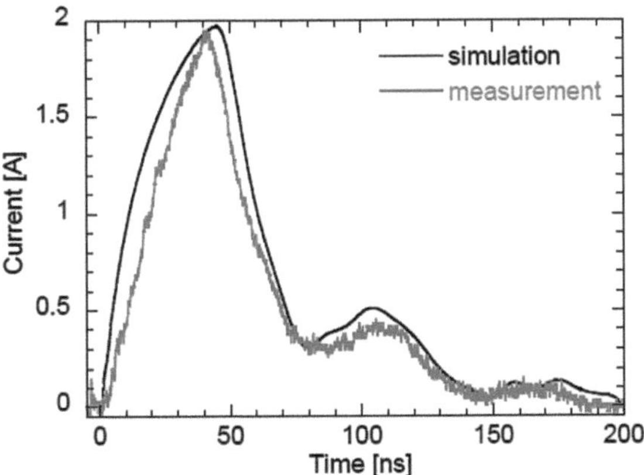

Fig. 5.36 Comparison of the simulated and measured current waveforms through nLDMOS SCR Clamp3 included into the Clamp-TVS test case for 1 kV HMM pulse

Table 5.4 Parasitic of a DIL IC package pin

R$_{pckg}$ (mΩ)	L$_{pckg}$ (nH)	C$_{pckg}$ (pF)
300	13.7	1.75

long trace was added to represent the connection between the TVS diode and the packaged IC in the simulation (see Fig. 5.37).

The simulated voltage waveforms for Clamp 3 within a DIL package have demonstrated that they are in correlation with the on-wafer measurements. The TVS turn-off time moment during the HMM stress is the function of the HMM stress level (Fig. 5.38).

The current through the ESD clamps can be simulated for different HMM stress levels to predict the failure level for the packaged devices with the Clamp-TVS test case tool. The comparison of the simulated waveforms for the three packaged Clamp-TVS test cases and the current waveforms measured for the standalone clamps is presented in (Fig. 5.39).

Because of the much smaller impedance of the IC package in comparison to the on-wafer setup none of the three configurations passes 8 kV HMM stress. Thus additional protection measures are required. For example the current into the on-chip protection can be simply limited by an additional isolation resistor [124] between the TVS diode and the IC input. Using Kirchhoff's laws the appropriate isolation resistance value can be estimated by the expression (5.2 and 5.3):

$$\frac{G_{ISO,ONClamp}}{G_{ISO,ONclamp} + G_{ONTVS}} = \frac{I_{safe,clamp}}{I_{30\ ns,HMM,8\ kV}}, \qquad (5.2)$$

$$R_{ISO} = \frac{1}{G_{ISO,ONclamp}} - R_{ONclamp}, \qquad (5.3)$$

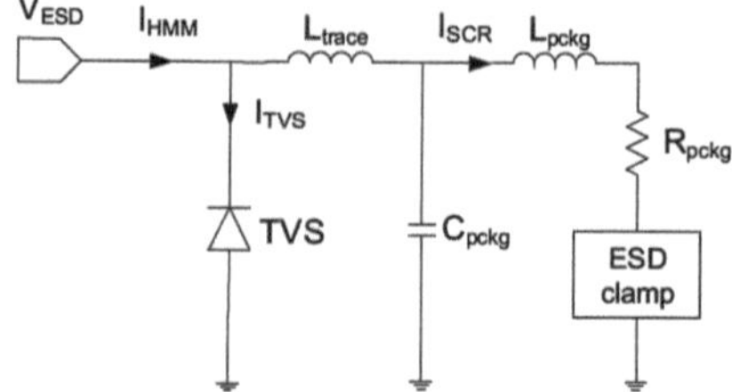

Fig. 5.37 Simulation setup with IC package parasitic components: I_{HMM} ESD stress current, I_{TVS} current through TVS diode, I_{Clamp} current through ESD clamp, L_{trace} board trace inductance, L_{pckg} package pin inductance, R_{pckg} package pin resistance, C_{pckg} package pin capacitance

Fig. 5.38 Simulated voltage across TVS diode with Clamp 3 connected, application case with DIL package; $V_{BD\ TVS}$ breakdown voltage of TVS diode

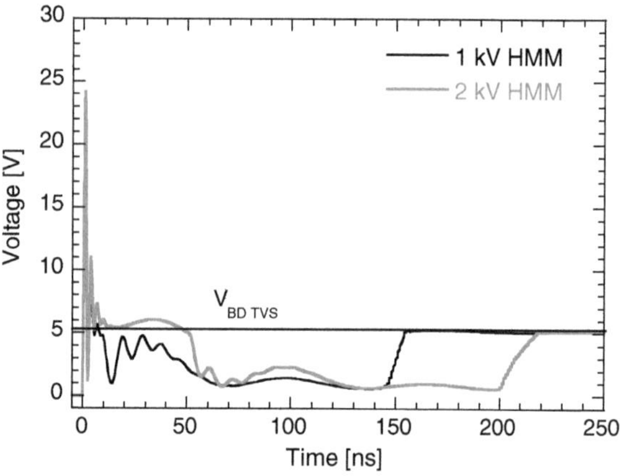

where R_{ISO} is the isolation resistance, $G_{ISO,ONclamp}$ the conductance of the isolation resistor and the ESD clamp in on-state, G_{TVS} the conductance of the TVS diode in on-state, $I_{safe,clamp}$ the safe current level through the ESD clamps and $I_{30ns,HMM,8kV}$ the current after 30 ns at 8 kV HMM stress. The required isolation resistance value, calculated using 5.2 and 5.3 for the three Clamp-TVS test cases (Table 5.5) points on the smallest required isolation resistor in case of the LVTSCR Clamp 2.

Although all three clamps have a similar ESD robustness, the nLDMOS SCR protected IC pin is more sensitive to system-level ESD stress when the parallel TVS protection is in place due to the low clamping voltage.

The cases in this section showed the limitations imposed by the current. In addition to the critical current level the critical voltage level must be taken into account too. The voltage across the on-chip ESD protection must not exceed a safe limit determined by the component-level ESD protection design window. This aspect is discussed in details in the following section using a specific CMOS gate protection example.

Fig. 5.39 Measured standalone and simulated failure current for three Clamp-TVS test cases that include SNMOS Clamp 1 (**a**), LVTSCR Clamp 2 (**b**) and nLDMOS-SCR Clamp 3 (**c**)

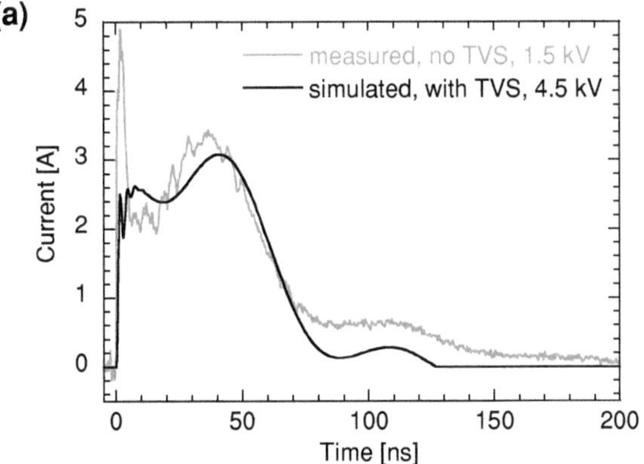

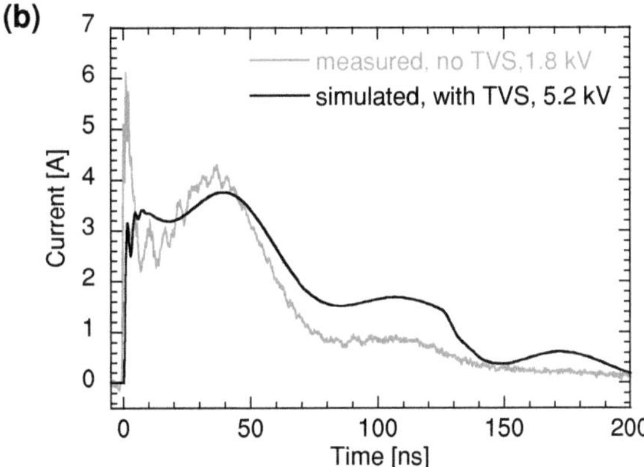

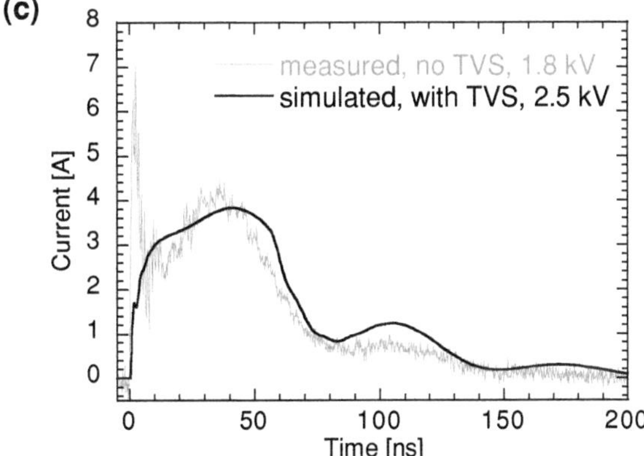

Table 5.5 Required isolation resistor—comparison for the three on-chip ESD clamps [127]

ESD clamp	R_{ISO} (Ω)
Clamp 1	3.1
Clamp 2	2
Clamp 3	3.7

5.5.3 Co-design in Advanced CMOS Technologies

Today high speed interface, RF, and other analog circuit blocks are leveraging the advantages of CMOS scaling down to lower operating voltage and feature dimensions. For example FinFET technologies [128] represent the next level of advanced CMOS scaling. With the scaling approach the robustness of the device is significantly reduced with the major "bottleneck" related to the thinning of the gate oxide (GOX). Reduction of the GOX significantly impacts the ESD protection design window. Several changes in the protection design were made to address the scaling trend, for example, using gated diodes as ESD protection device [129].

In this section, a local clamp protection design based on two bulk-FinFET gated diodes is used to protect a gate monitor structure. The diodes are designed with one single 80 µm wide fin found previously as the best configuration [129]. The gate length of each diode is 70 nm. The gate monitor is a 5×5 µm NMOS where drain, source and bulk contact are shorted to each other. It represents the input gate of e.g. an inverter (Fig. 5.40). The gate stack consists of 5 nm high-k material. This gives an effective oxide thickness of 1.6 nm.

The dependence of the breakdown voltage of the gate monitor upon the TLP pulse duration is presented in Fig. 5.41. The on-wafer HMM test for the two diode test cases demonstrates a significant reduction of the failure level from 1.9 kV without gate monitor (GM) to 0.7 kV when the GM is connected in parallel to the protection clamp. This indicates a gate oxide failure during system-level ESD stress. The quasi-static device parameter are extracted from TLP testing results on the two serial diodes and with the GM placed in parallel.

To design the circuit for system level compliance the on-chip clamp is combined with a selected off-chip TVS component with suitable characteristics (Table 5.6). The TVS has a very low breakdown voltage and a system-level ESD robustness of 8 kV IEC61000-4-2. The quasi-static device parameters show that the stress current is almost equally shared. Due to similar on-state resistance both on-chip and off-chip ESD components are engaged.

To design a suitable protection against system-level ESD stress, a design window is defined based on the component-level test results. It is defined by the GOX breakdown voltage during vfTLP stress less a safety margin as an upper limit and the operating or signal voltage plus a safety margin as a lower limit. For this case study the design window is defined in the 1.5–4.5 V range. The maximum current levels during HMM stress on the TVS diode are plotted over the maximum voltages (Fig. 5.42). Thus, even at low HMM stress level the voltage across the

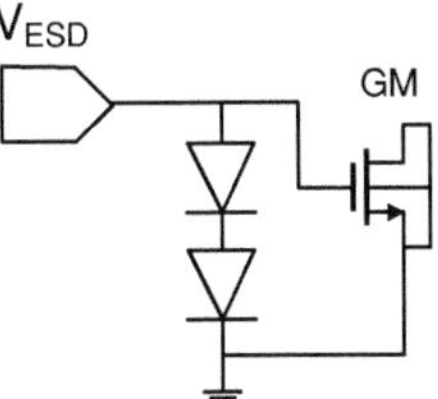

Fig. 5.40 Schematic of diode-based input protection (local clamping) with gate monitor (GM) in parallel

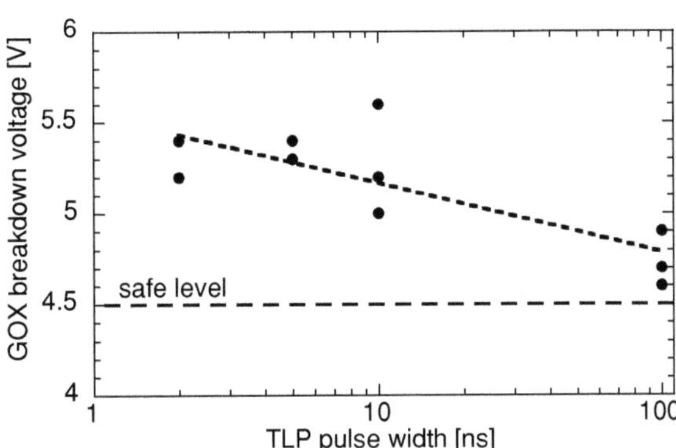

Fig. 5.41 Dependence of the breakdown voltage of the 1.6 nm high-k gate oxide monitor structure upon the TLP pulse width with the rise time 200 ps

TVS diode exceeds both the breakdown voltage of the GOX to be protected and the available design window.

Some measures for the limitation of the current and the voltage in the on-chip ESD protection during system-level ESD stress are required. The necessary information is extracted from TLP test results on the two FinFET diodes with the connected gate monitor structure in parallel. The safe current level should be below the TLP failure current I_{T2}, therefore a current level of 2.8 A was taken as a safe value. With this value and the data (Fig. 5.42) the value of the current limiting isolation resistor can be calculated as:

$$R_{ISO} = \frac{V_{max,TVS} - V_{BD,GOX}}{I_{safe,on-chip}},\qquad(5.4)$$

where R_{ISO} is the value of the isolation resistor, $V_{max,TVS}$ is the maximum voltage across the TVS at the target protection level, $V_{BD,GOX}$ the GOX breakdown voltage and $I_{safe,on-chip}$ is the safe current level through the on-chip protection during ESD stress. An isolation resistance value of 17.3 Ω is obtained using expression (5.4), a safe value for the GOX breakdown voltage of 4.5 V and a safe on-chip current of 2.8 A. A resistance value of 18 Ω is selected for design verification.

The designed ESD protection is simulated and verified with the calculated isolation resistor. The TVS model extraction was done similar to the above section. The on-chip protection stacked diode clamp was represented by the gated diodes that are faster components in comparison with the shallow-trench isolated (STI) diodes [130, 131]. Almost no overshoot occurs during triggering which justifies the

Table 5.6 Quasi-static device parameter obtained with TLP testing (pulse width: 100 ns, rise time: 200 ps); TVS diode: Vishay VESD01-02VG08

Device	R_{ON} (Ω)	V_{T1} (V)	I_{T2} (A)	V_{T2} (V)
2 diode + GM	0.8	1.5	3.0	4.9
TVS diode	1	2	n.a.	n.a.

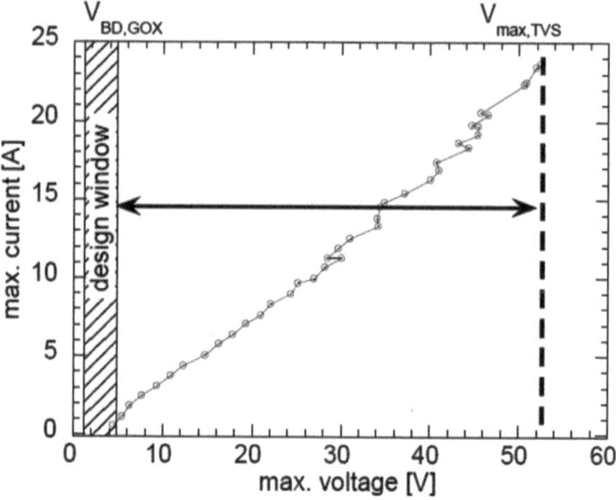

Fig. 5.42 HMM test results for standalone TVS (VESD01-02VG08): maximum current versus maximum voltage; $V_{BD;GOX}$ GOX breakdown voltage of gate monitor, $V_{max;TVS}$ maximum voltage at 8 kV HMM stress

use of a simple circuit model. To include the impact of a high bandwidth IC package, the equivalent model of a ball grid array (BGA) package was included in the circuit combining the on-chip protection and TVS. The simulated current through the stacked diode clamp in parallel with the GM at 8 kV HMM stress was compared to the current obtained from standalone HMM pass-fail test results (Fig. 5.43).

With the isolation resistor, no gate oxide breakdown would occur even at a HMM stress level of 8 kV. The calculated value for the necessary isolation resistor is much higher in comparison to the previous case study due to higher sensitivity of the GM structure to ESD stress.

5.5.4 Guidelines for Component-Level ESD Design

Component-level ESD clamps are often designed to provide a "strong snapback" with a low holding voltage and on-state resistance. The apparent purpose of such design is to limit the voltage at the protected IC pin at high ESD stress levels to the lowest possible limit. However, this design target may not be optimal when future IC pin interaction with a system level protected port is taken into account.

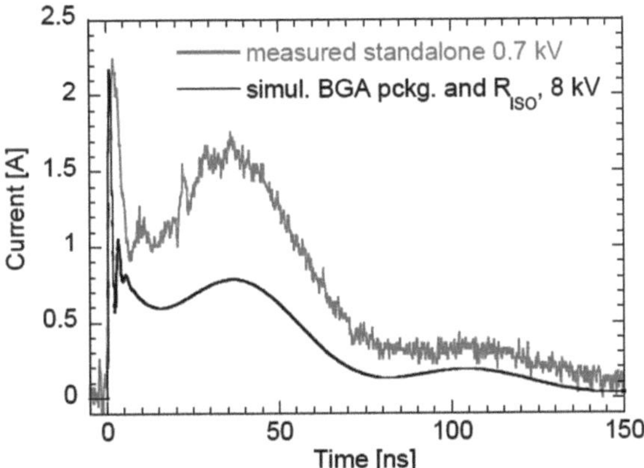

Fig. 5.43 Comparison of simulated current through on-chip ESD protection with GM (with BGA IC package and isolation resistor of 18 Ω) and measured current through standalone on-chip protection with GM in parallel (on wafer)

On the contrary, based on the examples above, the IC pins directly interfacing with the system level protection network may provide a better end result if the strong snapback capability is not in place. Due to the impedance of the PCB connections as well by the design itself the off-chip TVS components may have a higher on-resistance and higher clamping voltage than on-chip snapback clamps. To limit the current into the on-chip protection, a higher on-state resistance of the on-chip protection device is favorable. In the opposite case a second stage isolation resistor must be included in the design potentially impacting the IC performance in a negative way. On the other hand at proper matching of the on-chip and off-chip ESD network components the second stage isolation resistors or inductors can be completely avoided.

Figure 5.44 presents the impact of the ESD protection level on the on-resistance of typical on-chip ESD protection devices. The values are compared with the on-state resistance of the TVS diode used in the previous sections. If the on-state resistance of the on-chip component is higher than the on-state resistance of the TVS diode, more system-level ESD current is conducted by the TVS diode. An on-chip protection device with a higher on-state resistance is the preferred choice for a particular system-level protection design.

The measured high current capability of the non-silicided SNMOS is ~ 8 mA/μm. The capability of a larger device decreases significantly with the width and a larger isolation resistor between TVS diode and SNMOS is required to pass the desired level. Three recommendations can be formulated for the component-level ESD protection design of external IC pins: (i) avoid on-chip protection devices with low-holding voltage at the IC input, since the snapback into low holding voltage during ESD stress can take over the current path from the off-chip protection components making them passive; (ii) consider to avoid high HBM protection level in a single-stage ESD protection, since a lower HBM protection level likely is accompanied with higher on-state resistance of the on-chip ESD protection; (iii) avoid on-chip protection devices with too low intrinsic current capability since some second stage current conduction is still required.

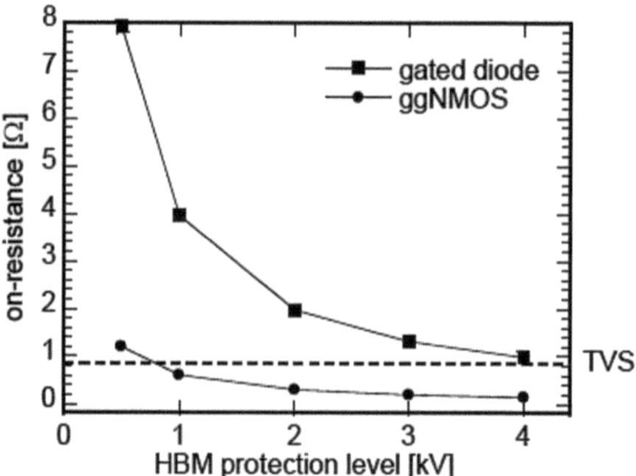

Fig. 5.44 Impact of ESD protection level on the on-resistance of on-chip protection devices: Comparison with the on-resistance of a TVS diode for 3.3 V applications, *SNMOS* no silicide blocking mask used

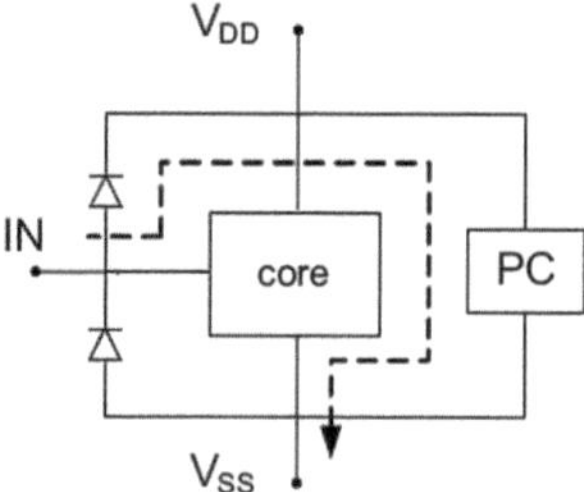

Fig. 5.45 Block diagram for ESD protection case study with the positive power supply V_{DD} and negative power supply pin V_{SS}, power clamp and added dashed line to show the current flow during ESD stress for INto V_{SS} pin combination

The application of these design recommendations are further discussed with the following test case. Two types of ESD clamps were comparatively evaluated: two ESD power clamps (PC) for the protection of ICs manufactured in a 130 nm CMOS technology. The PC includes the snapback type diode triggered SCR (DTSCR). The second PC is an active clamp. Both clamps are applied for the component-level ESD protection of supply pins. An ESD protection diode is added (Fig. 5.45) to simulate the worst case ESD stress in an IC.

The supply voltages for the 130 nm CMOS core device with a GOX of 2 nm (EOT) were 1.2 V. The measured GOX breakdown voltage was ∼7 V during a 1 ns long voltage stress [132]. Three component-level ESD design targets are evaluated to study the impact of the component-level ESD protection design on the behavior during the system-level ESD stress.

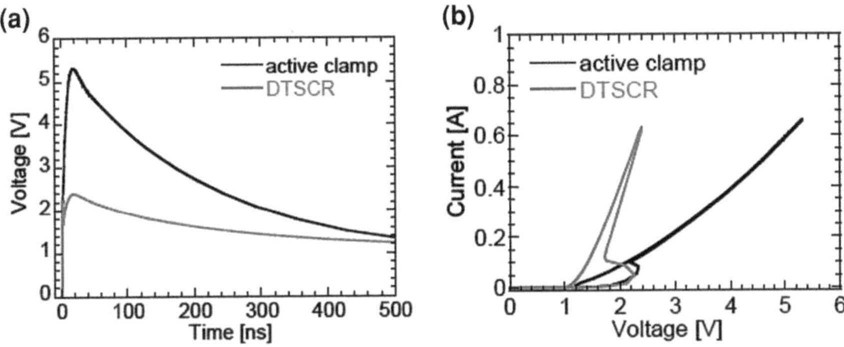

Fig. 5.46 Voltage waveforms at 1 kV HBM stress (**a**) and plotted HBM phase I-V character-istics (**b**) for the Active and DTSCR power clamps

Table 5.7 Clamp and device parameters of the test case components measured for three HBM protection levels

Device type	Intrinsic I_{T2} (mA/µm)	Device width (µm)		
		0.5 kV	1 kV	2 kV
diode-triggered SCR	50	7	13	25
active clamp	2	170	330	660
ESD diode	50	7	13	25

According to the 1 kV HBM voltage waveforms (Fig. 5.46a) and the plotted phase HBM I-V characteristics (Fig. 5.46b) the on-resistance of the active clamp is significantly higher. The protection level and the device parameters of the ESD clamps and the protection diode are summarized in Table 5.7. The ESD clamps operation during system-level ESD stress has been evaluated using mixed-mode simulations. The active clamp and DTSCR are implemented in the mixed-mode simulator with FEM parameters extracted from the original process simulation. The active clamp had conventional design with the RC-circuit and inverter forming the driver of the power NMOS clamping array. The diode-triggered SCR clamp included the reference circuit in form of three gated diodes to limit the overshoot during turn-on [43].

The input ESD protection diodes were added in the simulation setup as standard diode SPICE circuit models. A 1.5 V TVS diode was included in the circuit to represent the off-chip system-level ESD protection component (Fig. 5.47). The evaluation board and the IC package were simulated with a 5 mm long board trace and the parasitic of a DIL IC package.

The comparison of the simulated residual current after 30 ns during a 2 kV HMM pulse demonstrates that the residual current for both power clamp options does not exceed the design window determined by comparison with the HBM protection level (Fig. 5.48).

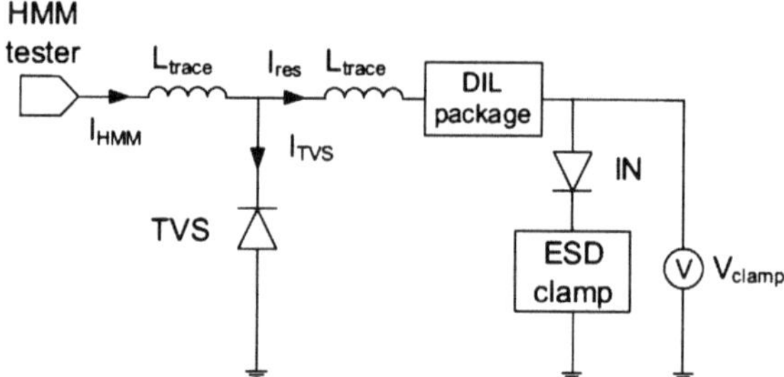

Fig. 5.47 Mixed-mode simulator circuit for the case study with TVS diode, input protection diode and ESD power clamp: I_{HMM} HMM stress current, I_{TVS} current through TVS diode, I_{res} residual current into IC, V_{clamp} voltage across ESD clamp and input protection diode, L_{trace} inductance of a board trace

Fig. 5.48 Simulated residual current after 30 ns into the on-chip ESD protection, HMM stress level: 2 kV

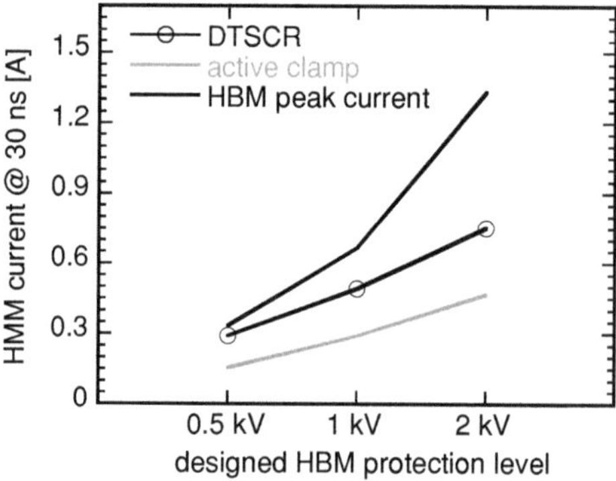

However, unlike for the 30 ns quasi-static regime, the simulated peak voltage at the beginning of the pulse is substantially different for the two clamp cases. The simulated peak voltage V_{clamp} across the ESD diode and the ESD clamp during 2 kV HMM stress is shown in Fig. 5.49. To avoid circuit failure, this voltage must not exceed the design window which is defined by the breakdown voltage of the GOX. This design window is exceeded significantly if the active clamp and the DTSCR are designed for a HBM protection level of 0.5 kV.

A significant voltage drop occurs right after turn-on of the clamps (Fig. 5.50). The voltage drop is caused by the residual HMM current peak and the higher on-resistance at low HBM protection level. The voltage across the DTSCR stays below the GOX breakdown voltage, and so within the design window, if the DTSCR is designed for a HBM protection level of 1 kV. Without additional off-chip protection devices, the active clamp can only be used if it is designed for a HBM protection level of 2 kV.

In this example the DTSCR is the preferred protection device. The higher on-resistance at lower HBM protection level is not a disadvantage during system-level ESD stress. The breakdown voltage and the on-state resistance of the input ESD

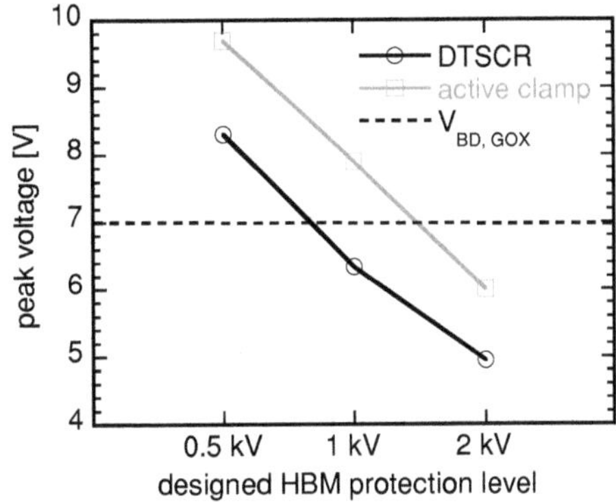

Fig. 5.49 Simulated peak voltage on-chip, HMM stress level: 2 kV, V_{BDGOX} breakdown voltage of gate oxide during a 1 ns long voltage stress

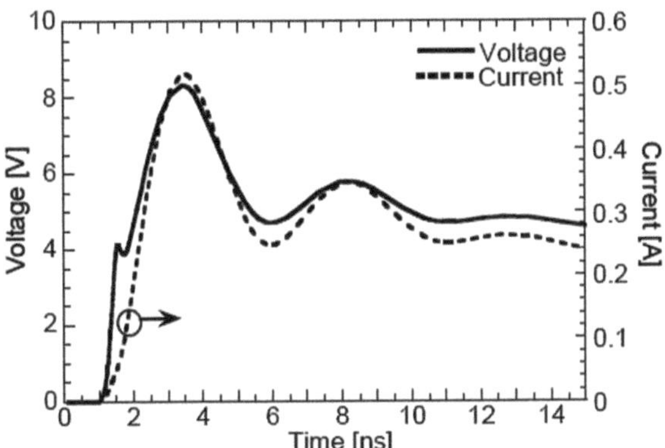

Fig. 5.50 Simulated voltage on-chip and residual current with the DTSCR as protection device, designed HBM protection level of DTSCR: 0.5 kV HBM, HMM stress level: 2 kV

protection diode maintain a clamping voltage level which is higher than the breakdown voltage of the TVS. Thus, the TVS is not turning-off during the system-level ESD stress. The low holding voltage prevents that after triggering the voltage exceeds the breakdown voltage of the GOX.

The active clamp has a much higher on-resistance during ESD stress due to its higher intrinsic on-state resistance and the much lower current capability. Thus, a significant larger layout area is required to obtain a sufficiently low on-resistance for the prevention of GOX failure during system-level ESD stress.

Above, both the protection against electro-thermal and gate oxide failure during system-level ESD stress have been demonstrated and verified with application examples. The main conclusion is that the transient behavior of off-chip and on-chip ESD protection devices needs to be included in the design process to prevent unexpected IC failures during the system-level ESD qualification. The presented simulation approach enables verification of the ESD protection design before IC packaging and mounting of the final system.

By adding HMM characterization to the design flow valuable information for both system- and component-level design is obtained. For the system-level design detailed information is given about the on-chip ESD protection and the available design window. This allows selection of the off-chip protection components based on a given on-chip ESD protection design. For the component-level design the transient device data and simulations generate a valuable input for the on-chip protection design. Board-level and IC package parasitic can be added to the simulation. The on-chip protection can be evaluated together with an off-chip protection long before the protected IC is packaged or mounted in a system.

The quasi-static device data obtained from TLP I-V curves, transient data like HMM failure current and the vfTLP failure voltages are required to design robust and effective ESD protection solutions for external IC pins subjected to the system-level ESD stress in an application board.

5.6 Comparison of System-Level ESD Co-design Methodologies

This final section brings the comparison between the two main system-level ESD design methodologies: *Datasheet Based Design* and *IC-System Co-design*. The goal of such comparison is to evaluate the suitability of each methodology for the system-level ESD protection design and to identify possible weaknesses in the design flow. Similarly to the previous material the comparison is done through the detailed analysis of a representative example.

At first, a functional and ESD robust RF system is designed with the available IC datasheet information demonstrating how an over-designed system is in reality obtained with this datasheet-based approach. In particular the design steps result in the application of an excessive amount for off-chip protection components. This design then is further optimized with TLP I-V characterization results taken into account and demonstrating that in this case the RF system requires less off-chip protection components. Finally, the RF input of the IC is evaluated with HMM characterization results which demonstrated that even less off-chip protection components are required followed by further optimization.

The test case is a wide band buffer amplifier designed for video, telecommunication and instrumentation applications. The maximum small signal bandwidth is 1.75 GHz for a supply voltage of 5 V and an output load impedance of 100 Ω is connected.

For electrical characterization of the amplifier the IC was mounted in a demonstrator board (Fig. 5.51). To ensure a constant permittivity over a wide frequency range, Rogers RO4003 was used as board material. The demonstrator board was laid out according to the recommendations of the amplifier manufacturer including a dedicated supply decoupling network and a dual DC bias connector. To guarantee the maximum bandwidth, the input capacitance of the amplifier was compensated with a matching board trace. The board layout includes also a 50 Ω resistor for the DC biasing of the amplifier input. This resistor reduces the gain of the amplifier by -6 dB.

Fig. 5.51 Top view of the
demonstrator board with
buffer amplifier when no
system-level ESD protection
components are added

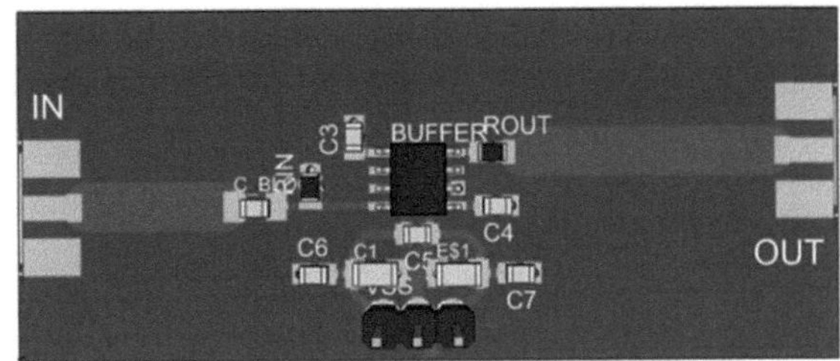

A TLP tester HANWA T-5000 is used as component-level ESD stress source to
extract the I-V characteristics, while the system-level ESD pulses were generated
with a HMM tester HANWA HED-W5000 M. That produces similar waveform to
the ones specified in the IEC6100-4-2 system-level ESD standard [126]. Both,
TLP and HMM tester are connected with SMA connectors to the demonstrator
board.

5.6.1 Design with Datasheet Information

The amplifier IC originally passes the standard component level 2 kV HBM stress
as well as 200 V MM stress at all pins with no further ESD related information
provided in the datasheet. Thus the type and electrical transient characteristics of
the IC ESD protection is assumed as unknown. Respectively, to design the system-
level ESD protection, a failure current of ~ 3.5 A can be used with some practical
confidence to determine the system level ESD design window for the off-chip ESD
protection components. The value of 3.5 A simply corresponds to the 200 V MM
peak current taken directly from the amplifier datasheet. In spite of that the MM
peak current may not correlate to the TLP failure current it is assumed that this 3.5
A current level can be used as a safe stress level to be further verified by the full
system test results.

According to the specification requirements the amplifier is designed to buffer
and transmit high frequency signals. Therefore to maintain the signal integrity the
off-chip system-level ESD protection components must be chosen with appropriate
low parasitic capacitance characteristics. Since the capacitance of the TVS inevitably causes a mismatch in the RF path an additional inductance was added to the
board to compensate for such mismatch. In this case study a simple micro strip line
was used to replicate an on-PCB inductor. The capacitance of each TVS device
and their compensating board traces must match the impedance the 50 Ω transmission line to realize a "clean" RF path through the board. The required
matching inductance is calculated as:

$$Z = 50 \ \Omega = \sqrt{\frac{L_{MATCH}}{C_{TVS}}}, \tag{5.5}$$

where Z is the matching impedance of the RF input, L_{MATCH} is the required matching inductance to match the TVS device capacitance C_{TVS}. To calculate the length of a board trace which has the required matching inductance, the matching inductance and the TVS capacitance in (5.5) are replaced by the parasitic inductance and capacitance of the board material:

$$Z = 50 \ \Omega = \sqrt{\frac{L_T}{C_{TVS} + C_T}}, \tag{5.6}$$

where L_T is the required PCB trace inductance to match the TVS device capacitance C_{TVS} and C_T the parasitic trace capacitance. The trace inductance and capacitance depend on the board material, the width and the length of the trace. They can be expressed as:

$$\begin{aligned} L_T &= L_0 \cdot l \\ C_T &= C_0 \cdot l, \end{aligned} \tag{5.7}$$

where L_0 is the parasitic inductance per unit length and C_0 the parasitic capacitance per unit length of the PCB trace. By solving the Eq. (5.6) with the expressions of the Eq. (5.7) the required trace length is calculated:

$$l = \frac{Z_0^2}{L_0 - Z_0^2 \cdot C_0} \cdot C_{TVS}, \tag{5.8}$$

where Z is the trace impedance and l the length of the matching trace.

The example for trace length dependence upon the desired compensated TVS capacitance is presented in Fig. 5.52. A few pF TVS capacitance already require a trace length of more than 1 cm. The required trace length increases even more with increasing trace width. Based on these calculations a very low capacitance bi-directional TVS device (TPD2USB30A) is selected as the on-PCB system-level ESD protection component. According to the specification this TVS provides 8 kV IEC61000-4-2 robustness (Fig. 5.53). According to the TVS datasheet, the on-state resistance is $\sim 1 \ \Omega$ and the capacitance C_{TVS} between each input pin to ground is ~ 0.7 pF.

Transient simulations are used to estimate the behavior of the amplifier and board components under system-level ESD stress. S-parameter simulations are used to verify the RF performance. Thereby, the system-level ESD protection design components were embedded in the RF design circuit and actual layout of the application board.

The system-level ESD protection was co-designed with the application board using transient, RF simulations and datasheet values in the RF design tool Agilent Design System (ADS). The TVS diode is implemented using the ADS diode model according to the model parameter and the TVS (Table 5.8). The TVS overshoot was represented by a series inductance. The inductance value is extracted from the TVS IBIS [133] model, provided by the TVS manufacturer.

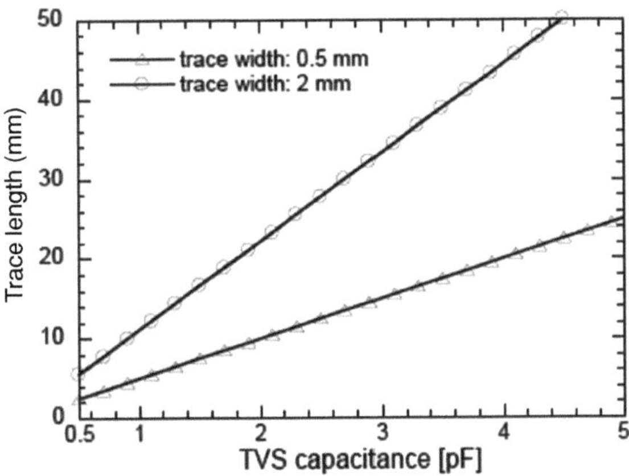

Fig. 5.52 Required trace lengths for matching a TVS capacitance with a microstrip line to 50 Ω for the board material RO4003 and the bandwidth 2 GHz

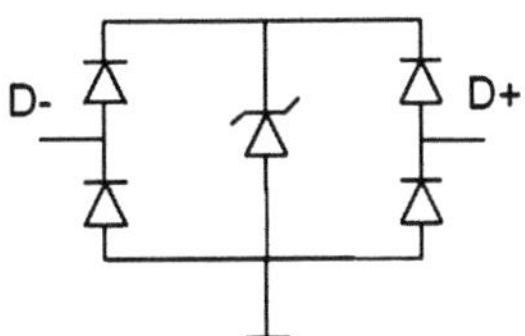

Fig. 5.53 The equivalent schematic of a TPD2USB30A TVS diode

Table 5.8 TVS diode modeling: ADS diode model parameter and their equivalent parameter extracted from the datasheet

ADS parameter	Parameter	Value
Ohmic resistance R_D	Dynamic resistance R_{ON}	1 Ω
Breakdown voltage B_V	DC breakdown voltage	4.5 V
Junction capacitance C_{J0}	Capacitance I/O to GND	0.7 pF

The amplifier itself was represented by the SPICE model provided by the IC supplier. Without any additional information for the on-chip ESD protection the system-level design can rely only on some assumptions. For example usually ESD protection diodes are used as the on-chip protection devices for RF inputs. In this case study an ESD protection diode with an on-resistance of 1.5 Ω is assumed as on-chip component. For the transient simulations, the diode on-state resistance was added to the ADS diode model, while the power clamp was not modeled. During system-level ESD stress a decoupling network is present at each supply pin and expected to sufficiently protect the power supply domain.

The passive components of the supply decoupling network are added using the part library of ADS and with the recommended capacitor values from the amplifier datasheet. The parasitic components of the capacitors equivalent circuit can be also extracted using HMM measurements unless they are available in the particular component library. This methodology has been presented earlier in this chapter Sect. 5.2.4. The board traces for the RF input and output of the amplifier are

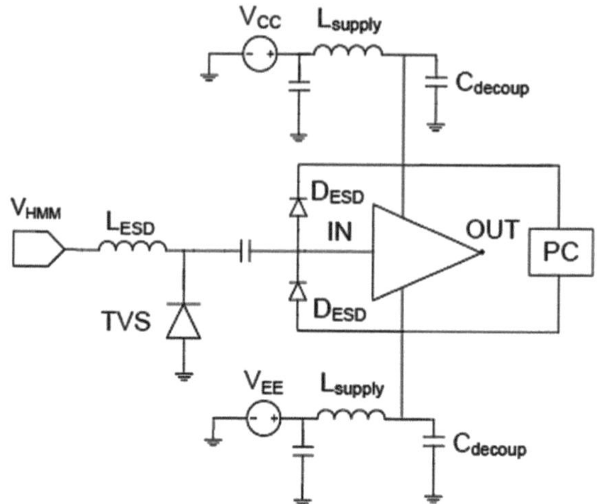

Fig. 5.54 Simulation circuit for the system-level ESD in ADS with the HMM tester model V_{HMM}, TVS device, the on-chip ESD protection diode D_{ESD}, the power clamp PC, the matching board traces L_{ESD}, the inductance of $L_{supply} = 1$ mH for DC source and the power supply decoupling capacitor C_{decoup}

modeled as micro strip lines. The PCB substrate model is extracted from the datasheet of the RO4003 board material used for the demonstrator board. The trace lengths are designed depending on the required matching inductance or as 50 Ω transmission lines.

The IEC61000-4-2/HMM stress source is represented with a lumped element model. The system-level ESD stress is applied to the RF input in the schematic (Fig. 5.54). The DC voltage sources are modeled with a parallel capacitor of 10 μF and a large series inductance of 1 mH to mimic the behavior of a more real voltage source, but with ideal output resistance and to block the flow of any transient current into the DC sources during the simulations.

The transient simulation approach is used to study the behavior of the ESD protection network operation during system-level ESD stress. The simulation results provide the residual current into the on-chip ESD protection for two different IEC61000-4-2 stress levels (Fig. 5.55).

For both stress levels the residual current into the on-chip protection exceeds the maximum current limit of 3.5 A in the design window. Based on this result, the system-level protection design requires significant improvement. However a simple approach with additional isolation resistor cannot be realized due to its impact on the RF signal and the expected degradation of the RF bandwidth. Therefore two TVS devices were used to reduce the on-resistance of the off-chip ESD protection and the residual current into the on-chip ESD protection.

Indeed, with an additional TVS component included in the design the residual current through the IC becomes acceptable (Fig. 5.56) for IEC61000-4-2 stress level of 4 kV. However to meet the 8 kV requirement an additional third TVS devices is required.

A significant difference in the residual current based on the simulation results is observed when the amplifier is powered up to 5 V (Fig. 5.57). In this case the residual current even for a stress of 8 kV is below the ESD design window limits

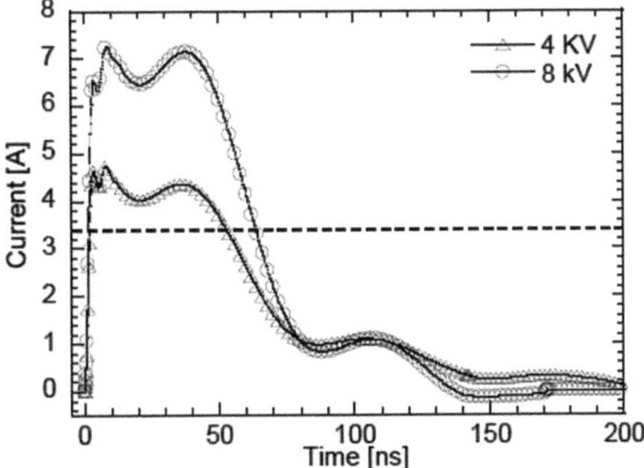

Fig. 5.55 Simulated residual current through the on-chip ESD protection when one TVS diode string is used, simulated for two different IEC61000-4-2 stress level and without supply voltage

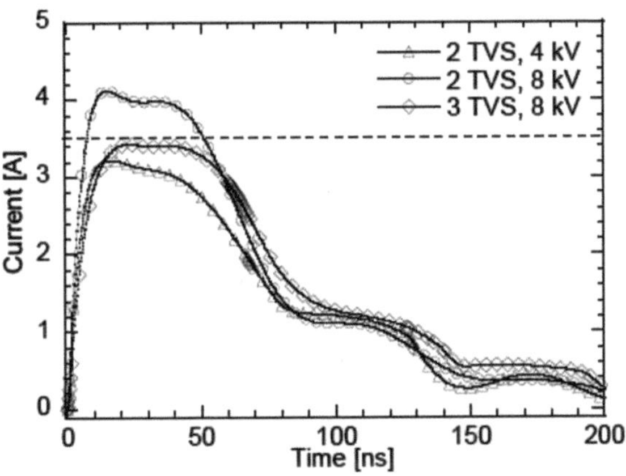

Fig. 5.56 Simulated residual current through the on-chip ESD protection without supply voltage, simulated for two different HMM stress level

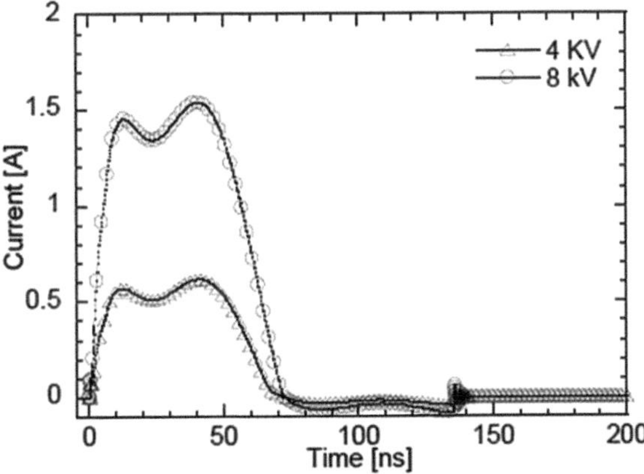

Fig. 5.57 Simulated residual current through the on-chip protection with supply voltage of 5 V

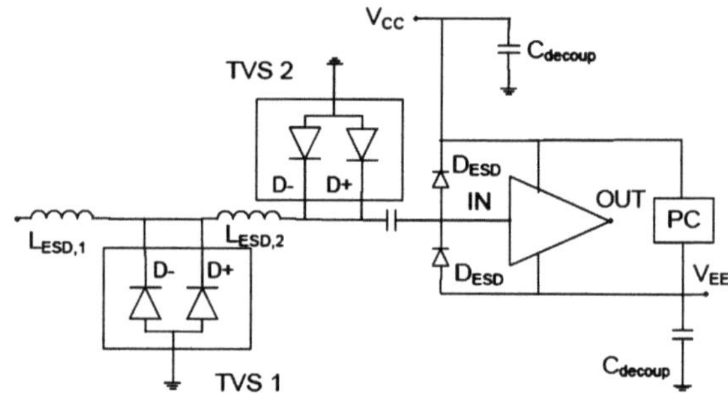

Fig. 5.58 Schematic of the system-level ESD protection design, where $L_{ESD;X}$ is the matching board trace; C_{decoup} is the supply decoupling network; TVS (TPD2EUSB30A); V_{CC} is the positive supply and V_{EE} is the negative supply

for the residual current level through the amplifier input. The two TVS devices increase the capacitance of the system-level ESD protection compared to only one diode string in one TVS device. The increased capacitance causes mismatching at the RF input of the demonstrator board. Impedance matching inductors (Fig. 5.58) are required to compensate for the increased capacitance. The concept of distributed ESD protection [134] is used to minimize the required inductance for each matching inductor.

Thus from the *Datasheet-Based* system ESD protection design approach two TVS components must be used to pass 4 kV system level pulse requirements.

The following section describes the design of the application board that includes the design and layout of the matching inductors for the compensation of the increased capacitance in the RF path due to the off-chip ESD protection devices.

Each TVS diode string has a capacitance of 0.7 pF. This results in total capacitance of 1.4 pF for each TVS device. The input capacitance of the amplifier during normal operation is 1.7 pF. Matching inductors are calculated using Eq. (5.8). Equivalent board traces are designed to compensate for these capacitances. The board material has an inductance of 6 nH/cm and a capacitance of 0.4 pF/cm for a bandwidth up to 2 GHz and a 0.5 mm wide trace. Based on those board properties the amplifier input is matched with a trace length of 8.5 mm and each TVS diode string with a trace length of 7 mm.

For the RF simulations, the calculated traces are added as micro-strip lines to the simulation setup. The HMM tester model is replaced by the ports required for the S-parameter simulation. There is only little impact on the input matching if the TVS devices are added to the design according to simulation results (Fig. 5.59) for input matching, with and without TVS.

The amplifier gain simulation in the demonstrator board shows minor impact of the ESD protection too (Fig. 5.60). The ripple, beginning around 1 GHz, is introduced by the equivalent transmission line for a bandwidth of 2 GHz.

This simulation-based design of the system-level ESD protection was experimentally verified using a demonstrator board (Fig. 5.61) for the characterization of both S-parameter and system-level ESD performance. The difference between the ESD protected board design and the original design is the transmission line from

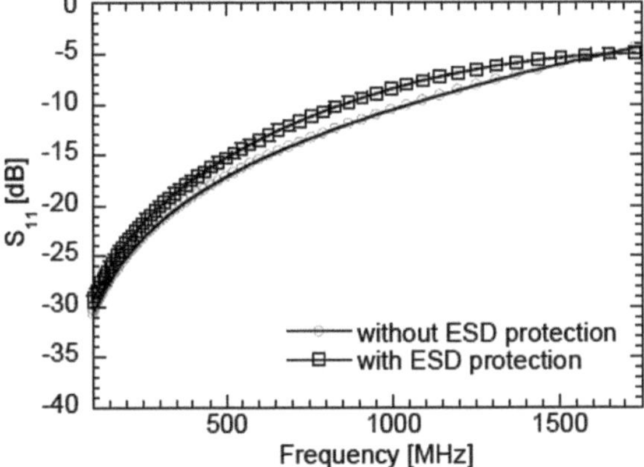

Fig. 5.59 Simulated input matching of the amplifier in the demonstrator board with and without added ESD protection at the supply voltage: ±5 V

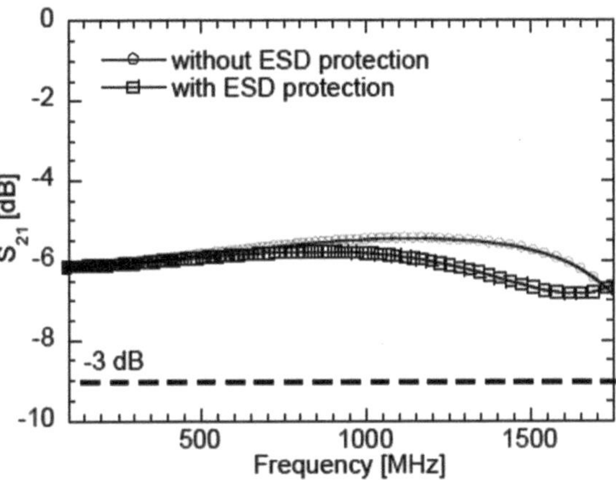

Fig. 5.60 Simulated gain of the amplifier in the demonstrator board with and without added ESD protection; supply voltage: ±5 V

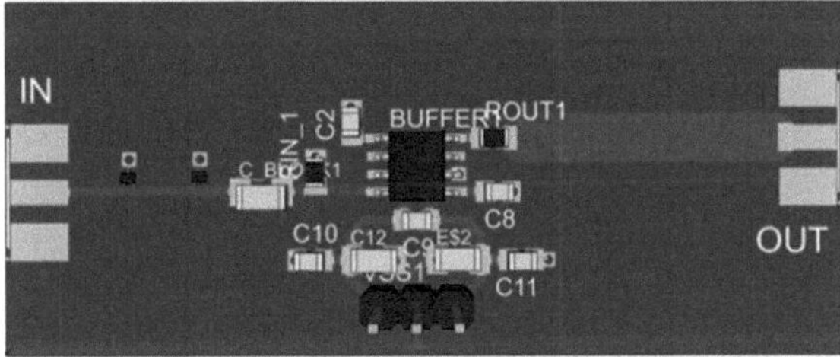

Fig. 5.61 Demonstrator board including system-level ESD protection; framed: matching board traces

the SMA connector to the decoupling capacitor at the amplifier RF input. It is replaced by an equivalent transmission line consisting of the TVS diodes and the required matching inductors.

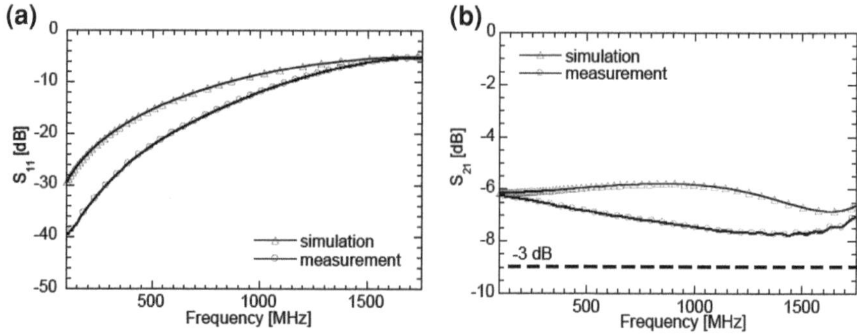

Fig. 5.62 Comparison of simulated and measured RF characteristics: **a** input matching with added ESD protection, **b** gain with added ESD protection at power supply 5 V

The S-parameters of the ESD protected design were measured with a network analyzer and compared to the simulation for input matching (Fig. 5.62a). According to the gain frequency dependence the added system level ESD protection does not significantly degrade the available bandwidth (Fig. 5.62b). At the same time the measurement results demonstrate a better RF performance than predicted by the simulations. The variations between the measurements and the simulations are attributed to the accuracy of the SPICE model of the amplifier and to a small difference between the modeled passives and the real application board.

According to the above transient simulation results the ESD passing level of the designed RF system input is below 4 kV HMM due to the excessive residual current in comparison with the amplifier datasheet estimation. On the contrary, according to the experimental data the HMM failure the passing level exceeds the 8 kV tester limit. At the test the failure criterion was set by monitoring the supply current above the datasheet limits. This unexpected result points out that a simple datasheet approach under the assumption of the on-resistance of the on-chip ESD protection may lead to an over-design. Therefore to improve the design the absolute failure level was evaluated by TLP characterization of the on-chip ESD protection to gain an insight regarding the type and characteristics of the IC component internal input protection.

5.6.2 *Design with Additional TLP Characterization*

TLP testing is used to characterize the on-chip and off-chip ESD protection devices. The TLP I-V characteristics are also used to improve the accuracy of the simulation models. 100 ns TLP testing is carried out on the unprotected standalone amplifier to obtain the TLP I-V curves of the on-chip ESD protection devices. The weakest pin combination during TLP testing is the power clamp between positive and negative supply pin. The TLP failure level is 5.5 A. This value is taken as a

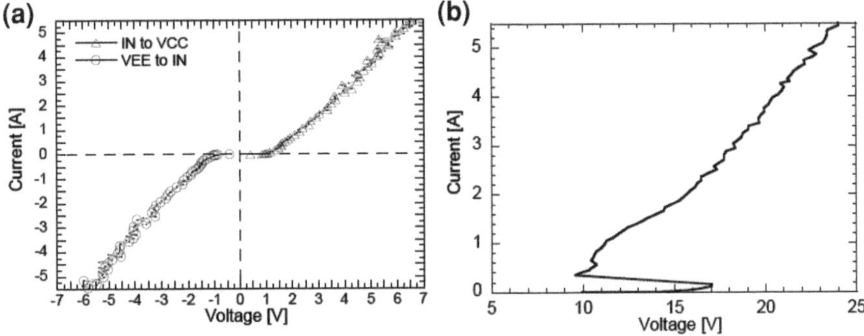

Fig. 5.63 100 ns TLP I-V curves of amplifier: **a** input ESD protection and **b** power clamp; *IN* RF input, *VCC* positive supply pin, *VEE* negative supply pin

Fig. 5.64 100 ns TLP I-V characteristics of the off-chip ESD protection TVS component (TPD2USB30A) measured up to 10 A TLP tester limit

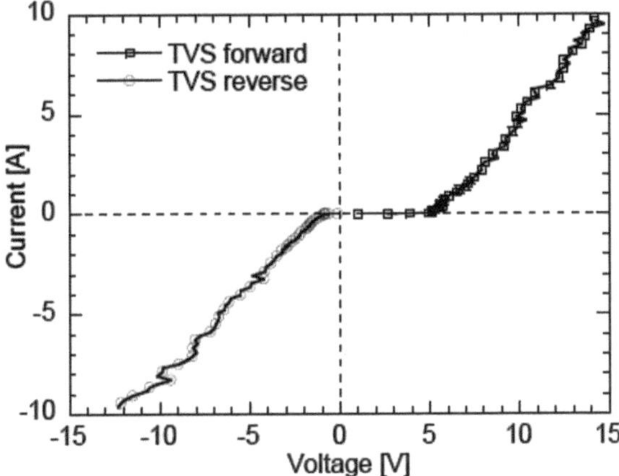

design window for the protection design since the exact discharge path in the IC is not known under system-level ESD stress conditions.

According to the TLP I-V characteristics (Fig. 5.63a) for the input combination with the positive and negative supply pins the RF input protection is based on ESD diodes with an on-state resistance of $\sim 1\ \Omega$. This value is lower than the assumed upon the datasheet information. Similarly from TLP characteristic for the power pins the clamp device can be identified as snapback n-p-n, NMOS or high holding voltage SCR device with the triggering voltage of ~ 17 V, minimum holding voltage of ~ 9.5 V and the on-state resistance for the entire current path including interconnects of $\sim 3\ \Omega$.

The same TLP approach for the discrete TVS component reveals the avalanche diode likely vertical device with a breakdown voltage of ~ 5.2 V and the clamping characteristics with an on-state resistance of $\sim 1\ \Omega$ both in reverse and forward bias modes (Fig. 5.64). In reverse mode the TVS simply provides the forward bias diode

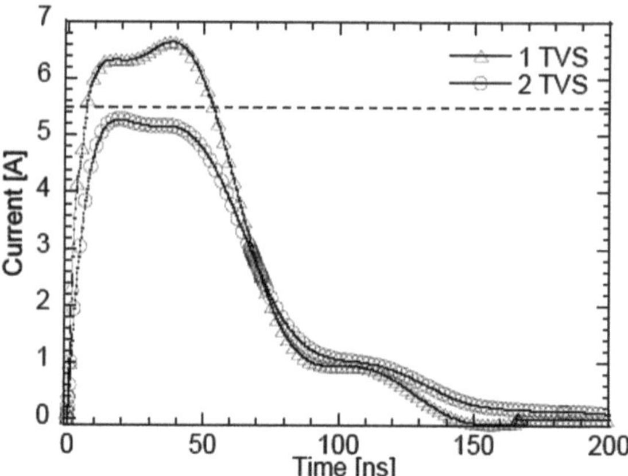

Fig. 5.65 Simulated residual current into the on-chip protection after TLP data taken into account with the dashed line marking the new design window according to the measured TLP current for the power pins

characteristics. The dynamic resistance of $\sim 0.9\ \Omega$ is similar to the value in the TVS datasheet.

This easily obtained information from the TLP characterization can be directly used to improve both the simulation models and achieve better design confidence. Indeed, according to the re-simulated results for the new safe current limit ~ 5.5 A (Fig. 5.65) two TVS on-PCB components are required to protect the IC input. To verify the simulation results the design option with only one TVS device was stressed with HMM pulse. However, surprisingly no failure was observed up to 8 kV HMM tester limit. Thus the failure level during this TLP-improved design was still not precise enough to deliver and optimize the protection solution that would avoid over-design and take at advantage both from lower cost and better signal integrity.

Thus, further optimization can be done to the input protection. Another insight from the simulation is provided by monitoring the voltage waveforms for different power supply conditions. Without applied supply voltage, the voltage during the stress at the supply pins is not reaching the 17 V triggering voltage. Based on the simulation results the proposed design is robust to 8 kV HMM stress. To verify the simulation result HMM stress is applied to the supply connector with the amplifier mounted in the test board. To monitor any turn-on of the power clamp the voltage between the stressed supply pin and the board ground is captured with a high-impedance voltage probe for each stress level. The amplifier power clamp does not turn on until a stress level of 8 kV and no device failure occurs. The off-chip decoupling capacitors protect the power clamp against system-level ESD stress.

HMM testing on the configuration with one TVS diode passes 8 kV although the simulations predicted a failure. The failure level used in the TLP-based design must be lower than the absolute ESD robustness of the on-chip ESD protection at the RF input. The following section shows that the design can be further optimized if HMM testing is carried out on the standalone IC.

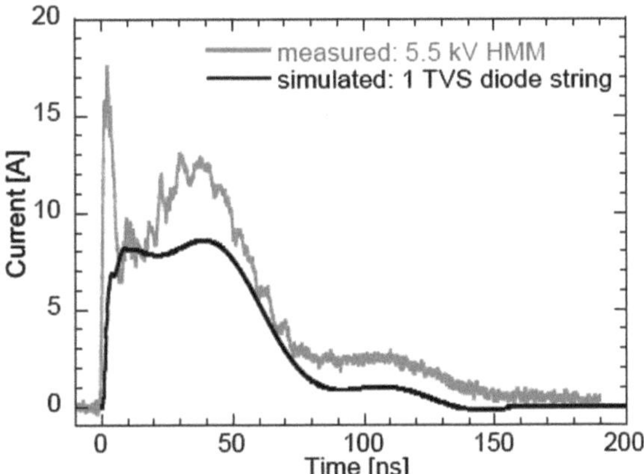

Fig. 5.66 Simulated residual current into the amplifier input, when protected with one TVS *diode string* comparison to measured failure current obtained with standalone HMM testing on amplifier input; HMM stress level during *simulation* 8 kV; no supply voltage applied

5.6.3 Design Optimization with HMM Testing

HMM pulse characterization of the RF input of the standalone amplifier IC was accomplished to obtain the absolute failure level during system-level ESD stress with no off-chip components connected. The HMM failure level of ~ 5.5 kV was recorded that corresponds to ~ 11 A of HMM pulse current at the time of 30 ns from the beginning of the pulse. The measurement and simulation HMM waveform comparison (Fig. 5.66) for the residual current with only one TVS used on the board is in correlation. This significantly optimized design has a reduced parasitic capacitance in the RF path down to only one TVS equivalent ~ 0.7 pF and requires only one single small board trace for the TVS capacitance matching. Hence, a much smaller board area is required to realize the system-level ESD solution.

5.6.4 Benchmarking and Comparison of the Designs

The standalone HMM measurements on the amplifier input ESD protection show a much higher ESD robustness then given by the manufacturers datasheet. Using only the datasheet ESD protection level leads directly to an overdesigned system-level ESD protection.

The available ESD protection windows based on the system with amplifier design for the three discussed design approaches are compared in (Fig. 5.67). Based on the ESD protection level in the IC datasheet, three TVS devices would be required to protect the amplifier against 8 kV HMM stress (Figs. 5.68a and 5.69a). Upon the TLP-based/SEED design approach the number of TVS components can be reduced to two. Finally only one TVS component is actually required if a more detailed analysis using HMM characterization is applied (Figs. 5.68b and 5.69b).

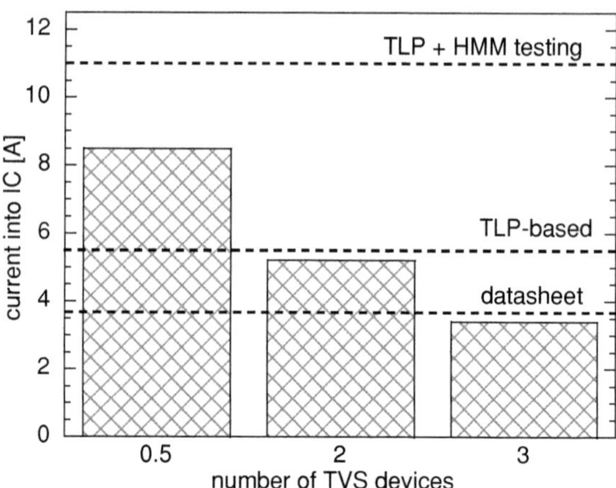

Fig. 5.67 Simulated amplitudes of the residual current through the IC amplifier input during HMM stress versus number of required TVS devices for the HMM stress level of 8 kV with dashed lines marking the design window for each methodology

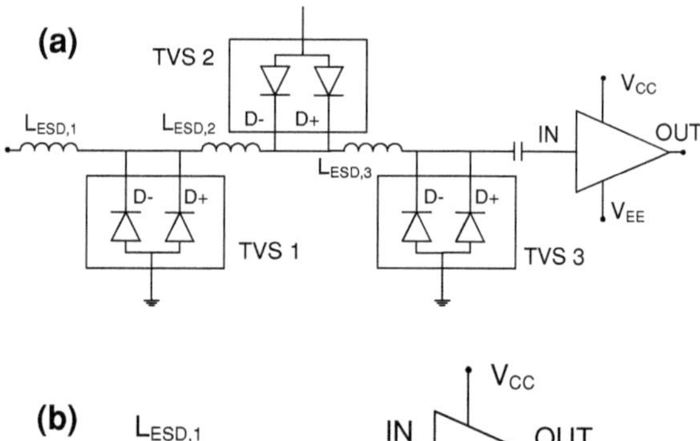

Fig. 5.68 Schematic of system-level ESD protection derived as a result of the datasheet based design (**a**) and IC-system co-design (**b**) with full TLP and HMM test data included in the analysis

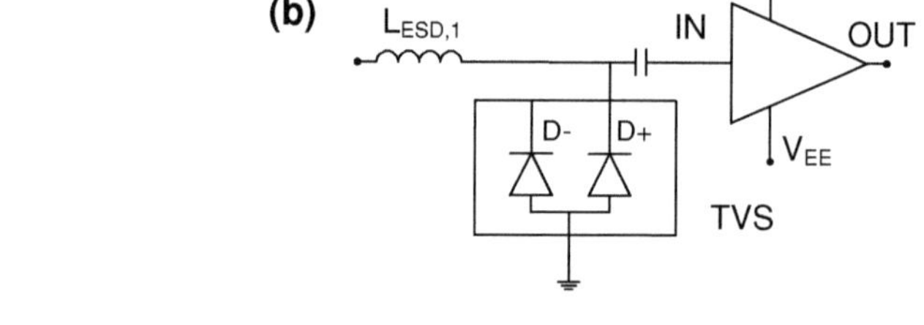

A sufficiently large safety margin above 20 % is obtained in case of the co-design solution which is suitable for the volume production where manufacturing related variations can occur. The improvement of the optimized design becomes even more visible when calculating the required board area for each design methodology (Fig. 5.70).

If the system design relies only on the datasheet based design methodology then five times more board area is required in comparison with the IC-system co-design based on the HMM testing data and transient simulations. Due to the lower number of TVS devices the capacitive load from the off-chip ESD protection is six times lower when using the co-design methodology. According to the simulated gain for two TVS diode configurations (Fig. 5.71) a ~ 80 MHz lower -3 dB frequency is

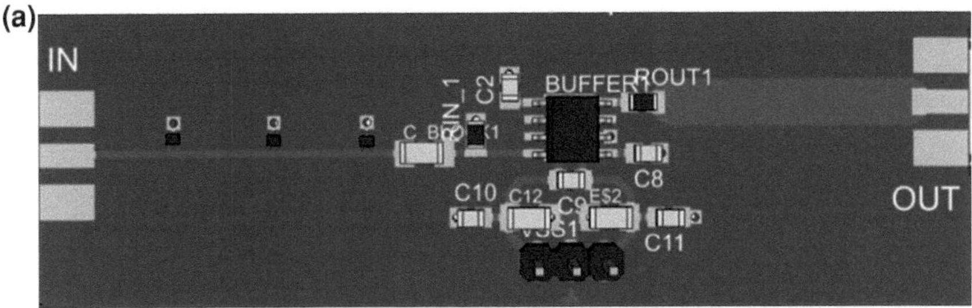

Fig. 5.69 Image of the demonstration board including system-level ESD protection after datasheet based design (**a**) and co-design (**b**)

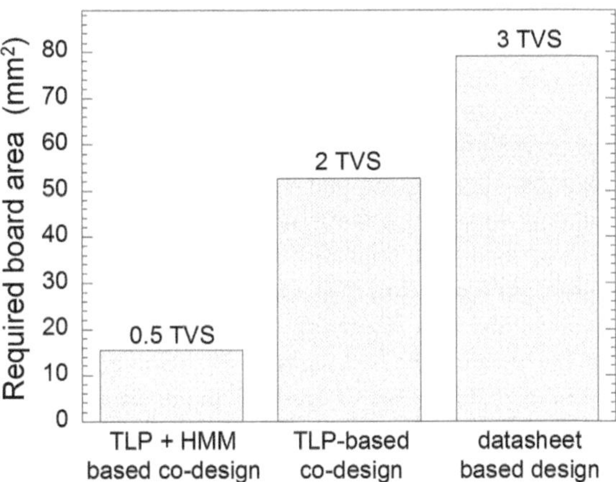

Fig. 5.70 Required board area for design according to the three different design methodologies output; minimum area: 3.1 × 5 mm for one TVS diode string and required matching inductor for 0.7 pF TVS capacitance

obtained when three TVS components are used. For both configurations the simulations are carried out with the required matching board traces in the simulation setup.

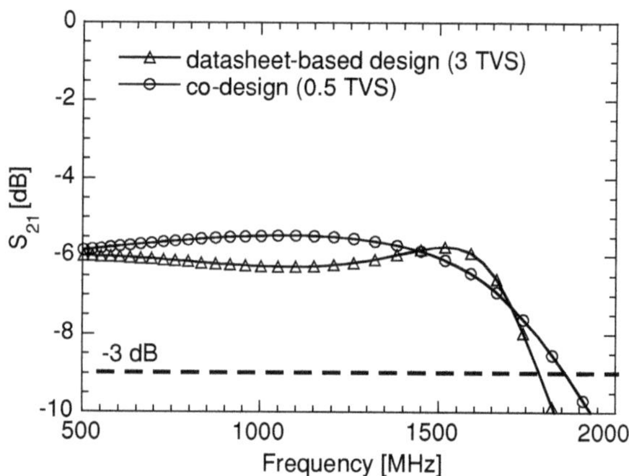

Fig. 5.71 Simulated gain for two different configurations: data sheet based design and co-design

5.7 Summary

Transient on-chip device and components characterization, composition of either simple or more complicated models followed by transient circuit simulation or mixed-mode analysis are essential key steps of the development of both IC and systems for the required ESD passing level target. This has been broadly demonstrated both in this chapter and across the entire book. In spite of that ESD device models can be implemented with a simplified approach to reduce the effort and the simulation time of large protection networks, the outcome from the analysis often reveals results that are intuitively unexpected. With a little bigger development investment the mixed-mode ESD simulations enable and deliver an opportunity to obtain higher accuracy of the analysis. This is often necessary to reproduce a complex behavior of the devices at the beginning of the ESD stress pulse to account for fast transient overshoots. Highly accurate FEM models can be created using the approach with parameterized templates and parameterized analytical profiles allow the simulation and extraction of the device behavior in the transient domain. These devices in combination with SPICE and compact models enable circuit-like simulations in a mixed-mode device-circuit approach under accurate calculation of the avalanche and injection transient effect thus greatly increasing the confidence of the IC-system blocks co-design. Case studies, including the presented in this chapter, can be used to analyze and benchmark system-level ESD co-design methodologies to compare and understand the advantages, limitations and drawbacks for each approach. Two major design methodologies have been evaluated in this chapter, the *Datasheet Based Design* and the *IC-System Co-design*.

The input components for the *Datasheet Based Design* are mainly the available information in the off-chip and IC component datasheets; expertise of the system designer and the phenomenological experience from similar prior designs. Today, this is a major approach across the industry. With this methodology even if the final design passes the ESD qualification it is hard to evaluate to what extend the final system performance is optimized and impacted by the trade-offs at the different design steps. This outcome is a product of the obvious fact that in most cases the system designer has no accurate knowledge about the device characteristics of the internal IC-component ESD protection solution implemented by the IC designers. Thus many system design steps are logically made in assumption of the worst-case scenario with a "black-box" approach. For example to achieve a robust ESD protection at board-level the TVS components can be chosen based on the datasheet containing incomplete information regarding the fast transient clamping voltage waveforms. Thus an overdesign of the system or lost time-to-market pace due to costly overdesign for ESD protection using additional TVS as a patch solution may occur as a direct reason of not accounting for the particular IC-level ESD protection implementation. According to the case study (Sect. 5.6) this can impact the performance of the protected system by increasing for example the capacitive loading of an RF path.

The preferred methodology is the *IC-System Co-design*. The meaning of this approach incorporates the steps of using experimental data or quasi-static pulsed characterization, HMM transient characteristics measurements with corresponding de-embedding, ESD model composition for the on-chip and off-chip components and through circuit or mixed-mode analysis to predict, optimize and design the most optimal system within the target design parameters. Transient simulations are the enabling factor for the advanced system design and for future generation IC design to implement the optimization for IC ESD protection. The TLP-based analysis combined with HMM characterization data obtained from the IC pins and added to the simulation design flow enables verification of the system ESD performance. A simple comparison of the residual HMM current and voltage transient waveforms in the ESD time domain as well as the failure level with the standalone IC HMM waveforms provide a confidence in understanding the margins of the system design. For example the residual current waveforms obtained at the IC pin as a part of the system can be compared with the standalone IC waveform to estimate the passing level.

5.8 Outlook

The system-level ESD co-design approach will likely remain a challenge for future SiP and SoC as the scaling of CMOS technologies continues. New materials (silicon-germanium) and new device architectures (3D transistors multi-gate-FET, FINFET) are rapidly emerging and will be used in the new systems design with the system level ESD requirements together with the new architectures. For example 3D integration will combine wafer die stacking to integrate different functionality like power management, logic and memory in one IC package.

Silicon interposers are currently the most promising technology. An interposer connects several active dies to each other, but also to the package substrate and to

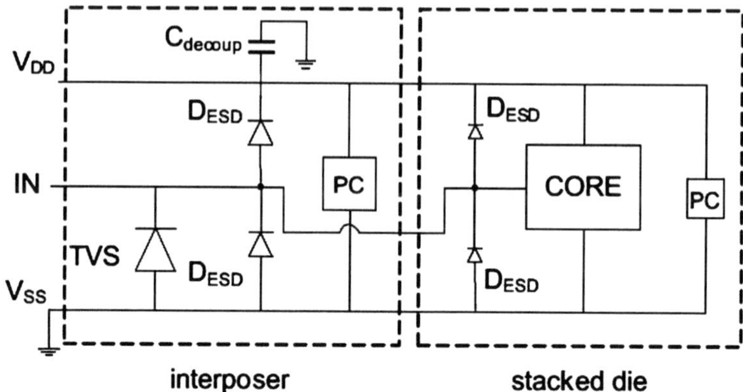

Fig. 5.72 Schematic of possible ESD protection design in interposer, C_{decoup} decoupling capacitor on interposer, D_{ESD} ESD protection diodes, *PC* ESD power clamps

the package pins. First products like next generation FPGAs have already entered into the market [135]. These large die size products use passive interposer to improve the yield of the whole product. A passive interposer contains only back-end of line structures like metal layers, via and through-silicon-via. Possible extension is the processing of front-end-of-line structures in the interposer. By adding implantation and well formation to the process flow, ESD protection devices can be created in the interposer. Both, component-level and system-level ESD protection devices can be made available to the designer.

This approach has two main advantages. The component-level ESD protection in the stacked dies can be reduced to a minimum. The stacking process is usually done in an environment with high ESD control measures. Thus manually handling of dies by operators is limited to an absolute minimum. Figure 5.72 illustrates a possible protection scenario. The interposer carries the system-level ESD protection device, component-level ESD protection and passive components like a decoupling capacitor. The interposer is manufactured in a less advanced low-cost technology. The stacked die is manufactured in a sub-20 nm CMOS technology. It has some component-level ESD protection which is dimensioned for low protection level like e.g. 100 V or 500 V HBM. This assumes that modern ESD control measures are in place during manufacturing which allow a safe handling of the not yet stacked dies.

The presented scenario will generate new challenges related to the ESD protection design, especially when the stacked dies are fabricated using advanced sub-20 nm CMOS technology. At the same time, the number of pins of the stacked dies can be even higher than in standalone high-pin count ICs which increases the equivalent capacitance. During the stacking process a single die can be charged and discharged while mounted. Thus, CDM pulse susceptibility can become one of the challenges. The HBM protection targets will be limited to a minimum level of e.g. 100 V. Most likely the single dies are only handled in automatic equipment and in areas with advanced ESD control based on the requirements of a well-established ESD control program. For the 3D and 2.5D type of systems some "ESD questions" arise.

In general it is hard to predict the ESD conduction paths in stacked die SiP's. What is the real cost saving if the ESD protection is shared between the interposer and the stacked die? Can the interposer be used to provide ESD protection, for example by co-packaging TVS devices?

ESD endurance is also amongst the future challenges. System-level ESD is integrated to the final products. Unlike the low-current component passing level demands the system ESD protection is designed to guarantee the passing level during the entire life time of the product rather than just to pass several pulses at the qualification of a representative sample pool. If the system design is implemented without verification of how many times the product can withstand ESD stress and is there any degradation of the passing level over the lifetime product endurance? To take this into account an additional ESD endurance testing and design is required. The authors in [136] carried out the experiments and statistical calculations to determine the spread and reduction of the failure level during repeated ESD stress on a TVS diode. The study generates a number of important questions. For example how can we take into account the endurance of ESD protection devices during the ESD protection design? Which safety margin should be used during the ESD protection design when taking into account both the endurance of the off-chip and the on-chip protection devices? Is there an impact on the long term reliability of gate oxides and interconnects?

These questions are not answered yet. They likely will generate an entire cycle of future research studies focused on advanced CMOS technology nodes with significant reduced current carrying capability and thin sensitive gate dielectric, robustness of high voltage medical, automotive, industrial and other emerging systems that the future will inevitably bring for us.

The authors of this book believe that the current and future challenges must be efficiently addressed using the IC-system co-design approach hopefully motivated through this book.

References

1. ANSI/ESD, *S20.20-2007—Protection of Electrical and Electronic Parts, Assemblies and Equipment* (Excluding Electrically Initiated Explosive devices), 2007
2. Industry Council on ESD target levels, *White Paper 1: A Case for Lowering Component-level HBM/MM ESD Specifications and Requirements*, 2007
3. California Micro Devices, *White Paper, January 2008, The Changing ESD Land-scape. ESD Protection Architecture Design Considerations for Next Generation Devices*, 2008
4. V.A. Vashchenko, V.F. Sinkevitch, *Physical Limitations of Semiconductor devices* (Springer, New York, 2008)
5. V.A. Vashchenko, A.A. Shibkov, *ESD Design for Analog Circuits* (Springer, New York, 2010)
6. M. Eherton, M. Khazhinsky, J. Miller, et al., in *EDA Tool for Checking ESD specific I/O ring integration rules, International ESD Workshop*, 2009
7. V.A. Vashchenko, W. Kindt, P. Hopper, M. ter Beek, Implementation of 60V Tolerant Dual Direction ESD Protection in 5V BiCMOS Process for Automotive Application, in *Proceedings of EOS/ESD Symposium*, 2004
8. IEC 61000-4-2:2008, Electromagnetic compatibility (EMC) - Part 4-2: Testing and measurement techniques - electrostatic discharge immunity test, 2008
9. S. Sze, *Physics of Semiconductor Devices* (Wiley, New York, 1981)
10. James Colby Protecting Electronic Devices Against ESD Littelfuse, Inc James Colby Board-Level Design Considerations for ESD Circuit Protection Littelfuse, Inc. colby@littelfuse.com
11. S. Joshi, E. Rosenbaum, Compact modeling of vertical ESD protection NPN transistors for RF circuits, in *Proceedings of EOS/ESDSymposium*, 2002, pp. 289–295
12. V. Vassilev, G. Groeseneken, S. Jenei, H. Maes, Modeling and 3extraction of RF performance parameters of CMOS Electrostatic Discharge Protection Devices, in Proceedeedings of EOS/ESD *Symposium*, 2002, pp. 111–118
13. C. Russ, K. Verhaege, K. Bock et al, A compact model for the grounded gate NMOS behavior under CDM ESD stress, in *Proceedings of EOS/ESD Symposium*, 1996, pp. 302–315
14. M. Mergens, W. Wilkening, S. Mettler, et al., Analysis and Compact modeling of lateral DMOS power devices under ES stress conditions, in *Proceedings of EOS/ESD Symposium*, 1999, pp. 1–10
15. V. Vassilev, et al., Analysis arid Improved Compact Modeling of the Breakdown Behavior of sub-0 25 micron ESD Protection ggNMOS Device, in *Proceedings of EOSESD Symposium*, 2001, pp. 62–70
16. V. Vassilev, V.A. Vashchenko, P. Jansen, et al., ESD circuit model based protection network optimisation for extended-voltage NMOS drivers. Microelectron. Reliab. **45**, 1430–1435 (2005)

V. Vashchenko and M. Scholz, *System Level ESD Protection*,   311
DOI: 10.1007/978-3-319-03221-4, © Springer International Publishing Switzerland 2014

17. B. Aliaj, V. Vashchenko, Q. Cui, J. Liou, A. Tcherniaev, M. Ershov, D. LaFonteese, 2.5-Dimensional Simulation for Analyzing Power Arrays Subject to ESD Stresses, in *Electrical Overstress/Electrostatic Discharge Symposium Proceedings,* 2009, pp. 3A.2.1–3A.27

18. C. Duvvury, S. Ramaswamy, A. Amerasekera, R.A. Cline, B.H. Andresen, V. Gupta, Substrate pump NMOS for ESD protection applications, in *Electrical Overstress/ Electrostatic Discharge Symposium Proceedings2000,* 26–28 Sept 2000, pp. 7–17

19. DECIMM^{TM} Angstrom Design Automation, Release 6.0, www.analogesd.com

20. V.A. Vashchenko, A.A. Shibkov, *TCAD Methodologies for Industrial ESD Design,* Seminar 4, IEW 2013

21. IEC 61000-4-2:2008, Electromagnetic compatibility (EMC)—Part 4-2: Testing and measurement techniques—electrostatic discharge immunity test, 2008

22. TESEQ surge tester, http://www.teseq.com/

23. ISO 10605:2008—Road vehicles—Test methods for electrical disturbances from electrostatic discharge, 2008

24. G. Boselli, A. Salman, J. Brodsky, H. Kunz, The relevance of long duration TLP stress on system level ESD design, in *EOS/ESD Symposium 2010,* Oct 2010, pp. 1–10

25. IEC 61000-4-5Electromagnetic compatibility (EMC)—Part 4-5: Testing and measurement techniques—Surge immunity test

26. D.E. Powell, B. Hesterman, Introduction to Voltage Surge Immunity Testing, in *IEEE Power Electronics Society Denver Chapter,* 2007

27. S. Marum, W. Kemper, Y-Y. Lin, P. Barker, Characterizing Devices Using the IEC 61000-4-5 Surge Stress

28. HANWA Tester, http://www.hanwa-ei.co.jp/english/seihin_11.html

29. HPPI Tester, http://www.hppi.de/

30. ANSI/ESDA SP5.6-2009—ESD testing—Human Metal Model (HMM)—Component Level, 2009

31. K. Hall, D. McCarthy, Tests with different IEC 801.2 ESD simulators have different results depending on product sensitivities, in *IEEE International Symposium on Electromagnetic Compatibility, 1995. Symposium Record,* Aug 1995, pp. 280–284

32. J. Barth, D. Dale, K. Hall, H. Hyatt, D. McCarthy, J. Nuebel, and D. Smith, Measurements of ESD HBM events, simulator radiation and other characteristics toward creating a more repeatable simulation or what simulators should simulate, *in EOS/ESD Symposium 1996,* Sept 1996, pp. 211–222

33. K. Wang, D. Pommerenke, R. Chundru, J. Huang, K. Xiao, P. Ilavarasan, M. Schaffer, Impact of ESD generator parameters on failure level in fast CMOS system, in *IEEE International Symposium on Electromagnetic Compatibility, 2003,* vol. 1, Aug. 2003, pp. 52–57

34. S. Frei, D. Pommerenke, An analysis of the fields on the horizontal coupling plane in ESD testing, in *EOS/ESD Symposium 1997,* Sept 1997, pp. 99–106

35. E. Grund, K. Muhonen, N. Peachey, Delivering IEC 61000-4-2 current pulses through transmission lines at 100 and 330 Ohm system impedances, in *EOS/ESD Symposium 2008,* Sept 2008, pp. 132–141

36. T. Maloney, N. Khurana, Transmission line pulsing techniques for circuit modeling of ESD Phenomena, in *EOS/ESD Symposium 1985,* 1985

37. A. Shibkov, V.A. Vashchenko, *ESD Design for Analog Circuits* (Springer, New York, 2010)

38. E. Grund, R. Gauthier, TLP systems with combined 50 and 500 Ohm impedance probes and KELVIN probes, in *EOS/ESD Symposium 2003,* Sept 2003, pp. 1–10

39. T. Daenen, S. Thijs, M.I. Natarajan, V. Vassilev, V. De Heyn, G. Groeseneken, Multilevel Transmission Line Pulse (MTLP) tester, in *EOS/ESD Symposium 2004,* Sept 2004, pp. 1–6

40. H. Gieser, M. Haunschild, Very-fast transmission line pulsing of integrated structures and the charged device model, in *EOS/ESD Symposium 1996,* Sept 1996, pp. 85–94

41. Ansi/ESD SP5.5.2-2007–ESD Testing Very Fast Transmission Line Pulse (vfTLP) Component Level, 2007
42. D. Tremouilles, S. Thijs, P. Roussel, M.I. Natarajan, V. Vassilev, G. Groeseneken, Transient voltage overshoot in TLP testing—Real or artifact?, in *EOS/ESD Symposium 2005*, Sept 2005, pp. 1–9
43. M. Scholz, S. Thijs, D. Linten, D. Tremouilles, M. Sawada, T. Nakaei, T. Hasebe, M.I. Natarajan, G. Groeseneken, Calibrated wafer-level HBM measurements for quasi-static and transient device analysis, in *EOS/ESD Symposium 2007*, Sept 2007, pp. 2A.2-1–2A.2-6
44. D. Linten, P. Roussel, M. Scholz, S. Thijs, A. Griffoni, M. Sawada, T. Hasebe, G. Groeseneken, Calibration of very fast TLP transients, in *EOS/ESD Symposium 2009*, Sept 2009, pp. 1–6
45. D.C. Smith, Current probes, more useful than you think, in *IEEE International Symposium on Electromagnetic Compatibility 1998*, vol. 1, 1998, pp. 284–289
46. S.-L. Jang, M.-S. Gau, J.-K. Lin, Novel diode-chain triggering SCR circuits for ESD protection. Solid State Electron. **44**(7), 1297–1303 (2000)
47. M. Mergens, C.C. Russ, K.G. Verhaege, J. Armer, P.C. Jozwiak, R. Mohn, B. Keppens, C.S. Trinh, Diode-triggered SCR (DTSCR) for RF-ESD protection of BICMOS SiGe HBTs and CMOS ultra-thin gate oxides, in *Electron devices Meeting (IEDM) Technical Digest. IEEE International*, Dec 2003, pp. 21.3.1–21.3.4
48. D. Linten, V. Vashchenko, M. Scholz, P. Jansen, D. Lafonteese, S. Thijs, M. Sawada, T. Hasebe, P. Hopper, G. Groeseneken, Extreme voltage and current overshoots in HV snapback devices during HBM ESD stress, in *EOS/ESD Symposium 2008*, Sept 2008, pp. 204–210
49. M. Mergens, W. Wilkening, S. Mettler, H. Wolf, A. Stricker, W. Fichtner, Analysis of lateral DMOS power devices under ESD stress conditions. IEEE Trans. Electron Devices **47**(11), 2128–2137 (2000)
50. M.H. Song, T. Smedes, J.C. Tseng, T.H. Chang, R. Derikx, R. Velghe, A contribution to the evaluation of HMM for I/O design, in *EOS/ESD Symposium 2011*, Sept 2011, pp. 1–7
51. M. Scholz, D. Linten, S. Thijs, M. Sawada, T. Nakaei, T. Hasebe, D. Lafonteese, V. Vashchenko, G. Vandersteen, P. Hopper, G. Groeseneken, On-wafer Human Metal Model measurements for system-level ESD analysis, in *EOS/ESD Symposium, 2009*, 31 Sept 2009
52. Y. Cao, D. Johnsson, B. Arndt, M. Stecher, A TLP-based human metal model ESD-generator for device qualification according to IEC61000-4-2, in *Asia-Pacific Symposium on Electromagnetic Compatibility (APEMC), 2010*, Apr 2010, pp. 471–474
53. V.M. Vashchenko, M. ter Beek, W. Kindt, P. Hopper, ESD protection of the high voltage tolerant pins in low-voltage BiCMOS processes, in *Proceedings of BCTM*, 2004, pp. 277–280
54. V.A. Vashchenko, M. ter Beek, W. Kindt, P. Hopper, Implementation of 60V Tolerant Dual Direction ESD Protection in 5V BiCMOS Process for Automotive Application, in *Proceedings of ESD/EOS Symposium*, 2004, pp. 117–124
55. O. Semenov, H. Sarbishaei, M. Sachdev, *ESD Protection Device and Circuit Design for Advanced CMOS Technologies* (Springer, Netherlands, 2008)
56. A. Amerasekera, C. Duvury, *ESD in Silicon Integrated Circuits* (Wiley, West Sussex, 1995)
57. S. Dabral, T.J. Maloney, *Basic ESD and I/O Design* (Wiley, Chichester, 1998)
58. K. Estmark, H. Gossner, W. Stadler, *Advanced Simulation Methods for ESD Protection* (Elsevier, Amsterdam 2003)
59. S.H. Voldman, *ESD RF Technology and Circuits* (Wiley, Chichester, 2006); *ESD Physics and Devices* (Wiley, Chichester, 2004); *ESD Circuits and Devices* (Wiley, Chichester, 2005)
60. A. Chatterjee, T. Polgreen, A low voltage triggering SCR for on-chip ESD protection at output and input pads. IEEE Elec. Dev. Lett. **12**, 21–22 (1991)
61. M. Mergens, C. Russ, K. Verhaege, et al, Diode-triggered SCR (DTSCR) for RF-ESD protection of BiCMOS SiGe HBTs and CMOS ultra-thin gate oxides, in *International Electron Devices Meeting*, 2003, pp. 515–518

62. J. Di Sarro, V.A. Vashchenko, E. Rosenbaum, P. Hopper, A dual-base triggered SCR with very low leakage current and adjustable trigger voltage, in *30thElectrical Overstress/ Electrostatic Discharge Symposium, 2008, EOS/ESD 2008*, 7–11 Sept 2008, pp. 242–248
63. J.Z. Chen, A. Amerasekera, T. Vrotos, Bipolar SCR ESD protection for high speed submicron Bipolar/BiCMOS frequency integrated circuits. IEDM 337–340 (1995)
64. V.A. Vashchenko, P. Hopper, Bipolar SCR ESD devices. Microelectron. Reliab. J **45**, 457–471
65. V.A. Vashchenko, A. Concannon, M. ter Beek, P. Hopper, High holding voltage cascoded LVTSCR structures for 5.5 V tolerant ESD protection clamps. IEEE Trans. Device Mater. Reliab. **4**(2), 273–280 (2004); A. Concannon, V.A. Vashchenko, M. ter Beek, P. Hopper, A device level negative feedback in the emitter line of SCR-structures as a method to realize latch-up free ESD protection, in *Proceedings of 41st Annual IEEE International Symposium on Reliability Physics*, 30 Mar to 4 Apr 2003, pp. 105–111
66. V.A. Vashchenko, V. Kuznetsov, P. Hopper, Implementation of Dual-Direction SCR Devices in Analog CMOS Process, in *Proceedings of EOSESD Symposium*, 2007, pp. 75–79
67. V.A. Vashchenko, D. LaFonteese, Lateral PNP BJT ESD protection devices, in *BCTM 2008*, pp. 53–56
68. A. Gendron, P. Renaud, S.C. Besse, M. Bafleur, N. Nolhier, Area-Efficient Reduced and No-Snapback PNP-based ESD Protection in Advanced Smart Power Technology, in *Proceedings of EOS/ESD Symposium*, Sep. 2006, pp. 69–76
69. P. Renaud, A. Gendron, M. Bafleur, N. Nolhier, High robustness PNP-based structure for the ESD protection of high voltage I/Os in an advanced smart power technology, in *Proceedings of BCTM*, 2007, pp. 226–229
70. V.A. Vashchenko, D. LaFonteese, System Level and Hot Plug-in Protection of High Voltage Transient Pins, in *EOS/ESD Symposium*, 2009
71. V. Vashchenko, A. Shibkov, Non-Linear Response of Self-Protected CMOS Drivers, RCJ Symposium 2013
72. B. Keppens, M. Mergens, J. Armer, et al., Active-Area-Segmentation (AAS) technique for compact ESD robust, fully silicided NMOS design, in *Proceedings of EOS/ESD Symposium*, 2003, pp. 250–258
73. C. Russ, Non-uniform triggering of gg-nMOSt investigated by combined emission microscopy and transmission line pulsing, in *Proceedings of EOS/ESD Symp*osium, 1998, pp. 177–186
74. B. Caillard, STMSCR: A New Multi-Finger SCR-Based Protection Structure Against ESD, in *EOS/ESD Symposium*, 2003
75. B. Aliaj, T. Mitchell, V. Vashchenko, Overcoming Multi Finger Turn-on in HV DIACs using Local Polyballasting, accepted for publication at EOS/ESD Symposium 2014
76. T. Smedes, Y. Li, ESD phenomena in interconnect structures, in *Proceedings of EOS/ESD Symposium*, 2003
77. K. Banerjee, Characterization of VLSI circuit Interconnect Heating and Failure under ESD conditions, in *Proceedings of the International Reliability*, 1996
78. W.R. Anderson, Metal and Silicon Burnout Failures from CDM ESD Testing, in *EOS/ESD Symposium*, 2009
79. G. Notermans, Pitfalls when correlating TLP, HBM and HMM testing, in *Proceedings of EOS/ESD Symposium*, vol. 6, 1998, p. 170
80. S. Voldman, High-Current Transmission Line Pulse Characterization of Aluminum and Copper Interconnects for Advanced CMOS Semiconductor Technologies, in *Proceedings of International Reliability Physics Symposium*, 1988, pp. 293–301
81. Y. Xi, S. Malobabic, V. Vashchenko, J. Liou, Correlation between TLP, HMM and System Level ESD Pulses for Cu Metallization, Device and Materials Reliability, IEEE Transactions on, 14(1) pp. 446–450, March 2014
82. A.J. Duncan, *Quality Control and Industrial Statistics*, 5th edn. (Irwin, Homewood, 1986)
83. http://www.itl.nist.gov/div898/handbook/pmc/pmc.htm

84. S. Kotz, N.L. Johnson, *Process Capability Indices* (Chapman & Hall, London, 1992)
85. A. Mutlu, M. Rahman, Statistical methods for the estimation of process variation effects on circuit operation. IEEE Trans. Electron. Packag. Manuf. **28** (4), 364–375 (2005)
86. D. Blaauw, K. Chopra, A. Srivastava, L Scheffer, Statistical timing analysis: from basic principles to state of the art. IEEE Trans. CAD ICs Syst. **27**(4), 586–607 (2008)
87. F. Roger, W. Reinprecht, R. Minixhofer, Process variation aware ESD Design Window considerations on a 0.18 μm Analog, Mixed-Signal High Voltage Technology, in *EOS/ESD Symposium*, 2011
88. M.X. Huo, K.B. Ding, Y. Han, S.R. Dong, X.Y. Du, D.H. Huang, B. Song, Effects of Process Variation on Turn-on Voltages of a Multi-finger Gate-Coupled NMOS ESD Protection Device, in *Physical and Failure Analysis of Integrated Circuits, IPFA*, 2009, pp. 832–836
89. M. Diatta, D. Trémouilles, E. Bouyssou, M. Bafleur, Investigation on Statistical Tools to Analyze Repetitive-Electrostatic-Discharge Endurance of System-Level Protections. IEEE Trans. DMRon **12**(4), 607–614 (2012)
90. A. Tazzoli, A. Shibkov, V. Vashchenkosubmitted, Effect of Process Technology Variation on ESD Clamp Parameters, accepted at EOS/ESD Symposium 2014
91. R. Myoung, ESD analysis on mobile device printed circuit boards, in *ANSYS*, 2012
92. Patent *RE38,608E, "low voltage punch-through transient suppressor employing a dual-base structure"*
93. JESD-78D—IC Latch-Up Test, JEDEC 2011
94. S. Voldman, Latchup, Wiley 2007
95. S. Gupta, J.C. Beckman, S.L. Kosier, Unbiased Guard Ring for Latch-up-Resistant, Junction-Isolated Smart-Power ICs, in *Proceedings of BCTM*, 2001, pp. 188–191
96. V. Vashchenko, B. Aliaj, J. Sheu and A. Shibkov, High Voltage Latchup Analysis, International ESD Workshop 2013
97. ESDA, ESDA Working Group 5.4 Transient-induced latch-up—Technical Report 5, 2012
98. S. Bargstadt-Franke, W. Stadler, K. Esmark, M. Streibl, K. Domanski, H. Gieser, H. Wolf, W. Bala, Transient latch-up: Experimental analysis and device simulation, in *EOS/ESD Symposium 2003*, Sept 2003
99. K. Domanski, S. Bargstadt-Franke, W. Stadler, U. Glaser, W. Bala, Development strategy for TLU-robust products, in *EOS/ESD Symposium 2004*, Sept 2004
100. K. Domanski, S. Bargstadt-Franke, W. Stadler, M. Streibl, G. Steckert, W. Bala, Transient-LU failure analysis of the ICs, methods of investigation and computer aided simulations, in *Proceedings of Reliability Physics Symposium*, April 2004, pp. 370–374
101. M.-D. Ker, S.-F. Hsu, Component-Level Measurement for Transient-Induced Latch-up in CMOS ICs Under System-Level ESD Considerations. IEEE Trans. Device Mater. Reliab. **6**(3), 461–472 (2006)
102. T. Brodbeck, W. Stadler, C. Baumann, K. Esmark, K. Domanski, Triggering of Transient Latch-up by System-Level ESD. IEEE Trans. Device Mater. Reliab. **11**(4), 509–515 (2011)
103. M. Scholz, S. Chen, S. Thijs, D. Linten, G. Hellings, G. Vandersteen, M. Sawada, G. Groeseneken, System-level ESD Protection Design using On-Wafer Characterization and Transient Simulations, Device and Materials Reliability *IEEE Transactions on*, **14**(1), pp. 104–111, March 2014
104. T. Suzuki, J. Iwahori, T. Morita, H. Takaoka, T. Nomura, K. Hashimoto, S. Ichino, A study of relation between a power supply ESD and parasitic capacitance. J. Electrostat. **64**(11), 760–767 (2006)
105. ISO 11898, Road vehicles-Controller area network (CAN), 2003–2013
106. CAN: W. Chen, N. Trichy, K. Soundarapandian, S. Pendharkar, J. Carpenter, J. Kohout, Experiences and challenges of CAN transceivers in upintegrated system basis chips, in *ICC 2005*, pp. 09-1-6
107. IEC/TS 62228 ed1.0, Integrated circuits—EMC evaluation of CAN transceivers, 2007

108. Hardware Requirements for LIN, CAN and FlexRay Interfaces in Automotive Applications, rev 1.3, 2012
109. A.A. Salman, F. Farbiz, A. Concannon, H. Edwards, G. Boselli, Mutual Ballasting: A Novel Technique for Improved Inductive System Level IEC ESD Stress Performance for Automotive Applications, in *EOSESD Symposium 2013*, paper 3B.1
110. W. Stadler, T. Brodbeck, R. Gartner, H. Gossner, Cable discharges into communication interfaces, in *Electrical Overstress/Electrostatic Discharge Symposium, 2006. EOS/ESD '06*, Sept 2006, pp. 10–15, 144, 151
111. R. Chundru, Z. Li, D. Pommerenke, K. Kam, C.-W. Lam, F. Centola, R. Steinfeld, An evaluation of TVS devices for ESD protection", in *IEEE International Symposium on Electromagnetic Compatibility (EMC), 2011*, Aug 2011, pp. 62–67
112. K. Shrier, T. Truong, J. Felps, Transmission line pulse test methods, test techniques and characterization of low capacitance voltage suppression device for system level electrostatic discharge compliance, in *EOS/ESD Symposium 2004*, Sept 2004, pp. 1–10
113. D. Johnsson, H. Gossner, Study of system ESD co-design of a realistic mobile board, in *EOS/ESD Symposium 2011*, Sept 2011, pp. 1–10
114. Industry Council on ESD target levels, *White Paper 3 - System Level ESD, Part I*, 2011
115. K. Verhaege, P.J. Roussel, G. Groeseneken, H. Maas, H. Gieser, C. Russ, P. Egger, Analysis of HBM ESD testers and specifications using a forth-order lumped element model, in *EOS/ESD Symposium 1993*, 1993
116. A. Amerasekera, C. Duvvury, W. Anderson, H. Gieser, S. Ramaswamy, *ESD in silicon integrated circuits* (Wiley, Chichester, 2002)
117. K. Esmark, H. Gossner, W. Stadler, *Advanced Simulation Methods for ESD Protection Development* (Elsevier Science & Technology, Amsterdam, 2002)
118. L. Lou, C. Duvvury, A. Jahanzeb, J. Park, SPICE simulation methodology for system level ESD design, in *EOS/ESD Symposium 2010*, Oct 2010, pp. 1–10
119. J. Willemen, A. Andreini, V. De Heyn, K. Esmark, M. Etherton, H. Gieser, G. Groeseneken, S. Mettler, E. Morena, N. Qu, W. Soppa, W. Stadler, R. Stella, W. Wilkening, H. Wolf, L. Zullino, Characterization and modeling of transient device behavior under CDM ESD stress, in *EOS/ESD Symposium 2003*, Sept 2003, pp. 1–10
120. Y.-C. Liang, V.J. Gosbell, Diode forward and reverse recovery model for power electronic spice simulations. IEEE Trans. Power Electron. **5**(3), 346–356 (1990)
121. Industry Council on ESD target levels, *White Paper 3—System Level ESD, Part II*, 2012
122. S. Bertonnaud, C. Duvvury, A. Jahanzeb, IEC System Level ESD challenges and effectice protection strategy for USB2 interfaces, in *EOS/ESD Symposium 2012*, Sept 2012, pp. 388–395
123. T. Li, J. Maeshima, H. Shumiya, D.J. Pommerenke, T. Yamada, K. Araki, An application of utilizing the system-efficient-ESD-design (SEED) concept to analyze an LED protection circuit of a cell phone, in *IEEE EMC Symposium, 2012*, Aug 2012, pp. 346–350
124. S. Marum, C. Duvvury, J. Park, A. Chadwick, A. Jahanzeb, Protecting circuits from the transient voltage suppressor's residual pulse during IEC61000-4-2 stress, in *31st EOS/ESD Symposium, 2009*, Sept 2009, pp. 1–10
125. G. Notermans, S. Bychikhin, D. Pogany, D. Johnsson, D. Maksimovic, HMM and TLP correlation, in *International ESD Workshop*, 2011
126. IEC 61000-4-2:2008, Electromagnetic compatibility (EMC)—Part 4-2: Testing and measurement techniques—electrostatic discharge immunity test, 2008
127. IEC 60063 Edition 2.0—Preferred number series for resistors and capacitors, 1963
128. S. Thijs, C. Russ, D. Tremouilles, A. Griffoni, D. Linten, M. Scholz, N. Collaert, R. Rooyackers, M. Jurczak, M. Sawada, T. Nakaei, T. Hasebe, C. Duvvury, H. Gossner, G. Groeseneken, Design methodology of FinFET devices that meet IC -Level HBM ESD targets, in *EOS/ESD Symposium 2008*, Sept 2008, pp. 294–302

129. S. Thijs, A. Griffoni, D. Linten, S.-H. Chen, T. Hoffmann, G. Groeseneken, On gated diodes for ESD protection in bulk FinFET CMOS technology, in *EOS/ESD Symposium 2011*, Sept 2011, pp. 1–8
130. M. Scholz, S. Thijs, D. Linten, D. Tremouilles, M. Sawada, T. Nakaei, T. Hasebe, M.I. Natarajan, G. Groeseneken, Calibrated wafer-level HBM measurements for quasi-static and transient device analysis, in *EOS/ESD Symposium 2007*, Sept. 2007, pp. 2A.2-1–2A.2-6
131. J.-R. Manouvrier, P. Fonteneau, C.-A. Legrand, P. Nouet, F. Azais, Characterization of the transient behavior of gated/STI diodes and their associated BJT in the CDM time domain, in *EOS/ESD Symposium 2007*, Sept 2007, pp. 3A.2-1–3A.2-10
132. A. Ille, W. Stadler, T. Pompl, H. Gossner, T. Brodbeck, K. Esmark, P. Riess, D. Alvarez, K. Chatty, R. Gauthier, A. Bravaix, Reliability aspects of gate oxide under ESD pulse stress, in *EOS/ESD Symposium 2007*, Sept 2007, pp.6A.1-1–6A.1-10
133. IBIS Open Forum, *IBIS (I/O Buffer Information Specification) Version 5.0*, 2008
134. B. Kleveland, T.J. Maloney, I. Morgan, L. Madden, T.H. Lee, S.S. Wong, Distributed ESD protection for high-speed integrated circuits. IEEE Electron device Lett. **21**(8), 390–392 (2000)
135. K. Saban, *White paper WP380: Xilinx stacked silicon interconnect technology delivers breakthrough FPGA capacity, bandwidth, and power efficiency*, 2011
136. M. Diatta, D. Tremouilles, E. Bouyssou, R. Perdreau, C. Anceau, M. Bafleur, Understanding the failure mechanisms of protection diodes during system level esd: toward repetitive stresses robustness. IEEE Trans. Electron Devices **59**(1), 108–113 (2012)

Index

V. Vashchenko and M. Scholz, *System Level ESD Protection*,
DOI: 10.1007/978-3-319-03221-4, © Springer International Publishing Switzerland 2014